Do Democracies Win Their Wars?

International Security Readers

Strategy and Nuclear Deterrence (1984)

Military Strategy and the Origins of the First World War (1985)

Conventional Forces and American Defense Policy (1986)

The Star Wars Controversy (1986)

Naval Strategy and National Security (1988)

Military Strategy and the Origins of the First World War, revised and expanded edition (1991)

—published by Princeton University Press

Soviet Military Policy (1989)

Conventional Forces and American Defense Policy, revised edition (1989)

Nuclear Diplomacy and Crisis Management (1990)

The Cold War and After: Prospects for Peace (1991)

America's Strategy in a Changing World (1992)

The Cold War and After: Prospects for Peace, expanded edition (1993)

Global Dangers: Changing Dimensions of International Security (1995)

The Perils of Anarchy: Contemporary Realism and International Security (1995)

Debating the Democratic Peace (1996)

East Asian Security (1996)

Nationalism and Ethnic Conflict (1997)

America's Strategic Choices (1997)

Theories of War and Peace (1998)

America's Strategic Choices, revised edition (2000)

Rational Choice and Security Studies: Stephen Walt and His Critics (2000)

The Rise of China (2000)

Nationalism and Ethnic Conflict, revised edition (2001)

Offense, Defense, and War (2004)

New Global Dangers: Changing Dimensions of International Security (2004)

Primacy and Its Discontents: American Power and International Stability (2009)

Going Nuclear: Nuclear Proliferation and International Security in the 21st Century (2010)

Contending with Terrorism: Roots, Strategies, and Responses (2010)

Do Democracies Win Their Wars? (2011)

—published by The MIT Press

Do Democracies Win Their Wars?

AN *International*
Security READER

EDITED BY
Michael E. Brown
Owen R. Coté Jr.
Sean M. Lynn-Jones
and Steven E. Miller

THE MIT PRESS
CAMBRIDGE, MASSACHUSETTS
LONDON, ENGLAND

Library of Congress Cataloging-in-Publication Data

Do democracies win their wars? / edited by Michael E. Brown . . . [et al.].
    p. cm. — (International security readers)
  Includes bibliographical references.
  ISBN 978-0-262-51590-0 (pbk. : alk. paper)
  1. War. 2. Democracy. I. Brown, Michael E. (Michael Edward), 1954– II. International security.
JZ6385.D63 2011
355.02—dc22
                    2010047085

10 9 8 7 6 5 4 3 2 1

Contents

vii
The Contributors

ix
Acknowledgments

xi
Preface   Sean M. Lynn-Jones

PART I: VICTORIOUS DEMOCRACIES: THE
CASE FOR A DEMOCRATIC ADVANTAGE

3
Search for Victory: Why Democracies Win   Dan Reiter and
Their Wars   Allan C. Stam

38
Powerful Pacifists: Democratic States and War   David A. Lake

---

PART II: THE DEMOCRATIC VICTORY DEBATE

67
Making Military Might: Why Do Democracies   Risa A. Brooks
Succeed? A Review Essay

94
Democracy and Victory: Why Regime Type   Michael C. Desch
Hardly Matters

137
The Power of Democratic Cooperation   Ajin Choi

149
Fair Fights? Evaluating Theories of Democracy   David A. Lake
and Victory

163
Understanding Victory: Why Political Institutions   Dan Reiter and
Matter   Allan C. Stam

175
Democracy and Victory: Fair Fights or Food Fights? Michael C. Desch

190
How Smart and Tough Are Democracies?   Alexander B. Downes
Reassessing Theories of Democratic Victory in War

233
Correspondence: Another Skirmish in the Battle over Dan Reiter and
Democracies and War   Allan C. Stam
                                                             Alexander B. Downes

---

PART III: DEMOCRACY AND DECEPTION:
SELECTION EFFECTS AND DEMOCRATIC
VICTORY

249
The Deception Dividend: FDR's Undeclared War   John M. Schuessler

282
Correspondence: FDR, U.S. Entry into World War II, Dan Reiter
and Selection Effects Theory   John M. Schuessler

The Contributors

MICHAEL E. BROWN is Dean of the Elliott School of International Affairs and Professor of International Affairs and Political Science at the George Washington University.

OWEN R. COTÉ JR. is Co-Editor of *International Security* and Associate Director of the Security Studies Program at the Massachusetts Institute of Technology.

SEAN M. LYNN-JONES is Co-Editor of *International Security* and a research associate at the Belfer Center for Science and International Affairs, John F. Kennedy School of Government, Harvard University.

STEVEN E. MILLER is Editor-in-Chief of *International Security*, Co–Principal Investigator with the Project on Managing the Atom, and Director of the International Security Program at the Belfer Center for Science and International Affairs, John F. Kennedy School of Government, Harvard University.

RISA A. BROOKS is Assistant Professor of Political Science at Marquette University.

AJIN CHOI is Associate Professor in the Graduate School of International Studies at Yonsei University.

MICHAEL C. DESCH is Professor and Chair of the Department of Political Science at Notre Dame University.

ALEXANDER B. DOWNES is Assistant Professor of Political Science at Duke University.

DAVID A. LAKE is the Jerri-Ann and Gary E. Jacobs Professor of Social Sciences and Distinguished Professor of Political Science at the University of California, San Diego.

DAN REITER is Professor and Chair of the Department of Political Science at Emory University.

JOHN M. SCHUESSLER is Assistant Professor in the Department of Leadership and Strategy at the Air War College.

ALLAN C. STAM is Professor of Political Science at the University of Michigan.

Acknowledgments

The editors gratefully acknowledge the assistance that has made this book possible. A deep debt is owed to all those at the Belfer Center for Science and International Affairs, Harvard University, who have played an editorial role at *International Security*. Special thanks go to Diane McCree and Katherine Bartel Gordon for their invaluable help in preparing this volume for publication. We thank Risa Brooks, Dan Reiter, and Allan Stam for revising and updating their previous work for this volume.

Preface

Sean M. Lynn-Jones

How are democracies different from states with other forms of government? Do democracies have a distinctive approach to governance at home and foreign policy abroad? For centuries, numerous writers have argued that democracy makes a difference. Most have celebrated its virtues as a form of government that offers citizens liberty and the opportunity to participate in politics. When it comes to international politics, however, many writers have offered more mixed evaluations.

Traditionally, there was a widespread belief that democracies were at a distinct disadvantage in international politics. This belief goes back as far as Thucydides, who wrote in *The Peloponnesian War* of how democratic Athens struggled and blundered in its long conflict with authoritarian Sparta.[1] Subsequent generations of political realists suggested that democracy hindered a state's attempts to conduct a successful foreign policy and to wage war effectively. Among other liabilities, democracies appeared to have trouble maintaining a consistent policy in pursuit of their national interests. Domestic political pressures and the vagaries of public opinion constantly distracted their leaders. Public opposition and moral qualms might dissuade them from using the ruthless measures employed by their authoritarian adversaries. For all their virtues, a free press, public debate, and an open society frequently made it hard for democracies to maintain the secrecy often necessary for successful diplomacy. Unlike dictatorial states, they could not demand that their societies make the sacrifices necessary to devote enormous human and material resources to their military establishments. Their people might demand peace when a threatening international environment made war necessary. During the Cold War, even U.S. leaders and staunch anticommunists wondered whether the United States could compete effectively with the Soviet Union.

In the 1980s, many scholars identified a more hopeful feature of democracies in international politics: the absence of war between democratic states. This finding, which became known as the democratic peace, actually was a rediscovery of a claim that the German philosopher Immanuel Kant had made almost two centuries earlier in his *Perpetual Peace*.[2] When Kant wrote, however, there were few democracies. With sufficient data and sophisticated statistical methods, social scientists constructed a more persuasive quantitative case for Kant's argument. They also attempted to explain why democracies did not fight one another, advancing hypotheses that attributed the democratic peace

---

1. Thucydides, *The Peloponnesian War*, trans. Rex Warner (Middlesex, U.K.: Penguin, 1954).
2. Immanuel Kant, "Perpetual Peace," in Immanuel Kant, *Kant: Political Writings*, 2d ed., Hans Reiss, ed., trans. H.B. Nisbet (Cambridge: Cambridge University Press, 1991), pp. 93–130.

to factors that included the political philosophy of liberalism, the constraints that domestic institutions impose on leaders in democracies, and the tendency of democratic publics to oppose war. Realist scholars and other skeptics articulated many critiques of the existence and logic of the democratic peace, and a lively debate emerged.[3]

More recently, academic research has highlighted another distinctive feature of the foreign policies of democracies: the tendency of democracies to win a large proportion of the wars they fight. Several scholars have found that democracies have been victorious in most of their wars since 1815. This finding, which flies in the face of the long-standing belief that democracies are at a disadvantage in international politics, has stimulated a vigorous debate. As in the case of the democratic peace debate, there is some agreement on the key empirical finding—that democracies often win their wars—but no consensus on why democracies win or whether democracy is actually the cause of victory. Proponents and critics of the "democratic victory" proposition have clashed on these questions.

This volume consists of contributions to the democratic victory debate. It begins with two essays that find that democracies tend to win the overwhelming majority of the wars they fight. Those essays offer multiple and contrasting explanations for why democracies seem to have an advantage in war. The remainder of the volume is a debate between proponents and critics of the democratic victory proposition. The critics question the statistical significance of the finding that democracies win most of their wars. They also challenge the claim that democracy is the cause of victory. Advocates of the democratic victory proposition advance vigorous responses to these empirical and logical critiques.

In "Search for Victory: Why Democracies Win Their Wars," Dan Reiter and Allan Stam present a comprehensive case for why democracies are usually victorious in war. They argue that democracies have won more than three-quarters of their wars since 1815. Reiter and Stam contend that the underlying cause of democratic victory is that elected leaders are accountable to the voters in their countries. This characteristic leads to two explanations for why democracies win wars.

---

3. Many contributions to the debate over the democratic peace appear in Michael E. Brown, Sean M. Lynn-Jones, and Steven E. Miller, eds., *Debating the Democratic Peace* (Cambridge, Mass.: The MIT Press, 1996). See also Sebastian Rosato, "The Flawed Logic of Democratic Peace Theory," *American Political Science Review*, Vol. 97, No. 4 (November 2003), pp. 585–602.

The first explanation of why democracies win most of their wars is the "selection effects" mechanism: democracies start wars only when they are likely to win. In general, leaders are more likely to go to war if they estimate that they have a high probability of winning. Reiter and Stam argue that elected leaders fear being voted out of office and that losing a war is often a cause of electoral defeat. Although leaders in nondemocracies may be ousted if they lose a war, elected leaders are far more likely to be removed by a dissatisfied public. Leaders who must face the voters are thus risk-averse and select wars that their countries are very likely to win while avoiding those that probably would end in defeat. In addition, democracies are better than autocracies at estimating the probability of victory. Open public debate, vigorous discussion, and the free flow of information enable democracies to make better decisions. In democracies, military officers are more likely to be promoted on the basis of merit, not political loyalties, so political leaders will tend to receive honest and useful advice from military leaders. Unlike in autocracies, militaries in democracies also are free from domestic security functions, so they can focus on preparing to fight international wars.

Reiter and Stam contend that the selection effects explanation predicts that democracies are especially likely to win wars that they initiate. Following Jack Snyder, they suggest that mixed regimes that share democratic and autocratic characteristics (e.g., Wilhelmine Germany before World War I) are particularly likely to engage in "foolish military and colonial ventures" and will therefore be most likely to lose the wars they start.[4] Such states make highly inaccurate estimates of their chances of victory, even if their leaders face some risk of being removed from power if they lose a war. Dictatorial regimes should be more likely to win than mixed regimes, but less likely than democracies. Reiter and Stam look at all wars from 1816 to 1985 and find that democratic states that initiated wars won 93 percent of the time, compared with 60 percent for dictatorial regimes and 58 percent for mixed regimes. The same pattern emerged after controlling for military-industrial capabilities, terrain, strategy, troop quality, allies, and distance.

Prominent twentieth-century wars illustrate how democracies are more likely to achieve victory. In 1971, India carefully lined up diplomatic and military support as the conflict between East and West Pakistan escalated. India's democratic institutions enabled it to assess when it would have the best chance

---

4. Jack Snyder, *Myths of Empire: Domestic Politics and International Ambition* (Ithaca, N.Y.: Cornell University Press, 1991).

of victory and to adopt an effective war-winning strategy. When India finally went to war, it decisively defeated Pakistan. The 1971 Indian case contrasts starkly with Japan's 1941 decision to attack Pearl Harbor. Japan believed its strategic situation was deteriorating. It was bogged down in China and faced an economic embargo from the United States. Japan's leaders, who faced little domestic criticism or public pressure, underestimated the willingness of the United States to fight and decided to gamble that Japan could win a quick victory. The result was a catastrophe for Japan.

Reiter and Stam conducted more in-depth tests to determine if democracies won wars because they successfully calculated the odds of victory. They found that democracies were more likely to initiate disputes when they estimated that they had a greater chance of winning a potential war. They also found that democracies reduced the costs of war by choosing to fight shorter wars, even though this meant that democracies sometimes had to settle for a draw instead of victory. Elected leaders, not surprisingly, are more likely to want to limit the costs that war imposes on the public. This dynamic operated for the United States in the Vietnam War and the 1991 Gulf War. In the former case, the United States hoped to be able to fight a short war, but could not. It ultimately ended the war to limit its costs and abandoned the goal of victory. In the Gulf War, the United States employed overwhelming force to achieve a rapid victory at low cost and then quickly ended the war after the liberation of Kuwait.

The second explanation of democratic victory is that democracies win wars because their militaries are more effective on the battlefield. In contrast to thinkers who claim that democracies do not foster a martial spirit in their citizens, Reiter and Stam argue that democracies produce better armies, for four reasons. First, soldiers who fight for elected governments will have higher morale, because they know that they are fighting for a government that upholds their interests and that deserves their loyalty and trust. Second, the individualistic political culture of liberal democracies produces soldiers who will take initiative on the battlefield. This trait is particularly important in modern warfare. Third, soldiers of democracies will probably hesitate to surrender to authoritarian enemies, who have a reputation for not treating prisoners fairly. On the other hand, soldiers of authoritarian regimes will be more likely to surrender to democratic opponents because they probably expect to be treated humanely. Fourth, the armies of democracies are more likely to have good leaders. Democracies promote leaders on the basis of merit, whereas military promotions in authoritarian countries depend on political factors. Reiter and Stam find that there is empirical support for most of these propositions, but it is not clear whether democracies always have an advantage in morale; authoritarian states often can use nationalism to bolster the morale of their militaries.

Reiter and Stam also consider and reject two other potential explanations of why democracies tend to win their wars: (1) democracies form more powerful alliances; and (2) democracies generate more military-industrial power.

Although logic suggests that democracies should ally with one another—especially against autocracies—and that they should be cooperative and effective allies, there is little empirical support for the argument that democracies win their wars because they form democratic alliances. Democracies often ally with nondemocracies. Some wars in which democracies have allied with one another (e.g., Korea and Vietnam) have ended in draws or defeats, not victory. Democracies do not seem to attract statistically significant additional amounts of aid from their democratic allies. Nor are they more likely to intervene on behalf of their fellow democracies. When democracies ally with other democracies, they often do so because they are motivated by realist concerns about power and security.

Reiter and Stam also find that there is little evidence for the claim that democracies win wars because they have larger economies and can devote more resources to war. They note that this claim includes arguments that democracies have larger economies and that the political legitimacy of democracies enables them to extract more resources from their populations. The evidence suggests, however, that there is no relationship between democracy and the size of a state's economy. Moreover, democracies are not better at extracting resources during wartime. Reiter and Stam find, for example, that compared to authoritarian regimes democracies mobilize a smaller proportion of their populations during wartime. Democracies actually have lower levels of tanks, artillery pieces, and air sorties. The share of gross domestic product that they devote to military spending is about the same as in nondemocracies. Reiter and Stam suggest that the domestic political constraints on leaders in democracies actually make it harder to mobilize resources for war.

Looking at recent history, Reiter and Stam point out that the United States easily won the conventional phases of its wars in Afghanistan in 2001–02 and Iraq in 2003. These rapid victories lend support to the claim that democracies win their wars. On the other hand, the marketplace of ideas in the United States did not prevent the George W. Bush administration from misunderstanding the character of the postconventional phase of each war. Nor did the existence of open debate in the United States reveal the facts about Iraq's ties to terrorists or weapons of mass destruction. U.S. forces also were far less militarily effective in the postconventional phases of the wars in Iraq and Afghanistan. Democracies do not seem to enjoy any special advantages in counterinsurgency operations. When wars drag on and impose high costs, the democratic publics call on their leaders to stop fighting, even if this means forgoing victory.

In "Powerful Pacifists: Democratic States and War," David Lake offers a different explanation for the tendency of democracies to win their wars. Lake's explanation for democratic victory begins with the assumption that states can be regarded as profit-maximizing firms that trade services for revenues. Protection from external threats is the most important service provided by the state. The demand for protection will increase as the level of external threat rises. States have a monopoly over the provision of protection. They have an incentive to seek rents—profits above the normal level—by exploiting this monopoly position and by creating or inflating foreign threats. In democracies, however, citizens have a greater opportunity to participate in politics, monitor the activities of the state, and even emigrate if they are dissatisfied. These factors constrain the ability of democratic states to seek rents. In autocracies, the state will not be subject to these societal constraints. Autocratic states will thus be able to seek higher rents. Autocracies are also more likely to engage in expansionist behavior, which can increase rents by eliminating competitors, increasing the external threat level and thus the demand for protection, and acquiring territories that can be exploited for revenue.

Lake contends that his theory of the rent-seeking state offers a better explanation of the international behavior of democracies than the liberal theories of the democratic peace. His theory predicts that democracies will be less expansionist and less war-prone than autocracies, that rent-seeking autocracies will go to war against democracies, and that democracies will engage in expansion only when the benefits outweigh the costs. In addition, Lake predicts that democracies are "significantly more likely to win the wars they fight against autocracies." Democracies emerge victorious because they have fewer economic distortions and thus generate more national wealth, can extract more resources for security because their societies support state policies, and form overwhelming coalitions against autocracies.

Although Lake cannot empirically test all of the propositions that emerge from his theory of the rent-seeking state, he finds strong evidence that democracies win the wars they fight. Since 1816, "democratic states, either singly or in combination with other states, have won four times as many wars as autocratic states." Whether a state is a democracy is a far better predictor of victory than other factors, including levels of military manpower and industrial production.

The next section of essays features contending perspectives on democratic victory. There is a logical and empirical case for the claim that democracies generally win wars, but critics have disputed the proposition. The essays in this section include challenges to the argument that democracies win most of their wars, as well as vigorous responses from proponents of the democratic victory argument.

In "Why Do Democracies Succeed? A Review Essay," Risa Brooks offers a critical analysis of the arguments of Dan Reiter and Allan Stam. She assesses their book *Democracies at War*, from which Reiter and Stam's contribution to this volume was derived. Although Brooks praises *Democracies at War* for making a rigorous, stimulating, and ambitious attempt to explore an important question, she concludes that the book "is not fully convincing in its claims about democracies and states' military effectiveness."

After summarizing *Democracies at War*, Brooks offers three sets of criticisms. First, she contends that Reiter and Stam do not offer a compelling explanation for why democracies win their wars. In her view, they substitute facts for arguments. Reiter and Stam rely on the existing literature for potential explanations that range from bargaining theory to political philosophy, but they do not offer an integrated deductive argument that shows why some of those explanations should be more powerful than others. She argues that *Democracies at War* simply uses empirical findings to create a central argument about the importance of public consent and accountability to voters in democratic countries. In addition, she posits that the book's emphasis on "superior leadership and more aggressive soldiering" as causes of democratic success in war neglects other factors that influence battlefield outcomes, including weapons handling, intelligence, logistics, training, and unit cohesion. Democracies might excel in some of these areas. Brooks also faults *Democracies at War* for paying little attention to the role of leaders' decisions about strategic and operational war plans. She recognizes that Reiter and Stam may be correct in claiming that democracies win wars because they have better soldiering and leadership, but wonders if they have given adequate attention to the many other potential causes of democratic victory.

Second, Brooks criticizes the logic of *Democracies at War* and claims that the authors rely on "questionable assumptions and a crude conceptualization of the relationships among key variables." She argues that the book conflates culture with institutions. Reiter and Stam claim that individuals have greater freedom and prerogatives in democracies, but these are attributes of a liberal political culture, not of democratic institutions. A country can have democratic institutions, including elections, without having a liberal political culture that favors individualism. Moreover, a country's military could have an organizational culture antithetical to individual initiative and improvisation. Some scholars argue, for example, that the British military in World War I was too rigidly hierarchical and stifled innovation.

Brooks also contends that *Democracies at War* lumps together a diverse group of autocratic states. Reiter and Stam claim that autocratic states often promote military officers for political reasons, whereas democracies tend to promote on the basis of merit. Brooks argues that autocracies have different patterns of civil-military relations and that the degree to which their militaries are politicized varies widely. North Vietnam, for example, selected and promoted talented military leaders much more often than Iraq, the authoritarian country often cited by Reiter and Stam.

Third, Brooks evaluates the empirical tests and methods used in *Democracies at War* and concludes that Reiter and Stam "poorly measure key variables and inadequately test key hypotheses." She argues that the book does not use a good measure of liberal political culture. Instead, it relies on measures of procedural democracy and the competitiveness of elections included in the POLITY data set. This problem is important because Reiter and Stam argue that liberal political culture endows soldiers in democracies with greater morale and initiative. Brooks also criticizes *Democracies at War* for relying on data from the Historical Evaluation and Research Organization (HERO) for measurements of initiative. She contends that HERO defines initiative in terms of which side attacks first in battle, whereas Reiter and Stam regard initiative as a trait of individual soldiers. Brooks suggests that the military forces of democracies may often take the initiative, as defined by HERO, because they often win the wars they fight and are therefore on the operational offensive. Winning is a cause of taking the initiative, not the other way around.

Further, Brooks faults *Democracies at War* for not controlling for variables that might be part of alternative explanations for the tendency of democracies to win their wars. Examples include the quality of a state's human capital, which depends on levels of education, literacy, and technological skills. Brooks notes that human capital may be a more powerful predictor of victory than democracy. Although many democracies will have high levels of education and human capital, others will not. One other study finds that democratic institu-

tions actually degrade military effectiveness, but human capital has the opposite effect.[5]

Brooks also criticizes the ways in which Reiter and Stam code democracies
and autocracies. They compare the relative degree of democracy in conflicts
and code a victory of a slightly more democratic (or less autocratic) state as a
win for democracy. Brooks questions whether small steps toward liberalization
(e.g., the decision of an authoritarian state to hold limited elections) cause authoritarian states to become more militarily effective.

Although she does not attempt her own large-scale empirical analysis,
Brooks points to several examples that may undermine the claim that democratic armies display more initiative and fight more effectively. The major example is Nazi Germany, whose soldiers consistently fought more effectively
than their democratic opponents. Other examples include North Vietnam and
the Zulu forces that fought British armies in the late nineteenth century.

Michael Desch's "Democracy and Victory: Why Regime Type Hardly Matters" offers another critique of the claim that democracies win their wars.[6]
Desch notes that there has been a long-standing debate between "democratic
defeatists" and "democratic triumphalists." The former believe that democracies are at a disadvantage in the struggle against nondemocracies. The latter
argue that democracies are more likely to win the wars they fight. After summarizing the arguments of the triumphalists, Desch offers three criticisms.

First, Desch examines the evidence that provides the basis for the claim that
democracies win a large proportion of their wars against nondemocracies. He
finds that there are multiple problems with the data. For example, democracies
sometimes are on the winning side of a war in which a nondemocratic ally
bore the brunt of the fighting and was responsible for the victory. World War II,
in which the Soviet Union fought a massive land war against Nazi Germany, is
a prime example. In other cases, states coded by the triumphalists as democracies turn out to be not very democratic. Desch also claims that wars in which
the democratic state enjoyed a huge advantage in overall power or a much
greater stake in the outcome of the conflict should be excluded. After making

---

5. Stephen Biddle and Stephen Long, "Democratic Effectiveness? Reassessing the Claim That Democracies Are More Effective in Battle," paper prepared for delivery at the annual meeting of the
American Political Science Association, Boston, Massachusetts, August 29–September 1, 2002; and
Stephen Biddle and Stephen Long, "Democracy and Military Effectiveness: A Deeper Look," *Journal of Conflict Resolution*, Vol. 48, No. 4 (August 2004), pp. 525–546.
6. For a longer version of Desch's arguments, see Michael C. Desch, *Power and Military Effectiveness:
The Fallacy of Democratic Triumphalism* (Baltimore, Md.: Johns Hopkins University Press, 2008).

these adjustments and focusing only on what he regards as fair tests, Desch finds that democracies may win a slim majority of their wars, but democracy is not a statistically significant explanation of these outcomes. Desch also points out that three countries—Britain, Israel, and the United States—account for many of the democratic victories, that preconditions for democracy (e.g., high levels of wealth, a liberal political culture, and geographical security) may explain why democracies often win wars, and that whether a government is consolidated or not may be an important explanation of how a state performs in war.

Second, Desch offers a critique of the selection effects argument—the claim that democracies win wars because they start wars they are likely to win. He argues that of all the factors that may determine whether a country wins or loses a war, democracy has one of the smallest marginal effects. In addition, the logic of the selection effects argument is flawed. Leaders of autocracies also fear that they will lose power if they start an unsuccessful war. Futhermore, Desch notes that Israel has a mixed record in the wars it has initiated, achieving a clear victory in 1967 but having much less success in 1956 and 1982.

Third, Desch questions whether democracies are better at fighting wars. He disputes claims that various causal mechanisms—wealth, alliances, sound strategy, public support, fighting proficiency—make democracies more militarily effective.

In Desch's view, democracies may be wealthier than other types of states, but democracy may not be the cause of their wealth. The causal arrow actually may point in the other direction. Desch disputes Lake's claim that democracies are not rent-seeking states, arguing that interest groups can seek rents by playing a large role in domestic politics. He also doubts that democracies are better at transforming their economic strength into military power.

Desch criticizes the claim that democracies win wars because they form strong alliances with other democracies. He points out that alliances that contain only democracies are rare. Most alliances are mixed. The winning coalition in World War II is a leading example. There is also little reason to believe that democratic alliances will be more effective than mixed alliances. Desch disputes the argument that democracies will keep their international commitments to one another because leaders will be sensitive to audience costs—how the public in a democracy will react if its leaders do not support other democracies. He suggests that democratic publics often pay little attention to foreign affairs.

Democratic triumphalists claim that the open public debate in democracies produces better strategic decisions, but Desch argues that there is no systematic evidence to support this claim. Israel has made strategic blunders, despite its democratic political process. Open debate is no guarantee of sound strategy, because democracies may not be able to reconcile divergent views and create coherent policies.

Desch questions whether democracies are more likely to win wars because their governments enjoy greater public support. He suggests that the public may have a much shorter time horizon than leaders, which makes it less likely that the public will support a particular war. He also points to the Vietnam War as a case in which the public in a democracy did not support a war.

Desch counters claims that democracies produce more motivated and proficient soldiers by noting that many other factors account for leadership and initiative on the battlefield. Nationalism, in particular, enhances loyalty and fighting prowess. He questions the Combat History Analysis Study Effort (CHASE) data that Reiter and Stam use, pointing out that independent assessments found many errors in the codings that HERO made in this data. Regardless of the quality of the data, Desch notes that the three most formidable armies of the twentieth century were deployed by Germany in World War I, Germany in World War II, and Israel. All three were motivated by nationalism.

The next three essays defend the argument that democracies tend to win the wars they fight and reply to Desch's attempts to undermine this argument. Desch then responds and defends his critique of democratic triumphalism.

In "The Power of Democratic Cooperation," Ajin Choi focuses on the debate over whether democratic allies are more effective than nondemocratic allies. She advances a new explanation for why democracies make better wartime partners: democratic political systems have "veto players" that influence the decisionmaking process and ensure that commitments are kept. Democracies also cooperate more effectively because they have high levels of communication among them. Choi presents data showing that states are more likely to win wars if they have a higher number of democratic partners. Mixed coalitions are more likely to win wars than are nondemocratic coalitions. Choi supports the quantitative data with detailed case studies that compare how Britain and Austria-Hungary supported their allies in World War I and how Britain and the United States cooperated in the air war against Germany during World War II, whereas the Soviet Union did not.

David Lake, in "Fair Fights? Evaluating Theories of Democracy and Victory," responds to Desch's criticisms of the empirical support for the democratic victory proposition. Lake argues that Desch finds no statistically significant relationship between democracy and victory because he excludes so many cases and thus analyzes a very small *n*. He also rejects Desch's arguments that wars with gross mismatches and asymmetric interests should be excluded, because Lake's theory predicts that democracies will be more powerful and more motivated in wars against autocracies. Lake argues that Desch should not assume that more powerful states always win wars. Instead, wars between states with very different amounts of aggregate power should be included in the analysis so that the relationship between power and victory also can be tested. Lake points out that "Powerful Pacifists" was written to test his theory of the rent-seeking state and that the propensity for democratic victory was one testable prediction of that theory. The theory does not claim that democracy is the most important determinant of victory in war, only that it has a positive effect. He notes that his subsequent research has generated additional support for the theory, including the finding that democracies have higher rates of economic growth because they are better at creating human capital. Lake concludes that "theory and evidence combine to support strongly the hypothesis that democracies will tend to win the wars they fight and for the reasons I suggest."

In "Understanding Victory: Why Political Institutions Matter," Dan Reiter and Allan Stam claim that Desch makes "a crucial error in logic and research design" when he excludes what he regards as "unfair fights" from the data set used to test the democratic victory proposition. Reiter and Stam argue that the cases involving wars that pit powerful democracies against weaker nondemocracies are precisely the type of cases that their theory predicts: wars in which democracies choose to fight weaker opponents. The existence of such cases, they maintain, "proves our selection effects theory that democracies seek out gross mismatches and avoid conflicts where their chances of victory are lower."

Reiter and Stam also rebut Desch's other criticisms. They agree that disaggregation of the data to separate large wars into multiple conflicts might bias the tests of democratic victory theory, but most of the cases created via disaggregation oppose the theory instead of supporting it. In response to Desch's claim that democracies often win with the support of powerful nondemocratic allies, they contend that they controlled for alliance contributions and found that democratic countries with powerful allies are more likely to win than other regime types. They stand by their coding decisions, which were made on the basis of whether a country achieved its immediate political aims in war. To rebut Desch's claim that asymmetric interests may account for some democratic victories, they cite Stam's earlier book, *Win, Lose, or Draw,* which found that asymmetric interests did not have a statistically significant effect on war outcomes.[7] Reiter and Stam note that Desch questions whether some states coded as democracies are actually democracies, but he does not provide support for this claim.

Although they reject Desch's decision to exclude what he regards as "unfair fights" from the data set, Reiter and Stam replicate his analysis using the truncated data set. They add a control variable for a measure of military-industrial capabilities and find that democracy has a statistically significant effect on war outcomes. Reiter and Stam thus argue that "even if one accepts all of Desch's critiques about the composition of the data set, his finding of no relationship between democracy and victory evaporates with the most basic improvement of his statistical model."

Furthermore, Reiter and Stam point out that their work includes case studies that help to establish that democracy caused war outcomes. France's relatively poor performance in war reflects the fact that France was sometimes an autocratic state after 1815. They defend the statistical basis for their claim that democracy has a significant effect on war outcomes. In response to Desch's claim that Israel has not always made wise strategic decisions, they point out that Israel has won wars and survived in a hostile environment, whereas its nondemocratic neighbors have a much worse security record. After reiterating their argument that democratic armies fight with a high level of tactical effectiveness, they acknowledge that the HERO data set is imperfect and that further research is necessary.

Michael Desch replies to the criticisms of Choi, Lake, and Reiter and Stam in "Democracy and Victory: Fair Fights or Food Fights?" He notes that he does not deny that democracy plays some role in determining war outcomes, but "it appears to be modest compared with other factors." He indicates that he agrees with the democratic triumphalists on some points. There is a striking tendency for democracies to win most of their wars. He agrees that democracy definitely is not an obstacle to success in war. Both sides in the debate also agree that many factors explain victory in war.

Desch believes that there are two important disagreements between him and the democratic triumphalists: "(1) how scholars should gauge democracy's

---

7. Allan C. Stam III, *Win, Lose, or Draw: Domestic Politics and the Crucible of War* (Ann Arbor: University of Michigan Press, 1996), pp. 114–115.

role in explaining victory; and (2) what makes democracies more likely to select winnable wars and fight more effectively." He argues that the triumphalists have yet to provide a "convincing test of their hypothesis as well as a
compelling set of causal mechanisms."

In discussing the first disagreement, which is about how to test the relationship between democracy and war outcomes, Desch emphasizes that statistical
significance is not the same as practical significance. He reiterates his argument
that democracy appears to be a less significant cause of victory than many
other factors and reports that he found no significant relationship between war
initiation by democracies and victory when he ran Reiter and Stam's model of
the selection effects argument using his smaller data set. Desch defends his decision to exclude cases that were not "fair fights" on the grounds that "dropping cases where the winner had an overwhelming power advantage is akin to
running a controlled experiment in the natural sciences." He responds to Lake
by contending that it is impossible to test Lake's theory against the alternative
explanation that states win when they have a preponderance of power and by
pointing out that Lake acknowledges that the relationship between wealth and
democracy may be spurious.

With regard to the second disagreement, which revolves around the causal
mechanisms that may or may not explain why democracies win wars, Desch
argues that it is questionable whether there is enough data to conclude that the
selection effects mechanism explains the pattern of outcomes. Since 1815, there
have been only 16 cases of democracies starting wars and half of those cases involve Britain, Israel, and the United States. The coding of 6 of these cases is
questionable, in Desch's view. He suggests that the selection effects mechanism seems to be at work in only 5 of the remaining 10 cases. Turning to the
wartime effectiveness mechanism, Desch points out that Reiter and Stam agree
that the evidence that soldiers in the armies of democracies fight better is problematic. He notes that Choi presents a critique of the audience costs explanation of why democratic alliances are more powerful, and that Reiter and Stam
provide a critique of Lake's argument that democracies make better allies.
Desch acknowledges that Choi presents a new theory that posits that transparency and checks and balances enable democracies to form cohesive and reliable alignments, but he raises doubts about whether these facts actually make
democracies better allies. Transparency, for example, can confuse other states
by providing too much information. Desch again notes that there have been
few all-democratic alliances, which is puzzling if democracies actually make
superior alliance partners.

Alexander Downes presents another challenge to the democratic victory proposition in "How Smart and Tough Are Democracies? Reassessing Theories of Democratic Victory in War." He notes that Reiter and Stam rely on the selection effects and war-fighting arguments (discussed above) to explain the apparent tendency of democracies to win most of their wars. After explicating those two explanations and summarizing many critiques of the theory of democratic victory, Downes offers his own critique.

Downes reanalyzes the quantitative data that Reiter and Stam use to support their arguments. Disputing Reiter and Stam's classification of states involved in wars as either initiators or targets (a category that includes states that join the initiators' side), he contends that states engaged in war should be divided into three categories: initiators, targets, and joiners. Downes also faults Reiter and Stam for excluding wars that ended in draws from their analysis, which includes only wars that ended in victory or defeat. In his view, draws should be included because a costly stalemate in war might make the public in a democracy more likely to vote a leader out of office. After making these changes to the analysis and data, he finds that democracies "are not significantly more likely to win wars." It turns out that "increasing democracy makes victory more likely and lowers the probability of defeat," but the effect is not statistically significant. Downes concludes that his reanalysis of the data "raises doubts about the robustness of the finding that democracies are better at choosing and fighting wars."

Downes also offers an analysis of the United States' decision to start bombing North Vietnam in 1965 and to commit substantial U.S. forces to the ground war in South Vietnam later that year. He notes that Vietnam is an anomalous case because the United States initiated a war that ended in either a defeat or a costly draw. U.S. leaders also understood that the prospects for victory were slim when they made the key 1965 decisions to initiate the bombing of North Vietnam and send ground troops to South Vietnam. Their actions are at odds with the selection effects argument that the need for public support will induce democratic leaders to choose winnable wars.

On the basis of his analysis, Downes suggests that democratic politics may explain why President Lyndon Johnson chose to fight the Vietnam War even though the prospects for success seemed remote. Johnson believed that refusing to fight in Vietnam would jeopardize his domestic agenda—the Great Society programs. Downes writes, "He worked assiduously to avoid a fractious debate between guns and butter by escalating the United States' military role in Vietnam quietly and gradually."

Downes concludes by noting that the U.S. war in Iraq that began in 2003 reveals another mechanism by which domestic politics may lead a democracy into a costly quagmire. To ensure that there was support for going to war, the George W. Bush administration concealed the likely true cost of the war—particularly the cost of occupying, governing, and rebuilding Iraq. He recommends further research on historical cases in which democracies initiated wars despite leaders' doubts that victory would be easy or likely.

In their exchange "Another Skirmish in the Battle over Democracies and War," Dan Reiter and Allan Stam respond to the critique of Alexander Downes, and Downes replies. Reiter and Stam first note that even when draws are included, democracies win most of their wars and are more likely to win than mixed regimes or autocracies. Democracies are particularly likely to win the wars they initiate. Reiter and Stam offer two criticisms of Downes's analysis. First, they contend that Downes did not take into account their previous argument that mixed regimes should fare worst of all in war and that there is thus a curvilinear relationship between regime type and war outcomes. Second, they defend their decision to exclude draws from their analysis by pointing out that there is a much more complicated relationship between regime type and draws. The willingness of democracies to settle for draws in wars changes with the duration of the war. Reiter and Stam claim that Downes's research design is not appropriate for analyzing draws: a research design that accounts for the length of a war is required. Reiter and Stam reanalyze the data after correcting for the two errors they claim that Downes made. They find that there is still a statistically significant relationship between democracy and victory in wars initiated by democracies.

Downes replies that the curvilinear hypothesis is secondary to the central argument that democracies are more likely to win their wars. In any case, he finds no significant evidence for the hypothesis. He also reiterates that draws should be included in the analysis because one would expect prudent and forward-looking democracies to avoid them. He notes that in the Vietnam War the United States knowingly entered a war that it doubted it could win; it did not simply "guess wrong" and settle for a draw as the war dragged on. He also notes that draws are generally perceived as failures, so it is unclear why leaders of democracies would be willing to settle for draws and thus risk being removed from office.

The final set of contributions to this volume features a debate over the selection effects argument, which holds that democracies win their wars because the need to maintain public support motivates them to select wars that they can win.

In "The Deception Dividend: FDR's Undeclared War," John Schuessler criticizes the selection effects argument by specifying the conditions under which it breaks down. Schuessler posits that democracies sometimes must fight wars that they cannot win easily. He argues that the mechanisms behind the selection effects argument may break down under some international conditions. For realist reasons, democracies may have no choice but to go to war against a powerful adversary. Schuessler writes, "Democracies are not always free to pick on weak and vulnerable opponents but must, from time to time, take on formidable foes because their territorial integrity or political independence would be threatened otherwise." For example, it may be in the national interest of a democracy to launch a preventive war against another country that might threaten it in the future, even if it is hard to secure public support for such a war. Democracies also may have trouble obtaining public support for military interventions in distant areas.

In cases in which war against a challenging adversary seems necessary, democratic leaders may circumvent the process of obtaining public consent by deceiving the public about their policies and plans for war. When significant dissent seems likely and leaders want to avoid a contentious debate, they may attempt to blame the adversary for the onset of hostilities. First, they will conceal their preparations for war. This may include not revealing the existence of any binding alliance commitments, negotiating in bad faith, or engaging in a piecemeal military buildup. Second, they will wait for a crisis sufficient to justify escalation. The decision for war will precede the crisis, but the crisis will inflame public opinion and make war possible. Such crises create a climate of emergency in which public consent will emerge.

Schuessler uses the events surrounding U.S. entry into World War II to conduct a plausibility probe of his argument. He claims that "President Roosevelt welcomed U.S. entry into war by the fall of 1941 and attempted to manufacture events accordingly." The United States recognized that Nazi Germany posed a threat to U.S. national security. As Germany overran France and threatened Britain, U.S. political and military leaders realized that Germany could not be defeated without U.S. involvement in the war. Roosevelt recognized, however, that it would be hard to persuade the U.S. public to support a major land war in Europe. In 1941, the Neutrality Acts still enjoyed strong support in Congress, and polls revealed little support for a declaration of war. The undeclared war against Germany in the Atlantic and the oil embargo on Japan were designed to provoke an incident that would facilitate U.S. entry into World War II. In the Atlantic, U.S. naval patrols led to clashes with German U-boats. Although Schuessler rejects the charge that Roosevelt knew the Japanese would attack Pearl Harbor and deliberately left the U.S. fleet vulnerable, he argues that Roosevelt saw the Pacific as a "back door" to war in Europe. Roosevelt approved the oil embargo against Japan and made demands for Japanese withdrawal from China knowing—and perhaps hoping—that these policies would induce Japan to strike first. When Japan attacked U.S. (as well as British and Dutch) forces and possessions in the Pacific, Roosevelt was relieved. It was U.S. policy to declare war on Germany in the event of a Japanese attack, but Germany removed any need for such a declaration by declaring war on the United States first. Roosevelt eased the United States into war without a public debate and with strong public support. In short, deception worked.

In their correspondence "FDR, U.S. Entry into World War II, and Selection Effects Theory," Dan Reiter and John Schuessler debate whether the World War II case undermines the selection effects explanation for why democracies win most of their wars. Reiter makes three arguments. First, the case does not undermine the broader democratic victory argument, because the United States emerged victorious from World War II. The United States ultimately chose a war that it could win. Second, there is some historical evidence that Roosevelt was not trying to provoke war. The United States avoided some naval incidents with Germany. There is no clear evidence that Roosevelt actually wanted Japan to attack the United States. He argued against a total oil embargo on Japan and negotiated with Japan in good faith. Third, the case shows the importance of public opinion in a democracy. It supports selection effects theory because Roosevelt was acutely aware of the need for public support. Moreover, there actually was public support for aggressive U.S. naval actions in the North Atlantic and a hard-line policy against Japan.

Schuessler responds to each of Reiter's three arguments. First, he reiterates that he was not trying to discredit the selection effects argument, only to indicate the conditions under which it breaks down. The case of U.S. entry into World War II reveals that events did not unfold in accordance with the causal logic of the selection effects argument (i.e., democratic leaders deceived the public instead of being constrained by public opinion), even if the ultimate outcome—U.S. victory in World War II—is consistent with the selection effects argument. Second, Schuessler claims that Roosevelt's policies remained well ahead of public opinion. Roosevelt recognized that the United States would have to enter the war and wrote of the need to provoke Germany to attack U.S.

forces. What Reiter regards as attempts to avoid incidents actually were attempts to avoid endangering convoys by sending them into the path of German submarines. In the Pacific, Roosevelt avoided giving Japan an ultimatum when the oil embargo was imposed because doing so would attract unwarranted attention and make the ultimatum unproductive. Third, Schuessler recognizes that Roosevelt's hard-line policy toward Japan was popular, but he argues that the public was not attentive to events in Asia and definitely was not ready to go to war over China.

The essays in this volume do not resolve the debate over democratic victory. In a debate that will probably continue for many years, there has been a remarkable and reassuring convergence thus far on what the key issues are. On both sides, scholars have focused on the overall record of war outcomes and a similar set of potential explanations for democratic victory. The debate has unfolded clearly and coherently with head-to-head clashes over research designs, data, and logic. The two sides may not agree on every issue, but they are not talking past each other.

There are many areas for potential further research by proponents and critics of the democratic victory proposition. Although many studies have analyzed the overall record of apparent democratic success in modern wars, further quantitative analysis might resolve some of the questions about the statistical and practical significance of democracy as a cause of victory by shedding further light on whether and how the militaries of democracies fight more effectively. As several contributions to this volume recognize, the existing data sets on this question are inadequate. In addition, most research on this topic has focused on ground forces. Democracies often deploy highly effective air forces and navies. The relationship between democracy and the effectiveness of those forces merits further study. The existing research on the selection effects mechanism—including many of the contributions to this volume—should be complemented by much more research on the military effectiveness mechanism.

Historical case studies also might contribute to the debate over the democratic victory proposition. Such studies could trace the process of causation to determine whether democracy explains the propensity of democracies to win most of their wars. Research could determine, for example, if open debate on policy options in democracies actually produces better strategic decisions. Case studies also might identify the conditions under which the explanations for democratic victory do not apply or break down, as John Schuessler demonstrates in his contribution to this volume.

Part I:
Victorious Democracies:
The Case for a Democratic Advantage

Search for Victory
Why Democracies Win Their Wars

Dan Reiter and
Allan C. Stam

The twentieth century ended with near consensus among political leaders, populations, and academics alike on the virtues of democracy. Many base their devotion to democracy on the belief that it brings at least three important individual and collective virtues: freedom, prosperity, and peace. International relations scholars have been particularly interested in the last of these virtues, proposing that democracies almost never fight each other.[1]

Through these virtues, democracies seem to offer an elegant and just solution to the vagaries of the human condition, the perfect recipe for the organization of society, and even, in the words of one observer, "the end of history."[2] But what if, in their dealings with other nations, democracies prove vulnerable to predation? Is democracy a luxury that states can only afford during times of peace? Observers ranging from Alexis de Tocqueville to Gen. William T. Sherman to Abraham Lincoln to John F. Kennedy have expressed doubts about the abilities of democracies to conduct foreign policy and to protect themselves by winning their wars.[3] Conversely, do democracy's skeptics have it wrong? Do the attributes associated with democratic institutions, in the worst of times, also allow states to provide for their national security?

Largely underappreciated by scholars and political observers alike has been a fourth virtue of democracy: democracies win interstate wars. Since 1815, democracies have won more than three quarters of the wars in which they have participated. This is cause for cheer among advocates of democracy. It would appear that citizens of democratic nations not only enjoy the individual

---

Portions of this article originally appeared in Dan Reiter and Allan C. Stam, *Democracies at War* (Princeton, N.J.: Princeton University Press, 2002). This article is reprinted with permission from Princeton University Press.

---

1. Bruce Russett, *Grasping the Democratic Peace: Principles for a Post–Cold War World* (Princeton, N.J.: Princeton University Press, 1993).

2. Francis Fukuyama, *The End of History and the Last Man* (New York: Avon, 1992).

3. *The Federalist Papers*, No. 8, ed. Garry Wills (New York: Bantam, 1982); Alexis de Tocqueville, *Democracy in America*, trans. George Lawrence (New York: Harper Perennial, 1969), p. 228; and Sherman, quoted in Gerald F. Linderman, *Embattled Courage: The Experience of Combat in the American Civil War* (New York: Free Press, 1987), pp. 36–37. Secretary of State Dean Acheson agreed: "In the conduct of their foreign relations democracies appear to me decidedly inferior to other governments." Quoted in Michael D. Pearlman, *Warmaking and American Democracy: The Struggles over Military Strategy, 1700 to the Present* (Lawrence: University Press of Kansas, 1999), p. 10. See also Zbigniew Brzezinski, "War and Foreign Policy: American-Style," *Journal of Democracy*, Vol. 11, No. 1 (January 2000), pp. 172–178; Hans J. Morgenthau, *Politics Among Nations: The Struggle for Power and Peace*, 5th rev. ed. (New York: Alfred A. Knopf, 1978), p. 153; and Jean-François Revel, *How Democracies Perish*, trans. William Byron (Garden City, N.Y.: Doubleday, 1984).

and collective benefits of peace, prosperity, and freedom; they can also defend themselves against outside threats from autocrats and oligarchs.

In this chapter, we explore why democracies win wars, summarizing the arguments and evidence we presented in our 2002 book, *Democracies at War*.[4] We will show why history has repeatedly proven the pessimists wrong. We also discuss how it is that the nation-states most capable of safeguarding freedom also exhibit prowess on the battlefield by, paradoxically to some, putting governance in the hands of the people.

Our central argument is that democracies win wars because elected leaders are accountable to voters. The notion that those who govern should be accountable to the consent of the people is at the core of all democratic systems. In democracies, leaders who act without the consent of the voters do so at considerable political risk. This commitment to consent, contrary to the negative declarations of observers such as Tocqueville, George Kennan, and Walter Lippmann, offers democracies a set of peculiar advantages that enable them to prevail in war.

We outline two explanations of why democracies win their wars. The first explanation, the "selection effects" explanation, proposes that democracies only start wars when their chances of winning are very high. Elected leaders know that there are few greater political disasters than wasting the lives of their citizens in a losing cause, and starting a war that goes poorly will curtail their ability to pursue their political agenda and increase their chances of being removed from office. The explicit threat of electoral punishment and the need to generate consent of the governed at the time of action pushes democratic leaders to be particularly cautious when starting wars, and typically, to start only those wars which they will go on to win. Although Tocqueville and

---

4. Dan Reiter and Allan C. Stam, *Democracies at War* (Princeton, N.J.: Princeton University Press, 2002). Much of this essay is excerpted from that book, and in some places we have slightly updated the discussions and citations, though we have shied away from reviewing the voluminous post-2002 literature on the relationship between democracy and war outcomes. Our contributions to the post-2002 literature on this question include Dan Reiter and Allan C. Stam, "Understanding Victory: Why Political Institutions Matter," *International Security*, Vol. 28, No. 1 (Summer 2003), pp. 168–179; Dan Reiter and Allan C. Stam, "Correspondence: Another Skirmish in the Battle over Democracies and War," *International Security*, Vol. 34, No. 2 (Fall 2009), pp. 194–200; Dan Reiter, "Correspondence: FDR, U.S. Entry into World War II, and Selection Effects Theory," *International Security*, Vol. 35, No. 2 (Fall 2010), pp. 176–185; Allan C. Stam and Dan Reiter, "Democracy and War Outcomes: Extending the Debate," paper presented at the annual meeting of the International Studies Association, Chicago, Illinois, March 2, 2007; and Dan Reiter, "A Closer Look at Case Studies on Democracy, Selection Effects, and Victory," unpublished manuscript, Emory University, June 21, 2010.

others feared this caution would paralyze democratic leaders, we demonstrate the opposite. Democratic leaders are prepared to use military force, but, except when facing total conquest, unlike their autocratic counterparts they are unwilling to risk decisive defeat. Beyond consent, because democratic societies enjoy greater freedom of speech and press as well as healthier civil-military relations, there is more thorough discussion of the likely cost and outcome of a potential war before fighting begins, both in and out of government. A freer and more robust marketplace of ideas means that elected leaders will enjoy more accurate estimates of their chances of winning, further enabling them to avoid military disasters. We present this argument here, and then also consider the implications of this theory for why democracies initiate wars, and why democracies tend to fight shorter wars.

A second advantage, which we term the warfighting explanation, emerges from the effects of liberalism on soldiers fighting on the battlefield itself. What kinds of soldiers, sailors, and airmen might we expect a society based on popular consent to produce? General Sherman worried that the fighting men of liberal societies stand to be defeated, as most men, if given the choice, will resist the rigors of military discipline necessary for victory on the battlefield. We turn Sherman's worry inside out: the soldiers produced by consent-based societies will in fact enjoy certain advantages. Specifically, the emphasis on the individual and their concomitant rights and privileges in democratic societies commonly produces better leaders and fighters more willing to take the initiative on the battlefield. Rather than empowering individuals at the expense of the collective, democratic institutions are associated with states filled with individuals more capable of serving the state's needs in times of duress.

In the remainder of this chapter, we develop the logic and some of the evidence supporting these two explanations. We also consider two other possible reasons why democracies win the wars they fight, that democracies win because they enjoy more powerful alliances and greater military-industrial power. We propose that the historical evidence generally does not support either of these explanations of democratic victory.

Democracy, Initiation, and Victory

States pick their fights deliberately, and war begins when at least one state chooses to attack another. In social science terms, we mean that war initiators select themselves into the population of war participants. Before attacking, a state's leaders make some guess as to the potential war's likely outcome.

This estimate of the chances of winning can approach 0 percent if the leader thinks his side has essentially no chance to win or approach 100 percent if he is almost certain of victory. Sometimes leaders are quite precise in these calculations. During the 1911 Morocco Crisis between France and Germany, the high-ranking French civilian leader Joseph Caillaux asked French Marshall Joseph Joffre point blank: "General, it is said that Napoleon did not give battle except when he thought his chances of success were 70 out of 100. Do we have a 70 in 100 chance of victory if the situation forces us into war?" When Marshall Joffre replied in the negative, Caillaux stated, "in that case we will negotiate."[5] The U.S. government during the Vietnam War brought quantification of wartime prospects to new heights, making very precise calculations of the chances of different outcomes given different policy inputs. An infamous 1965 policy memo by U.S. Undersecretary of State John McNaughton laid out subjective estimates of U.S. chances of success and collapse in Vietnam over the next three years for a variety of policy options. McNaughton estimated, for example, that the United States had a 70 percent chance of victory by 1968 if the United States escalated its ground commitment to several hundred thousand troops.[6]

When a leader considers launching a war, she will do so if her estimate of the chances of winning is high enough. Each leader has in mind a minimum acceptable threshold of victory; if her estimated chance of victory is above this threshold, she will order an attack, and if it is below, she will not. Of course, the level of acceptable risk will vary from state to state and leader to leader. A real gambler might be willing to attack if she thinks there is only a 55 percent chance of victory, whereas a conservative leader averse to risk might attack only if he thinks there is at least a 90 percent chance of victory. All else being equal, we expect that as a leader's estimate of his state's chances of winning goes up, he will become more likely to approve an attack.[7] If states consider whether or not they will win a war when they make the decision to initiate the use of force, then understanding the decision to attack sheds light on who wins wars. More conservative, risk-avoidant states will win more of the wars they start, as they will only attack when they are very confident they will win.

---

5. Joffre, quoted in David G. Herrmann, *The Arming of Europe and the Making of the First World War* (Princeton, N.J. Princeton University Press, 1996), pp. 152–153.
6. Larry Berman, *Planning a Tragedy: The Americanization of the War in Vietnam* (New York: Norton, 1982), p. 140.
7. The model we present here is conceptually similar to that of Scott Sigmund Gartner and Randolph M. Siverson, "War Expansion and War Outcome," *Journal of Conflict Resolution*, Vol. 40, No. 1 (March 1996), pp. 4–15.

Such states are like a boxing champion who fights only weak opponents to safeguard his grip on his title.

Elected leaders tend to be risk-avoidant actors due to the expected political costs of anticipated or actual failure. Typically, they avoid starting high-risk wars, and as a result are particularly likely to win the wars they initiate. The essence of democracy is popular control of the government. One way or another, the leadership of a democracy must answer to its people, usually through elections. Undemocratic governments, on the other hand, need not hold regular, competitive elections and are not ultimately answerable to the will of the people. Displeased publics are far more likely to oust democratically elected governments than nonelected governments. We may become cynical of contemporary leaders obsessed with public opinion polls, but the undistracted focus of a democratic leader on what the public thinks of his or her performance is the central spirit of democracy, government as an expression of the will of the people. Even Henry Kissinger, one of the twentieth century's most influential realist figures, recognized that democratic publics punish leaders who incur foreign policy disasters.[8]

One of the gravest policy failures a nation can confront is defeat in war. Defeat damages the pride of the nation, needlessly expends its human and financial resources, and may endanger the very survival of the nation. As one historian put it, societies demand cheap victory: "The American public has wanted only one thing from its commanders in chief: quick wars for substantial victories with minimal costs."[9] Governments that lead their nations into unsuccessful wars are especially likely to confront an angered citizenry. Democratic governments have much to fear from an angered public, as their hold on power is particularly dependent on the continuing pleasure of the people. U.S. President George H.W. Bush recognized this on the eve of the Gulf War in 1990, stating the possibility in the bluntest terms: "I'll prevail or I'll be impeached."[10] Repressive governments, less vulnerable to the displeasure of their peoples, are less likely to be myopically concerned with defeat. They need not face the public in elections, and they can violently repress opposition if the need arises. The historical evidence bears out these expectations linking regime type, war outcomes, and exit from power.[11]

---

8. Henry Kissinger, *Years of Renewal* (New York: Simon and Schuster, 1999), p. 1068.
9. Pearlman, *Warmaking and American Democracy*, p. 13.
10. Bush, quoted in Colin L. Powell, with Joseph E. Persico, *My American Journey* (New York: Random House, 1995), p. 499.
11. Bruce Bueno de Mesquita and Randolph M. Siverson, "War and the Survival of Political Leaders: A Comparative Study of Regime Types and Political Accountability," *American Politi-*

Because democratic executives know they risk ouster at the worst or policy-making impotence at best if they lead their state to defeat, they are especially unwilling to launch risky military ventures, and will require a greater confidence in victory before launching war. Democratic France was unwilling to risk war with Germany in 1911 with a less than 70 percent chance of victory. This was not an isolated incident. France's leaders had similarly backed down in the 1898 Fashoda Crisis with Britain when they realized the considerable military inferiority France faced on the ground in Egypt and at sea against the British navy.[12] The prediction that then follows the selection effects explanation is that democracies are especially likely to win wars that they initiate. This does not imply that they win because they are necessarily more powerful, rather that they more consistently avoid wars they would have gone on to lose had they actually fought them.

Conversely, autocratic leaders are more willing to initiate high-risk wars, as they are more likely to retain political power even if the war turns badly. Saddam Hussein initiated a disastrous and bloody stalemate against Iran in the 1980s, blundered into one of the greatest military defeats in modern history against the United States in the 1990s, and yet remained the leader of Iraq. Had he been a democratically elected leader, he most certainly would have thought more carefully about accepting such grave gambles.

Several strands of empirical evidence support the assumptions of the selection effects explanation. Scholarship portrays American public opinion as essentially stable, rational, and prudent, meaning the public can be trusted to assess the risks, dangers, and potential benefits of war.[13] If democratic leaders were insensitive to public support and opposition, we might expect that in some wars, we would see significant opposition to the effort, and sometimes not. In reality, however, democratic governments have enjoyed a lack of political opposition at the outset of the wars they have initiated.[14] Except in cases of

cal Science Review, Vol. 89, No. 4 (December 1995), pp. 841–855; and H.E. Goemans, "Which Way Out? The Manner and Consequences of Losing Office," *Journal of Conflict Resolution,* Vol. 52, No. 6 (December 2008), pp. 771–794.

12. Darrell Bates, *The Fashoda Incident of 1898: Encounter on the Nile* (London: Oxford University Press, 1984), pp. 151–159.

13. Benjamin I. Page and Robert Y. Shapiro, *The Rational Public: Fifty Years of Trends in Americans' Policy* (Chicago: University of Chicago Press, 1992); Bruce W. Jentleson, "The Pretty Prudent Public: Post Post-Vietnam American Opinion on the Use of Military Force," *International Studies Quarterly,* Vol. 36, No. 1 (March 1992), pp. 49–73; and Bruce W. Jentleson and Rebecca L. Britton, "Still Pretty Prudent: Post–Cold War American Public Opinion on the Use of Military Force," *Journal of Conflict Resolution,* Vol. 42, No. 4 (August 1998), pp. 395–417.

14. Gad Barzilai, "War, Democracy, and Internal Conflict: Israel in a Comparative Perspective," *Comparative Politics,* Vol. 31, No. 3 (April 1999), p. 318.

existential wars, democracies are casualty sensitive, as they adopt military and foreign policies that enable them to suffer significantly fewer civilian and military casualties when they do fight wars.[15] Support for war in democracies slides as casualties mount and the prospects for victory fade.[16] Notably, potential opponents of democracies have tried to exploit democratic casualty sensitivity, for example trying to bluff democracies into staying out of war by threatening high casualties. In July 1990, before the invasion of Kuwait, Saddam Hussein brusquely told U.S. Ambassador April Glaspie, "Yours is a society which cannot accept ten thousand dead in one battle."[17]

Beyond the dependence of elected leaders on public consent, there is a second factor that produces the selection effect: democracies produce better estimates of the probability of victory than do their autocratic counterparts. That is, the estimates of winning that democratic institutions produce are more accurate representations of their actual probabilities of victory than the information provided to authoritarian leaders. Not only do democracies start wars they believe they will win, but also their beliefs about outcomes are less biased than are the outcome estimates produced by autocracies.

How is it that democracies are better at forecasting war outcomes and associated costs? Democratic governments receive more and higher quality information, both from within their own bureaucracies, as well as from public debate associated with open opposition parties and a free press. As a result they are more likely to make better policy choices and typically only initiate winnable wars. The proposition that the vigorous discussion of alternatives and open dissemination of information in democratic systems produce better decisions is an idea at the core of political liberalism, traceable to thinkers such as John Milton, Thomas Jefferson, and John Stuart Mill. Unfettered opposition parties and a free press both work to expose flaws in produced policies, generating vigorous debate. Further, military officers in democracies are more likely

---

15. Benjamin A. Valentino, Paul K. Huth, and Sarah E. Croco, "Bear Any Burden? How Democracies Minimize the Costs of War," Journal of Politics, Vol. 72, No. 2 (April 2010), pp. 528–544. See also Michael Horowitz, Erin Simpson, and Allan Stam, "Domestic Institutions and Wartime Casualties," *International Studies Quarterly*, Vol. 55, No. 4 (December 2011).

16. John E. Mueller, *War, Presidents and Public Opinion* (Lanham, Md.: University Press of America, 1985); Scott Sigmund Gartner and Gary Segura, "War, Casualties, and Public Opinion," *Journal of Conflict Resolution*, Vol. 42, No. 3 (June 1998), pp. 278–300; Scott Sigmund Gartner and Gary Segura, "Race, Casualties, and Opinion in the Vietnam War," *Journal of Politics*, Vol. 62, No. 1 (February 2000), pp. 115–146; and Christopher Gelpi, Peter D. Feaver, and Jason Reifler, *Paying the Human Costs of War: American Public Opinion and Casualties in Military Conflicts* (Princeton, N.J.: Princeton University Press, 2009).

17. Janice Gross Stein, "Deterrence and Compellence in the Gulf, 1990–91: A Failed or Impossible Talk?" *International Security*, Vol. 17, No. 2 (Fall 1992), p. 175.

to be promoted on the basis of merit, whereas in autocracies military officers are more likely to be promoted on the basis of political reliability, so as to reduce the risks of coups d'état. Elected leaders will enjoy higher quality advice from superior military minds, whereas unelected leaders will hear what they want to hear from toadies and yes-men. Fearful military subordinates probably prevented Saddam Hussein from getting an accurate picture of American military power in 1990, which in turn led Iraq into an utterly disastrous military confrontation.[18] Armies of democracies are also more likely to be focused solely on external defense whereas the armies of autocratic states often are more concerned with maintaining domestic order and stability.[19]

So far, our discussion has framed democracy in a simple, rather one-dimensional fashion: states are either more or less democratic. An alternative approach is to relax this assumption and enrich our typology of states. Consider instead three different kinds of states: democracies (such as the United States), mixed (i.e., oligarchic or cartelized) regimes that share democratic and autocratic characteristics (such as Wilhelmine Germany), and unitary dictatorships that are highly repressive and undemocratic (such as Stalinist Russia or Saddam Hussein's Iraq).

Although these three kinds of states can be placed on a spectrum with democratic states being the freest, mixed regimes being less free, and dictatorships being the least free, regime type does not necessarily vary linearly with foreign policy behavior. One scholar predicted that democracies are least likely to experience imperial overexpansion, dictatorships are somewhat more likely to overexpand, and mixed regimes are the most likely to overexpand. He proposed that mixed regimes are especially susceptible to logrolling coalitions. Rather than trying to settle on the single best policy, political opponents join together to produce a single overambitious policy that to the detriment of the national interest offers something for everyone. This tendency toward logrolling pushes mixed regimes into foolish military and colonial ventures. Such systems are also more likely to fall prey to imperial mythmaking that makes expansion seem falsely appealing. The unitary nature of dictatorships makes them less likely to fall prey to logrolling or mythmaking, but foregoes democratic advan-

18. Kenneth M. Pollack, "The Influence of Arab Culture on Arab Military Effectiveness," Ph.D. dissertation, Massachusetts Institute of Technology, 1996, chap. 3; and Barry M. Blechman and Tamara Cofman Wittes, "Defining Moment: The Threat and Use of Force in American Foreign Policy," *Political Science Quarterly,* Vol. 114, No. 1 (Spring 1999), p. 14.
19. Stanislav Andreski, *Military Organization and Society* (Berkeley: University of California Press, 1968); and Stanislav Andreski, "On the Peaceful Dispositions of Military Dictatorships," *Journal of Strategic Studies,* Vol. 3, No. 3 (December 1980), pp. 3–10.

tages of the marketplace of ideas which provide broad checks on a single leader.[20]

These arguments indicate that among initiators, mixed regimes are least likely to win, followed by dictatorships, followed by the relatively more victory-prone democracies. Although the leaders of both mixed regimes and dictatorships face relatively less risk of a domestic political threat following defeat in war than do democracies, mixed regimes are more likely to suffer from imperial mythmaking and overestimate their chances of victory.[21] In our model, mixed regimes and totalitarian states are more risk acceptant than liberal democracies, hence both are willing to start wars with lower estimates of their chances for victory, but mixed, or oligarchic, regimes make the least accurate estimates of their chances of victory.[22]

If the selection effects explanation is correct, then democratic war initiators ought to be more likely to win their wars than other kinds of war initiators. Further, highly repressive initiators ought to be more likely to win than mixed regime initiators. To test the selection effects explanation, we look at all wars from 1816–1985, a war being defined as a military clash between two countries in which there are at least 1,000 battle casualties.[23] Table 1 sorts each war participant according to whether it was a democracy, a mixed regime, or a dictatorship, whether or not it initiated the war, and whether it won or lost.[24] Consistent with the notion that states consider carefully their chances of victory before starting a war, initiators do better than targets, winning 65 percent of the time compared to 41 percent for targets. Of the three types of states in our typology, democratic initiators do best, winning 93 percent of the time compared to dictatorial initiators winning 60 percent of the time, and mixed regime initiators winning 58 percent of the time. Just as democracies almost never fight each other, when they do fight, democracies almost never start wars they go on to lose. In more sophisticated tests controlling for factors such as military-industrial capabilities, terrain, strategy, troop quality, allies, and distance, the central result holds: democratic initiators are significantly more

20. Jack Snyder, *Myths of Empire: Domestic Politics and International Ambition* (Ithaca, N.Y.: Cornell University Press, 1991).
21. Ibid.
22. See also H.E. Goemans, *War and Punishment: The Causes of War Termination and the First World War* (Princeton, N.J.: Princeton University Press, 2000).
23. Because of space constraints, in this chapter we can present only brief summaries of the statistical tests discussed in full detail in *Democracies at War* and elsewhere.
24. Some wars, such as the Korean War, end in draws. Draws play a complicated role in our theory, as the theory predicts that democracies experience decisive outcomes in short wars but more often settle for draws in long wars. We explore draws in chapter 7 of *Democracies at War*.

Table 1. Winning Percentage for War Initiators and Targets by State Type

		Dictatorships	Mixed	Democracies	Total
War Initiators	Wins	21	21	14	56
	Losses	14	15	1	30
	Winning Percentage	60%	58%	93%	65%
Targets	Wins	16	18	12	46
	Losses	31	27	7	65
	Winning Percentage	34%	40%	63%	41%

likely to win their wars than other types of war initiators, and highly repressive initiators are more likely to win than oligarchic initiators.[25]

India's decision for war against in Pakistan in 1971 demonstrates how democratically induced caution helped set the stage for military victory. As conflict between East and West Pakistan escalated in 1971, India took several actions to build international support and bolster its chances for winning a war if one should come. It succeeded in persuading the World Bank to cut off economic assistance (minus humanitarian aid) to West Pakistan, convinced the United States to cut off military aid to West Pakistan, signed a bilateral treaty with the Soviet Union that gave India protection from Security Council censure if the United Nations decided to enter the conflict, and secured a steady supply of Soviet military supplies.[26]

While the Indian political leadership pursued diplomatic goals, it simultaneously trained and armed East Pakistani guerrilla forces to harry the West Pakistani authorities. Once the Indian leadership had achieved its diplomatic goals it simply waited for the proper military circumstances to escalate the conflict. On December 3, 1971, the Pakistanis attacked Indian Air Force bases near the West Pakistani–Indian border, providing the excuse the Indians needed to engage in open warfare with Pakistan.[27] In just a few weeks, the Indians decisively defeated the West Pakistanis using a combined-arms, maneuver-based military strategy, reaching the capital of East Pakistan on the eastern front, and thwarting Pakistani attacks on the western front.[28] India

---

25. See Reiter and Stam, *Democracies at War,* chap. 2.
26. Sumit Ganguly, *The Origins of War in South Asia: Indo-Pakistani Conflicts since 1947* (Boulder, Colo.: Westview, 1986), p. 120.
27. Ibid., p. 129.
28. John J. Mearsheimer, *Conventional Deterrence* (Ithaca, N.Y.: Cornell University Press, 1983).

was victorious over Pakistan, and East Pakistan achieved independence as Bangladesh.

The accouterments of democracy helped India win its war with Pakistan. Prime Minister Indira Gandhi wanted victory, not just war for its own sake. In April 1971, her chief of staff dismissed calls for immediate war, recognizing that the Indian Army was not yet ready for a two-front war (that is, simultaneously intervening in East Pakistan and parrying a likely counterattack launched from West Pakistan), and that India needed an additional six months to prepare for such a conflict. An autocracy might not have accepted such prudent delay.[29] Pausing several months before striking substantially increased India's chances for victory, as it allowed for the fighting to begin after the monsoon season ended in September, afforded the opportunity to train the Bangladeshi force Mukhti Bahini in guerrilla and conventional tactics enabling them to aid regular Indian forces, and permitted improvement in Indo-Soviet ties.[30] The Indian leadership's desire for a short and successful war pushed them to adopt a maneuver-oriented military strategy, which also helped them win. India also benefited from superior civil-military relations. During the war, Indian officers were more professional and more closely identified with the rank-and-file of the army, enabling them to manage the war more effectively. Conversely, the politicization of the military in Pakistan undermined its effectiveness.[31]

The attack of Japan, then a military oligarchy, on Pearl Harbor, Hawaii, on December 7, 1941, contrasts starkly with democratic India's careful consideration of war with Pakistan thirty years later. The Japanese leadership perceived itself to be in a dangerous bind in 1941, as its land war against China was bogging down and U.S. opposition to the war in China had led to a trade embargo that was becoming increasingly effective with the passage of time. The U.S. economic sanctions limited Japanese access to crucial raw materials, particularly oil, which was essential to Japan's continued campaign of imperial expansion. Japan saw an adventurous bid to capture the raw materials of

---

29. Compare the Indian behavior in 1971 to the Pakistanis' hasty response to India's 1998 nuclear weapons tests. Following Pakistan's quick tests of its own in response (a decision driven by pressure from the military), the United States was forced to cut off aid to Pakistan, doing substantial damage to the Pakistani economy. Being able to count on an impetuous response, India was able to weaken Pakistan without firing a single shot in against Pakistan. See Samina Ahmed, "Pakistan's Nuclear Weapons Program: Turning Points and Nuclear Choices," *International Security*, Vol. 23, No. 4 (Spring 1999), pp. 178–204.
30. Richard Sisson and Leo E. Rose, *War and Secession: Pakistan, India, and the Creation of Bangladesh* (Berkeley: University of California Press, 1990), especially pp. 198, 208–210.
31. See Ahmad Faruqui, "The Enigma of Military Rule in Pakistan," unpublished manuscript, Palo Alto, California, 2000.

Southeast Asia as one way to make its empire economically sustainable. It also recognized, however, that widening the war in Asia would bring Japan into conflict with the United States. Because the Japanese believed capturing Southeast Asia was essential to their nation's continued existence as an Asian great power, war with the United States was inevitable. Japan chose to launch a surprise attack against Hawaii, the Philippines, and elsewhere to maximize its chances for victory, recognizing beforehand that in the end their chances for victory in a general war against the United States were likely quite low.[32]

Japan's willingness to gamble aside, the absence of Japanese democratic institutions contributed to its poor estimate of its chances of victory. The hypernationalist military oligarchy in Japan drastically underestimated the American willingness to fight when attacked, and ignored the overwhelming latent power of the much larger American economy. An organized political opposition and a citizenry better informed about the low chances of victory against the United States and unafraid of government oppression might have spoken out against such a war. An elected Japanese government fearful of electoral backlash would have made the necessary concessions to avoid a catastrophic war. In the words of the Japanese historian Saburo Ienaga: "If the popular will had influenced national policies, the conflict might have been avoided or at least shortened. It was a vicious cycle: the weakness of democracy was one cause of the war, and the war further eroded freedom."[33]

We conducted other empirical tests that confirm our hypothesis that democracies win wars because they only initiate war when they are very confident they will win.[34] Recall that our theory proposes that democracies have a relatively high threshold of acceptable risk compared to other kinds of states, meaning they will avoid risky ventures, only initiating war when they are quite confident they will win. So far, our empirical analysis has tested this conceptualization indirectly, finding that democratic initiators are especially likely to win wars that they have initiated.

A more direct test would look at the actual decision to initiate wars. Our theory predicts that for all states, both democratic and nondemocratic, the

---

32. Scott D. Sagan, "The Origins of the Pacific War," in Robert I. Rotberg and Theodore K. Rabb, eds., The Origin and Prevention of Major Wars (Cambridge: Cambridge University Press, 1988), pp. 323–352; and Michael A. Barnhart, Japan Prepares for Total War: The Search for Economic Security, 1919–1941 (Ithaca, N.Y.: Cornell University Press, 1987).
33. Saburo Ienaga, The Pacific War, 1931–1945: A Critical Perspective on Japan's Role in World War II (New York: Pantheon, 1978), p. 97.
34. Reiter and Stam, Democracies at War, chap. 2. See also Bruce Bueno de Mesquita, James D. Morrow, Randolph M. Siverson, and Alastair Smith, "Testing Novel Implications of the Selectorate Theory of War," World Politics, Vol. 56, No. 3 (April 2004), pp. 363–388.

chances that a state will initiate war will go up as its estimated chance of victory rises. We would also predict, however, that as a state's estimate of its chances for victory rises, the chances of initiation increase greatly for democracies, whereas the chances of initiation increase only moderately for nondemocracies. That is, when the odds of victory are low to medium, democracies are significantly less likely to initiate war than are nondemocracies, but as the chances of victory become higher, the likelihood that a democracy will initiate war increases faster than the likelihood that a nondemocracy will initiate war. As a democracy becomes more confident it will win, the constraints on its decisions fall away and its foreign policy behavior resembles that of a nondemocracy.[35] In statistical tests of thousands of pairs of states going back to 1815, we found that democracies become increasingly willing to initiate a dispute as their estimated chances of victory increase. We also found the same result when we examined the decisions of states to escalate smaller scale disputes to war.[36] These results are consistent with selection effects theory and with the results in table 1. Though democracies are less willing to initiate the use of force when they are unsure they will win, as they become more confident they will win they shrug off these constraints and become willing to use force.

Selection effects theory predicts that democracies become more likely to initiate wars as they are more confident that they will win. This is part of a bigger picture that elected leaders strive to avoid foreign policy failures, which include wars that are unsuccessful or provide benefits incommensurate with their costs. That is to say, elected leaders strive to avoid costly wars as well as failed wars.

How do elected leaders avoid costly wars? Existing research indicates that democracies suffer fewer civilian and military casualties in their wars. We explored the hypothesis that democracies reduce costs by fighting shorter wars.[37] We conducted statistical analysis, controlling for a variety of factors which might affect the duration of wars, such as military-industrial capability, strategy, terrain, and other factors. We found that democracies fight significantly shorter wars than do other kinds of states. We also found that democracies' desires to fight short wars to some degree mediate their desires to fight winning

---

35. Note that this is not entirely consistent with some variants of democratic peace theory. In our view, if popular consent were generated within two disputing democracies, war would be possible.
36. Reiter and Stam, *Democracies at War*, chap. 2.
37. Ibid., chap. 7. See also Branislav Slantchev, "How Initiators End Their Wars: The Duration of Warfare the Terms of Peace," *American Journal of Political Science*, Vol. 48, No. 4 (October 2004), pp. 813–829; and D. Scott Bennett and Allan C. Stam, "The Duration of Interstate Wars," *American Political Science Review*, Vol. 90, No. 2 (June 1996), pp. 239–257.

Figure 1. Changes in Initiator's Chances of Victory over Time: Democracies and Nondemocracies

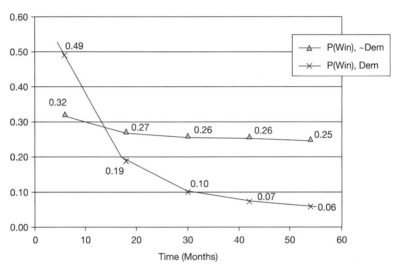

wars. Specifically, our results indicated that as war endures, eventually the costs of war become too much for a democracy to bear, and democracies begin to seek negotiated settlements, or draws, as a means to exit the war and stem the escalating costs.

Figures 1 and 2 demonstrate these results. Figure 1 tracks the likelihood of victory for democratic as compared to nondemocratic initiators as war duration increases. Although democratic initiators are initially much more likely to win than nondemocratic initiators, over time this advantage dissipates and then disappears. Figure 2 tracks the likelihood of democratic and nondemocratic belligerents reaching a negotiated settlement, or draw, over time. Although initially both democracies and nondemocracies share about the same likelihood of fighting to a draw, as the duration of the war increases, democracies become increasingly likely to seek out and accept negotiated settlements.

The democratic and autocratic perspectives on war duration are demonstrated well by the American experience in the Vietnam War. President Lyndon Johnson promised and hoped to fight a relatively short war. The strategy, however, was predicated on the assumption that North Vietnam would cave in to U.S. demands if the United States could make the war sufficiently costly for the communists. The plan failed, in part because of differing attitudes about

Figure 2. The Probability of Democracies and Nondemocracies Accepting a Draw over Time

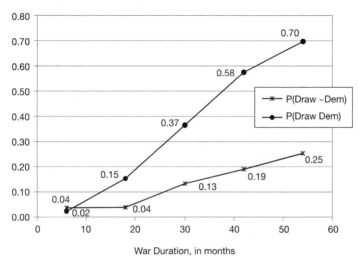

acceptable costs of war between the democratic United States and authoritarian North Vietnam.[38]

The 1991 Gulf War between Iraq and the U.S.-led coalition also demonstrates democracies' desire to fight short wars. From the outset of American war planning, "it was well understood within the American military that holding down casualties was a political prerequisite for launching a military offensive." Some plans, such as for a Marine invasion of Bubiyan Island, were dismissed as likely to incur too many casualties.[39] More generally, the desire for a short war with few American casualties meant a strategy of immediate and overwhelming force, forgoing the slow escalation strategy previously

38. Scott Sigmund Gartner, *Strategic Assessment in War* (New Haven, Conn.: Yale University Press, 1997); Andrew F. Krepinevich, *The Army and Vietnam* (Baltimore, Md.: Johns Hopkins University Press, 1986); John E. Mueller, "The Search for the 'Breaking Point' in Vietnam: The Statistics of a Deadly Quarrel," *International Studies Quarterly*, Vol. 24, No. 4 (December 1980), pp. 497–519; and A.F.K. Organski and Jacek Kugler, The War Ledger (Chicago: University of Chicago Press, 1980).
39. Michael R. Gordon and Bernard E. Trainor, *The Generals' War: The Inside Story of the Conflict in the Gulf* (Boston: Little, Brown, 1995), p. 133; U.S. News and World Report, *Triumph without Victory: The Unreported History of the Persian Gulf War* (New York: Times Books, 1992), p. 171; and General H. Norman Schwarzkopf, *It Doesn't Take a Hero* (New York: Bantam, 1992), p. 362.

pursued in Vietnam. As Gen. Colin Powell put it at the time, "I don't believe in doing war on the basis of macroeconomic, marginal-analysis models. I'm more of the mind-set of a New York street bully: 'Here's my bat, here's my gun, here's my knife, I'm wearing armor. I'm going to kick your ass.'"[40] This idea of massive escalation with overwhelming force evolved into what was later called the Powell Doctrine. President Bush's belief that victory could come quickly and cheaply bolstered his commitment to the war. In his memoir he wrote, "Briefing after briefing had convinced me that we could do the job fast and with minimum coalition casualties…I had no fear of making the decision to go."[41] The American concern to keep the war brief and casualties few shaped the coalition's military strategy in major ways, including encouraging an extremely intense phase of aerial bombing before the launching of the ground campaign, and providing rationale for a massive flanking operation around Kuwait rather than permitting large-scale set piece battles. Perhaps most important, these considerations pushed the United States to end the war after Kuwait had been liberated but before Baghdad had been captured, though the tremendous success in liberating Kuwait made the overthrow of Saddam's government at least militarily feasible. Ending the war after the liberation of Kuwait kept the war short and low in casualties. Ground operations stopped after 100 hours, during which time 146 Americans were killed and 467 wounded.[42]

Selection effects theory proposes that elected leaders become more likely to initiate war as public constraints diminish. Such constraints wane when a possible war promises to be successful, short, and low cost. Selection effects theory also describes two other sets of conditions which can make elected leaders more likely to initiate war. First, war initiation becomes more likely as the public perceives larger advantages of initiation (or perceives greater disadvantages of inaction). The public becomes more willing to absorb greater costs of war as the stakes become higher, for example if the public perceives a grave and direct threat to national security. In 1956 and 1967, for example, public constraints on the Israeli government's decision to initiate war were lower because in both cases the public perceived threats to national security from Egypt. Protection of national security aside, publics sometimes support

---

40. Powell, quoted in U.S. News and World Report, *Triumph without Victory*, p. 172.
41. Bush, quoted in George Bush and Brent Scowcroft, *A World Transformed* (New York: Alfred A. Knopf, 1998), p. 462.
42. Powell, *My American Journey*, p. 485; U.S. News and World Report, *Triumph without Victory*, pp. 275–276; Gordon and Trainor, *The Generals' War*, pp. 476–477; and John Mueller, *Policy and Opinion in the Gulf War* (Chicago: University of Chicago Press, 1994), p. 69.

wars of empire, as was the case in 1898 when the United States initiated the Spanish-American War.[43]

Another set of circumstances which can make the initiation of conflict more likely is if an elected government can act covertly. Although it is very difficult to initiate and wage interstate wars covertly, governments can sometimes engage in other, smaller scale military or paramilitary actions covertly, such as supplying rebels or attempting to overthrow a foreign government. Actions kept secret are unlikely to face constraints either from the public or other branches of government. Covert uses of force have a relatively high rate of failure, perhaps because such actions do not benefit from open debate and the scrutiny of the marketplace of ideas. Indeed, in the words of former National Security Adviser McGeorge Bundy, "The dismal historical record of covert military and paramilitary operations over the last 25 years is entirely clear."[44] The disaster of the Bay of Pigs invasion of Cuba in 1961 is a good example of the link between secrecy and policy failure. The plan as conceived began in the Eisenhower administration. President Kennedy gave his approval despite several deep flaws, which in all likelihood would have led to reconsideration of the venture if they had been exposed to public debate.[45]

Democracy and Battlefield Military Effectiveness

A second major proposition connecting democracy and war outcomes posits that democratic soldiers fight with higher levels of effectiveness in battle. Whereas the selection effects proposition argues that democracies win wars because they make better choices about what wars to enter, the battle effectiveness proposition focuses on what actually happens during war. We term this and other explanations connecting democracy to war outcomes which focus on intrawar dynamics—two of which we explore in the sections to follow—*warfighting* explanations.

The proposition that democratic soldiers fight better than their counterparts drawn from authoritarian polities gets at a fundamental disagreement over the

---

43. Reiter, "A Closer Look at Case Studies on Democracy, Selection Effects, and Victory"; and Gerald F. Linderman, *The Mirror of War: American Society and the Spanish-American War* (Ann Arbor: University of Michigan Press, 1974).

44. McGeorge Bundy, *New York Times*, June 10, 1985.

45. John Prados, *Presidents' Secret Wars: CIA and Pentagon Covert Operations since World War II* (New York: Morrow, 1986). On secrecy, see Daniel Patrick Moynihan, *Secrecy: The American Experience* (New Haven, Conn.: Yale University Press, 1998). On the failure of U.S. policy in Latin America, see Walter LaFeber, *Inevitable Revolutions: The United States in Central America* (New York: Norton, 1984).

merits of free versus command societies. A conventional view is that free societies excel at the arts and commerce, but lack the martial spirit that combat requires and authoritarian systems supposedly imbue their populations with. Some think of Athens, Britain, and the United States as three countries in their day classified as prosperous and culturally blessed free republics. In contrast, Sparta, Nazi Germany, and the Soviet Union are often perceived as bland, brutal, and powerful. Some have expressed the concern that liberalism imposes significant encumbrances on martial effectiveness. This is a fundamental controversy in political thought, whether or not the individualism unleashed within liberalism undermines collective action and respect for achieving communitarian goals. The twentieth century German thinker Carl Schmitt considered this idea and proposed that liberalism's emphasis on the individual precluded the state from calling on individuals to self-sacrifice on the battlefield.[46] In the 1930s, Nazi sympathizers such as the American aviator Charles Lindbergh were pessimistic about the Western democracies' chances of prevailing over the new fascist dictatorships in the event of war.[47] During the Cold War, observers of the Communist threat such as Russian author Aleksandr Solzhenitsyn worried that the devotion to individual, material pursuits in Western liberalism impeded the self-sacrifice demanded by national defense, so that the American Athens was systematically handicapped in its race for survival with the Soviet Sparta.[48]

An alternative to the view that oppression builds stoic societies stocked with warriors is that freedom produces better fighters. We make four propositions as to why democracies might produce armies that are more effective. First, soldiers serving popularly elected governments fight with higher morale, enabling them to overcome obstacles and setbacks that would otherwise lead them to quit the field. Soldiers are more likely to accept the dangers of the battlefield and place their lives at risk if they are serving in a military overseen by a government grounded in democratic political institutions. They are more likely to perceive the war effort and the leadership itself as reflecting their own interests if the need for popular consent constrains the government and can be removed from office if it fails to hold up its end of the social contract. Further, in this line of reasoning, soldiers are more confident that a democratically

---

46. Carl Schmitt, *The Concept of the Political*, trans. George Schwab (New Brunswick, N.J.: Rutgers University Press, 1976), p. 71; and John P. McCormick, *Carl Schmitt's Critique of Liberalism: Against Politics as Technology* (Cambridge: Cambridge University Press, 1997), p. 257.

47. On Lindbergh, see Peter Grose, *Gentleman Spy: The Life of Allen Dulles* (Boston: Houghton Mifflin, 1994), p. 130.

48. Aleksandr I. Solzhenitsyn, *A World Split Apart* (New York: Harper and Row, 1978), pp. 13–15.

elected government will obey the laws and abide by its promises, because failure to do so may result in its removal from power. In essence, the source of political legitimacy lies in the rule of law rather than the cult of personality, as is the case in many autocracies. Over the past millennia, a number of thinkers, including Sun Tzu, Herodotus, Pericles, John Locke, Thomas Paine, John Quincy Adams, Thomas Jefferson, and others, have made the point that soldiers fight harder for popular governments.[49]

Modern political scientists have built more formal and extensively specified theories expanding on these ideas. One focused on "contingent consent," examining the factors that are likely to determine the degree to which members of society comply with government military conscription policies. A key component of this model is the trustworthiness of government, a characteristic democratic governments are more likely than autocracies to exhibit. Importantly, it predicts that democratic governments are more likely to enjoy higher levels of societal consent in reaction to military conscription policies. Along similar lines, we propose that governments installed by popular election operating under the constraints of democratic political institutions are more likely to earn the trust and loyalty of their citizens, which will translate into greater consent and willingness to take the initiative on the battlefield and therefore in turn leading ultimately to higher levels of military effectiveness through greater morale.[50]

A second argument connecting regime type and battlefield military performance focuses on political culture. Democratic, liberal culture emphasizes the importance of the individual. Indeed, the very essence of liberal democracy is the guarantee that the individual is minimally fettered, whether in the spheres

---

49. Sun Tzu, *The Art of War*, trans. Samuel B. Griffith (London: Oxford University Press, 1963), p. 64; Herodotus, *The History*, trans. David Grene (Chicago: University of Chicago Press, 1987), book 5, para. 78; Thucydides, *History of the Peloponnesian War*, trans. Rex Warner (Harmondsworth, U.K.: Penguin, 1954), book 2, para. 39; John Locke, *Second Treatise of Government*, ed. C.B. Macpherson (Indianapolis: Hackett, 1980), p. 26; Stephen Holmes, *Passions and Constraint* (Chicago: University of Chicago Press, 1985), p. 19; Thomas Paine, *Prospects on the War and Paper Currency*, 1st American ed. (Baltimore, Md.: S. and J. Adams, 1794), p. 17; and *Bartlett's Familiar Quotations*, 16th ed. (New York: Little, Brown, 1992), p. 368.
50. On contingent consent, see Margaret Levi, *Consent, Dissent, and Patriotism* (Cambridge: Cambridge University Press, 1997). On the importance of trust within military organizations and how it influences battlefield efficiency or lack thereof, see Allan C. Stam III, *Win, Lose, or Draw: Domestic Politics and the Crucible of War* (Ann Arbor: University of Michigan Press, 1996). Stam hypothesized that ethnic and class cleavages, along with political institutions will influence trust within military organizations. See also Stephen Biddle and Robert Zirkle, "Technology, Civil-Military Relations, and Warfare in the Developing World," *Journal of Strategic Studies*, Vol. 19, No. 2 (June 1996), pp. 171–212; and Pollack, "The Influence of Arab Culture on Arab Military Effectiveness."

of politics, speech, art, religion, or commerce. Armies are microcosms of the societies they come from, and to a large extent, the military mirrors the qualities of the society from which it is drawn.[51] The emphasis on individual initiative in liberal political culture, therefore, spills over into superior initiative on the battlefield. Alexis de Tocqueville made this point, observing that democracies are likely to enjoy the tactical advantage of having soldiers acting on reason rather than on instinct, without threatening military discipline.[52] Superior initiative and inspired leadership demand just this kind of flexibility on the part of individual soldiers. Modern warfare requires the display of individual initiative on the battlefield to exploit best emergent but fleeting opportunities and to cope with unanticipated conditions. The exercise of individual initiative is especially important in the exercise of mobile and maneuver-based warfare, as its very premise is allowing lower ranked officers and troops the freedom to exploit a fluid battlefield.[53] During the Arab-Israeli wars, for example, the differences in fighter pilot performance between democratic Israel and its autocratic Arab neighbors emerged from the greater emphasis on individual initiative in liberal Israeli culture. The first prime minister of Israel, David Ben-Gurion, explained: "A look at our defence system, the most disciplined organization we have in the country, indicates just how much we predicate our very existence on democracy and on the popular consent that only this form of government can elicit. I think the citizen aspect of the IDF [Israeli Defense Force] makes for both imaginative thinking in military affairs and for exemplary conduct above and beyond the call of duty."[54] Just as democratic citizens exhibit personal initiative at home, so do democratic soldiers on the battlefield.

Third, soldiers fighting for authoritarian leaders are more likely to surrender to democratic enemies than vice versa. Liberal societies' emphasis on the fair and equal treatment of the individual can serve to erode the loyalty of opposing soldiers, inducing the surrender of enemy forces. Democracies will have an easier time convincing enemy soldiers that if captured, their captors will treat them well. Some may point to the egregious treatment inflicted since

51. Andreski, *Military Organization and Society*; and Samuel P. Huntington, *The Soldier and The State: The Theory and Politics of Civil Military Relations* (Cambridge, Mass.: Harvard University Press, 1981).

52. Tocqueville, *Democracy in America*, pp. 658–659.

53. John Stuart Mill, *On Liberty*, ed. Alburey Castell (New York: Appleton Century Crofts, 1947); Stephen Peter Rosen, *Societies and Military Power: India and Its Armies* (Ithaca, N.Y.: Cornell University Press, 1996); Eliot A. Cohen and John Gooch, *Military Misfortunes: The Anatomy of Failure in War* (New York: Free Press, 1990); Pollack, "The Influence of Arab Culture on Arab Military Effectiveness"; and Mearsheimer, *Conventional Deterrence*.

54. David Ben-Gurion, *Recollections*, ed. Thomas R. Bransten (London: Macdonald Unit Seventy-Five, 1970), pp. 95–100; and Pollack, "The Influence of Arab Culture on Arab Military Effectiveness."

September 11, 2001, by U.S. soldiers on detainees at places such as Abu Ghraib, but the backlash against those abuses and the changes in U.S. policy toward improved treatment of detainees demonstrates the real costs that democracies suffer when they treat prisoners poorly. Autocracies, on the other hand, will be hard pressed to convince enemy soldiers to accept capture and the ill-treatment it will likely entail. Enemy soldiers are less likely to believe that an autocracy will abide by its promises to fair treatment of prisoners of war (POWs), even those promises codified by international treaty, given how poorly autocratic governments treat their own populations and respect law. More generally, authoritarian states are more likely to thrive on war cultures that require the brutalization of POWs, World War II Japan and Germany being prime examples.

In many wars, concerns about ill treatment made soldiers less likely to surrender to autocratic enemies than to democratic enemies. In the Korean War, 38 percent of Americans taken prisoner died in captivity. American forces recognized this and resisted surrender. Despite three years of fighting with hundreds of thousands of American forces deployed in Korea absorbing tens of thousands of fatalities, only 7,190 Americans were taken prisoner.[55] During the Vietnam War, there were 661 recognized American POWs, the vast majority suffering torture, compared to over 50,000 North Vietnamese soldiers captured.[56] The Wehrmacht treated Soviet POWs brutally (nearly two-thirds of Soviets POWs captured by Germany were dead by 1944), and German military effectiveness suffered. Alfred Rosenberg, minister for the Occupied Eastern Territories who was put in charge of the treatment of Soviet POWs, felt that brutalization of Soviet POWs was responsible for bolstering the fighting spirit of the Red Army "and thereby also for the deaths of thousands of German soldiers. . . . An obvious consequence of this politically and militarily unwise treatment has been not only the weakening of the will to desert but a truly deadly fear of falling into German captivity."[57]

Conversely, soldiers often believe they will be treated well by democratic foes, and are more willing to surrender to them. During World War II,

55. John W. Dower, *War without Mercy: Race and Power in the Pacific War* (New York: Pantheon Books, 1986), p. 48; Joan Beaumont, "Protecting Prisoners of War, 1939–95," in Bob Moore and Kent Fedorowich, eds., *Prisoners of War and Their Captors in World War II* (Oxford: Oxford University Press, 1996), p. 279; Max Hastings, *The Korean War* (New York: Touchstone, 1987), p. 304; and Micheal Clodfelter, *Warfare and Armed Conflicts: A Statistical Reference to Casualty and Other Figures, 1618–1991*, 2d ed. (Jefferson, N.C.: McFarland,, 1992).
56. Stuart I. Rochester and Frederick Kiley, *Honor Bound: American Prisoners of War in Southeast Asia, 1961–1973* (Annapolis: Naval Institute Press, 2007).
57. Jürgen Förster, "The Dynamics of Volksgemeinschaft," in Williamson Murray and Allen R. Millett, eds., *Military Effectiveness*, Vol. 3 (Boston: Allen and Unwin, 1988), pp . 21–23; Albert Speer, *Inside the Third Reich: Memoirs by Albert Speer*, trans. Richard and Clara Winston (New York:

Germans were much more willing to surrender to U.S. and British forces in the West than to Soviet forces in the East. As one Panzer corporal put it, "In Russia, I could imagine nothing but fighting to the last man. We knew that going into a prison camp in Russia meant you were dead. In Normandy, one always had in the back of his mind, 'Well, if everything goes to hell, the Americans are human enough that the prospect of becoming their prisoner was attractive to some extent.'"[58] More recently, during the 1991 Gulf War between the U.S.-led coalition and autocratic Iraq, over 86,000 Iraqis surrendered in just a few days of ground combat.[59]

A fourth connection between regime type and battlefield military effectiveness concerns military leadership. Democratic armies are likely to enjoy superior military leadership. This is because civil-military relations within a democracy are likely to engender relatively more meritocratic militaries that promote the competent over the politically connected. Progress away from monarchy and aristocracy and toward democracy engenders social leveling and the general spread of egalitarian norms throughout society, the government, and the military. In turn, this means that military organizations are more likely to operate on meritocratic rather than class-based principles.

Conversely, authoritarian militaries are likely to be less meritocratic, and in turn on average will have relatively inferior leadership. Given the weaker or nonexistent institutionalized civilian control of the military in autocratic regimes, a nondemocratic political leader is more likely than a democratic leader to perceive that the armed forces may pose a domestic political threat. The tendency for autocrats to staff their militaries with politically dependable ethnic confreres rather than effective leaders is especially notable in the Middle East. Syrian strongman Hafez al-Assad stocked his military with Alawis; King Hussein of Jordan promoted to senior positions officers from the East Bank region; and Saddam Hussein of Iraq made sure his officer corps was dominated by minority Sunnis. Making matters worse, autocrats rotate officers relatively more frequently to prevent them from developing close ties to their troops, and squelch rather than listen to constructive criticism from within the officer corps.[60] The need for officers to be politically unthreatening in a nondemo-

MacMillan, 1970), p. 250n; and Leonidas E. Hill, ed., *Die Weizsacker-Papiere 1933–1950* (Frankfurt: Propylaen Verlag, 1974), pp. 327–328. The authors thank Christian Tuschhoff for assistance with translation.
58. Quoted in Stephen E. Ambrose, *Citizen Soldiers: The U.S. Army from the Normandy Beaches to the Bulge to the Surrender of Germany, June 7, 1944–May 7, 1945* (New York: Simon and Schuster, 1997), p. 33.
59. Adam Roberts, "The Laws of War in the 1990–91 Gulf Conflict," *International Security*, Vol. 18, No. 3 (Winter 1993/94), p. 160.
60. Risa Brooks, *Political-Military Relations and the Stability of Arab Regimes*, Adelphi Paper, No. 324

cratic state will generate lower effectiveness throughout the military. The propensity for Arab authoritarian leaders to install political toadies in high positions undercuts Arab military effectiveness.[61] Some scholars have argued that the disastrous performance of the Red Army in 1941 was due to the purges of the officers' ranks in the years immediately before the German invasion.[62]

We subjected three of these four propositions to quantitative empirical testing (we did not have sufficiently high quality data to test the surrender proposition).[63] We looked at a data set of all battles from 1800 to 1982, looking at measures of morale, initiative, and leadership. We found that democratic armies enjoy significantly higher levels of initiative and leadership than autocratic armies, but they do not enjoy higher levels of morale. This may be because autocratic regimes such as Nazi Germany and North Vietnam can sometimes use nationalism or ideology to inspire their troops to fight and die for their countries, political institutions aside.

Democracy, Alliances, and War Outcomes

A third major proposition linking democracy and victory in war concerns alliances. Some have proposed that democracies are more likely to win their wars because they enjoy more powerful alliances.[64] Three different strands of logic have been developed. First, democracies are more likely to form alliances among themselves, as democracies recognize each other as partners more likely to abide by formal commitments. Second, during wartime democracies see a common interest in fending off autocratic aggressors, so when one democracy is attacked others are especially likely to come to its aid. Third, democracies serve as better allies in wartime, as democratic allies are more likely to cooperate deeply with each other than are autocratic allies.

Though the logic behind each of these claims appears reasonable, there is lit-

(London: International Institute for Strategic Studies, 1998), pp. 32–33; Gordon Tullock, *Autocracy* (Boston: Kluwer , 1987); and Khidhir Hamza with Jeff Stein, *Saddam's Bombmaker: The Terrifying Inside Story of the Iraqi Nuclear and Biological Weapons Agenda* (New York: Scribner, 2000), pp. 174–175.
61. Pollack, "The Influence of Arab Culture on Arab Military Effectiveness"; and Risa Brooks, "Institutions at the Domestic/International Nexus: The Political-Military Origins of Strategic Integration, Military Effectiveness and War," Ph.D. dissertation, University of California, San Diego, 2000.
62. John E. Jessup, "The Soviet Armed Forces in The Great Patriotic War, 1941–5," in Murray and Millett, *Military Effectiveness*, pp. 256–276; and Antony Beevor, *Stalingrad* (New York: Viking, 1998), p. 23.
63. Reiter and Stam, *Democracies at War*, chap. 3.
64. See, for example, David A. Lake, "Powerful Pacifists: Democratic States and War," *American Political Science Review*, Vol. 86, No. 1 (March 1992), pp. 24–37; and Ajin Choi, "Democratic Synergy and Victory in War, 1816–1992," *International Studies Quarterly*, Vol. 48, No. 3 (September 2004), pp. 663–682.

Table 2. Target's Democracy and Defender Intervention in Wars, 1816–1992

	No Potential Defender Intervened	At Least One Defender Intervened	Total
Nondemocratic target	52 (80%)	13 (20%)	65 (100%)
Democratic target	8 (73%)	3 (27%)	11 (100%)
Total	60 (79%)	16 (21%)	76 (100%)

Notes: Pearson chi squared (1) = 0.30; P = 0.58.

tle empirical evidence that democracies win their wars because of superior alliance dynamics. Most rigorous studies find that before 1945 states of similar regime type were not any more likely to ally with each other, and others have found that democracies may actually be significantly less likely than other kinds of states to honor their alliance commitments in wartime.[65] Research reveals that after 1945 there was a tendency for similar states to ally with other, but the handful of wars fought by democratic coalitions have not all ended in democratic victory, as for example the Korea and Vietnam Wars ended in draws (as did the Suez War, for Britain and France).[66]

Beyond peacetime alliance behavior, the more directly relevant question is whether democracies are especially likely to provide substantial assistance to each other in times of war, specifically, sending military assistance to aid a democracy under attack. We begin with a simple question: are democratic targets more likely to attract assistance of any kind than are autocratic targets? Among the 76 states targeted in wars from 1816–1992, 11 of them were democracies.[67] Among these 76 targets, 18 attracted assistance from third parties. Table 2 provides a cross-tabulation of whether the target state was a democracy and whether it attracted a defender.

The results indicate that though democratic targets are slightly more likely

65. On democracies not allying with each other, see Brian Lai and Dan Reiter, "Democracy, Political Similarity, and International Alliances, 1816–1992," *Journal of Conflict Resolution*, Vol. 44, No. 2 (April 2000), pp. 203–227. On democracies not honoring their alliance commitments, see Erik Gartzke and Kristian Skrede Gleditsch, "Why Democracies May Actually Be Less Reliable Allies," *American Journal of Political Science*, Vol. 48, No. 4 (October 2004), pp. 775–795.
66. See, for example, Lai and Reiter, "Democracy, Political Similarity, and International Alliances, 1816–1992."
67. That is, states with Polity scores of 7 or above on a –10 to +10 scale. Setting the threshold at 5 or 6 does not change the results.

to attract assistance (27 percent of the time) than are nondemocratic targets (20 percent of the time), this difference is not statistically significant. Therefore, we must reject the proposition that democratic targets are systematically more likely than autocratic targets to attract allies. This in itself is important: there is little if any evidence to suggest that democratic targets are more likely to attract assistance of any kind during war than are nondemocratic states.

Consider another tack: are democracies especially likely to intervene in favor of targeted democracies? This gets back to the theoretical discussion above: democracies win wars because other democracies recognize an international liberal community, and are willing to rescue targeted democracies from attack. We explore this idea by asking the specific question: what causes bystanders to intervene on behalf of targeted states? If the arguments about the effects of liberalism discussed above are correct, then we would expect that among potential interveners, democracies should be especially likely to take action to defend democracies that find themselves under attack by autocracies. We conducted statistical analysis to analyze this question, and controlled for other factors that are likely to affect a state's decision to intervene.[68] The data we analyzed included all opportunities for states in the international system to intervene on behalf of all targeted states from 1816 to 1992. Although we found that factors such as geographical contiguity, trade in goods and services, international alliances, and national power may affect a state's decision to intervene to assist another state, democracies are not more likely to intervene to save targeted democracies. In short, the domestic political characteristics of the target state do not have any systematic effect on whether or not a potential intervener acts on the target's behalf.

We take a closer look at the instances of democracies fighting together in wartime, asking two questions: First, did democracies join the war because the targeted state was a democracy? Second, did the democracy's choice to join affect the war's outcome? To work toward a solution for this puzzle, consider the eight wars from 1815–2000 in which at least two democracies fought on a side. These cases are: the Boxer Rebellion, the two World Wars, the Korean War, the Suez War, the Vietnam War, the 1991 Gulf War, and the 1999 Kosovo War.

Of these eight wars, one (Korea) was clearly a draw (Suez was also probably a draw for Britain and France) and in another, the democratic coalition either lost or fought to a draw (Vietnam). Hence, democratic balancing cannot count as contributing to victory in these two cases. In the six or so cases where the

---

68. Reiter and Stam, *Democracies at War*, chap. 4.

democratic coalition won, the important question to ask is whether democracies entered the war to rescue a fellow democracy from defeat, or if the democracies were fighting on the same side for other reasons. This is important, because the theory predicting that democracies win wars because they aid each other during wartime forecasts that democracies will not join together to fight wars of empire but instead, will ally to defend each other. In two of these wars, the Boxer Rebellion and the Suez War, democracies are clearly not fighting to help each other, but rather are aggressors seeking to alter the status quo and to serve their state's individual national interest.

In another, the 1991 Gulf War, democracies came to the rescue of an autocratic state (Kuwait). Realist motives, such as access to oil and maintaining a regional balance of power, and not democratic fraternalism drove the democrats' coalition formation and subsequent intervention. The Korea and Vietnam Wars were similar: democracies were assisting autocratic targets out of geopolitical motivations. Further, the participation of multiple democracies in the Gulf War was unnecessary for victory, as the only two states truly necessary for military victory were the United States and Saudi Arabia, the latter of which was not a democracy. In the more recent Kosovo War, NATO intervened not to protect another democracy but rather to prevent ethnic cleansing of a stateless people. The leaders of the Kosovars shortly afterward demonstrated everything but democratic inclinations. In addition, the United States made the principal military contributions to the Kosovo operation.

This leaves two wars in which democratic coalitions won defensive wars, the two World Wars. Are these examples of democracies coming to each other's rescue to protect the liberal democratic community? The answer is at best mixed. In World War I, democratic Belgium and France became involved when attacked by oligarchic Germany. Democratic Britain joined the war for principally realist reasons: to protect the balance of power in Europe, to maintain its reputation, and to honor its commitment to Belgian neutrality, not Belgian democracy.[69] Democratic fraternalism was also absent on the other side of the Atlantic. Although President Woodrow Wilson has become the leader most closely associated with liberalism, a desire to rescue other beleaguered democracies is not the best explanation of Wilson's motives behind his decision to request American intervention into the First World War. The

---

69. Michael Howard, *The Continental Commitment: The Dilemma of British Defence Policy in the Era of the Two World Wars* (London: Ashfield, 1972); Donald Kagan, *On the Origins of War and the Preservation of Peace* (New York: Anchor, 1995), pp. 204–205; and Luigi Albertini, *The Origins of the War in 1914*, 3 vols., trans. and ed. Isabella M. Massey (London: Oxford University Press, 1957), especially Vol. 3, p. 409.

United States let its democratic confrères fight unassisted for three bloody years, while trying desperately to maintain its neutrality. Unable to generate significant (and sufficient) domestic consent for war, the Wilson administration sat on the side as literally millions of soldiers and civilians from European democracies perished. When U.S. intervention did come in the spring of 1917, Wilson favored it not because he feared an imminent German victory over the United States' European ideological counterparts, although this was a distinct possibility. Rather, Germany's unrestricted submarine warfare in the Atlantic played a substantial role in spurring U.S. popular outrage. German submarines threatened American commerce and maritime traffic throughout the Atlantic; this was unacceptable to both Wilson and, eventually, the U.S. public. Further fanning the smoldering enmity between the United States and Germany, the Germans proposed in the now infamous January 1917 Zimmermann Telegram that Mexico join Germany in a potential war against the United States. In exchange for assisting the Germans, Mexico was to receive back the territories of New Mexico, Arizona and Texas.[70] With popular consent obtained in reaction to the German's aggressive and provocative actions, Wilson subsequently guided the United States' entrance into the war.

World War II also does not demonstrate clearly the dynamic of democracies rescuing democracies. Britain and France declared war on Germany in September 1939, coming to the rescue of autocratic Poland. Earlier that year both had stood aside during the spring 1939 demise of the democratic Czechoslovak republic, in spite of (in the French case) formal alliance commitments to the Czechs. Once the war started, the small democratic neutrals in Europe did not voluntarily jump on the democratic bandwagon. Some— Ireland, Switzerland and Sweden—did not join the democratic coalition at all. Others—Belgium, the Netherlands, Luxembourg, Norway and Denmark— joined only after the Germans invaded. One—Finland—joined the Axis.

The United States did not enter the war when it broke out in 1939. It did not come quickly to the rescue of either democratic France or Britain, even in the summer of 1940 when France fell and Britain faced mortal danger during the Battle of Britain, the preparatory phase before Operation Sea Lion, Germany's planned invasion of Britain. The United States did not enter the war in Asia until Japan, at war principally with autocratic Nationalist China, attacked Pearl Harbor on December 7, 1941. The United States declared war against

---

70. Ernest R. May, *The World War and American Isolation, 1914–1917* (Cambridge, Mass.: Harvard University Press, 1959), p. 427; and Arthur S. Link, *Wilson: Campaigns for Progressivism and Peace* (Princeton, N.J.: Princeton University Press, 1965), pp. 342–345.

Japan immediately after the Pearl Harbor attack, but Roosevelt refused to request that Congress declare war immediately on Germany and Italy, due in part to divisions in public opinion. America entered the war in Europe only after Hitler and Mussolini first declared war against the United States on December 11, 1941.[71]

In sum, even anecdotal evidence to support the proposition that democracies win wars because they help each other in war is scant. In the eight wars in which two democracies fought on the same side and in defense against aggression, in only two did the democratic coalitions emerge victorious while protecting the interests of democratic targets—World Wars I and II. In both the world wars, however, realist motivations spurred the democratic belligerents to action, not liberal fraternalism. Some have argued that democracies serve as better allies than autocracies, though notably democratic coalitions failed to secure victories in Korea or Vietnam, instead accepting negotiated settlements best characterized as draws. The World Wars also saw instances of poor interdemocratic cooperation. There was poor Anglo-French cooperation throughout World War I, and shallow cooperation between U.S. and other Allied forces later after U.S. entry.[72] During World War II, Anglo-Belgian-French coordination both before and during the German invasion of the West in May–June 1940 was quite poor, as was Anglo-American coordination in Italy in 1943. Anglo-Franco-American cooperation during the 1991 Gulf War also had its bumps, as both Britain and France refused to accept their initial combat assignments in the ground campaign.[73]

Democracy, Military-Industrial Power, and Victory

A fourth explanation of democracy success in war is that democracies win wars because they can wield greater amounts of war materiel. Scholars have

---

71. Robert Dallek, *Franklin D. Roosevelt and American Foreign Policy, 1932–1945* (Oxford: Oxford University Press, 1979), p. 312; and Dan Reiter, *How Wars End* (Princeton, N.J.: Princeton University Press, 2009), chap. 4.

72. William James Philpott, *Anglo-French Relations and Strategy on the Western Front, 1914–18* (New York: St. Martin's, 1996), p. 50; John Keegan, *The First World War* (New York: Alfred A. Knopf, 1999), p. 350; David F. Trask, *The AEF and Coalition Warmaking, 1917–1918* (Lawrence: University Press of Kansas, 1993); and James F. Dunnigan and Austin Bay, *From Shield to Storm: High-Tech Weapons, Military Strategy, and Coalition Warfare in the Persian Gulf* (New York: William Morrow, 1992), pp. 60–61.

73. Brian Bond, *France and Belgium 1939–1940* (London: Davis-Poynter, 1975), pp. 96, 100; and Peter Calvocoressi, Guy Wint, and John Pritchard, *Total War: The Causes and Courses of the Second World War*, Vol. 1, rev. 2d ed. (New York: Pantheon, 1989), p. 138; and Ernest R. May, *Strange Victory: Hitler's Conquest of France* (New York: Hill and Wang, 2000), pp. 304, 394, 295.

made two different arguments on the material advantages of democracy: that democracies in general have larger economies, and that during wartime democracies can extract proportionately greater resources from their societies because democratic populations are more willing to make material sacrifices for victory.

The first claim is that democracies win because their economies are bigger. Notably, the broader literature is mixed on whether democracy causes development, development causes democracy, neither, or both.[74] A more straightforward question is whether democracies win wars because they have larger economies. Our statistical analysis, summarized previously, finds that both democracy and military-industrial capabilities make victory more likely, indicating that at the least democracy is contributing to victory in ways beyond economic strength. We conducted further analyses on the relationship between a state's level of democracy and its economic capabilities—in particular those capabilities that translate into war materiel on the battlefield. Looking at the belligerents in all wars from 1816 to 1990, we found there to be no statistically significant relationship between a state's political institutions and the size of its economy.[75]

Separate from overall economic prosperity is the question of how much a state can extract from its economy for the war effort. In large part, a nation's real power (as opposed to its theoretical potential) emerges as a product of its aggregate material resources and its political capacity for resource extraction during times of need.[76] Some have argued that democracies are able to extract more from their societies during wartime than dictators and oligarchs. Some make the simple argument that because democratic governments are more

---

74. Larry Diamond, "Economic Development and Democracy Reconsidered," *American Behavioral Scientist,* Vol. 35, No. 4/5 (March/June 1992), pp. 450–499; Philip Keefer and Stephen Knack, "Why Don't Poor Countries Catch Up? A Cross-National Test of an Institutional Explanation," *Economic Inquiry,* Vol. 35, No. 3 (July 1997), pp. 590–602; Ross E. Burkhart and Michael S. Lewis-Beck, "Comparative Democracy: The Economic Development Thesis," *American Political Science Review,* Vol. 88, No. 4 (December 1994), pp. 903–910; John F. Helliwell, "Empirical Linkages between Democracy and Economic Growth," *British Journal of Political Science,* Vol. 24, No. 2 (April 1994), pp. 225–248; Robert J. Barro, *Determinants of Economic Growth: A Cross-Country Empirical Study* (Cambridge, Mass.: MIT Press, 1997); Yi Feng, "Democracy, Political Stability and Economic Growth," *British Journal of Political Science,* Vol. 27, No. 3 (July 1997), pp. 391–418; Adam Przeworski, Michael E. Alvarez, José Antonio Cheibub, and Fernando Limongi, *Democracy and Development: Political Institutions and Well-Being in the World, 1950–1990* (Cambridge: Cambridge University Press, 2000); John B. Londregan and Keith T. Poole, "Does High Income Promote Democracy?" *World Politics,* Vol. 49, No. 1 (October 1996), pp. 1–30; and Matthew A. Baum and David A. Lake, "The Political Economy of Growth: Democracy and Human Capital," *American Journal of Political Science,* Vol. 47, No. 2 (April 2003), pp. 333–347.
75. Reiter and Stam, *Democracies at War,* chap. 5.
76. Organski and Kugler, *The War Ledger.*

popular, democracies will be able to extract more resources, either human or economic, from society for the war effort, increasing the chances of victory over autocratic states.[77] Others make a more sophisticated claim, arguing that in both peace and war, democracies are better able to convert the resources the state extracts from its society into public goods versus excludable private goods that authoritarian states' leaders rely on to maintain their hold on power. In war, this argument suggests that democracies' necessary focus on public goods drives publics to make greater material sacrifices for victory on the battlefield.[78]

History reveals, however, that democracies are not more effective at extracting resources during wartime than are other types of regimes. Two scholars found that political capacity is an important determinant of the outcomes of major wars since 1900, but that "differences in the form of government do not determine the degree of political effort."[79] We conducted further statistical tests using different data and confirmed these findings.[80]

Another way of thinking about economic extraction for the war effort is to look at the percentage of the population that is serving in the military. This is a useful measure of resource extraction, as putting more people in the armed forces, while obviously advantageous on the battlefield, can strain the national economy and create labor shortages, thereby imposing hardship on the civilian population. We conducted statistical tests to examine whether democracies have put more of their populations under arms since 1816. We looked at both the population of all countries whether at war or in peace as well as just those countries at war. The results indicate that for all states, democracies, in fact,

---

77. See Lake, "Powerful Pacifists," p. 30; Alan C. Lamborn, *The Price of Power: Risk and Foreign Policy in Britain, France, and Germany* (Boston: Unwin Hyman, 1991), p .84; and Levi, *Consent.* Kenneth Schultz and Barry R. Weingast's argument could be applied here, as it might imply that democracies can access greater resources in wartime by having more credit availability, enabling them to "finance larger and longer wars." Schultz and Weingast, "Limited Governments," in Randolph M. Siverson, *Strategic Politicians, Institutions, and Foreign Policy* (Ann Arbor: University of Michigan Press, 1998), p. 17. It is an interesting point, which they demonstrate nicely in the Dutch revolt against Spain and the seventeenth century Anglo-French rivalry. Its applicability to modern circumstances may be more limited, however; modern democracies might be less likely to exploit these financial advantages in a long war, because since 1815 they have tended to fight short wars to maintain support at home. Kenneth A. Schultz and Barry R. Weingast, "The Democratic Advantage: An Institutional Foundations of Financial Power in International Competition," *International Organization,* Vol. 57, No. 1 (Winter 2003), pp. 3–42.
78. Bruce Bueno de Mesquita, James D. Morrow, Randolph Siverson, and Alastair Smith, "Institutional Explanation of the Democratic Peace," *American Political Science Review,* Vol. 93, No. 4 (December 1999), pp. 791–807.
79. Jacek Kugler and William Domke, "Comparing the Strength of Nations," *Comparative Political Studies,* Vol. 19, No. 1 (April 1986), p. 66. See also Jacek Kugler and Marina Arbetman, "Relative Political Capacity: Political Extraction and Political Reach," in Arbetman and Kugler, eds., *Political Capacity and Economic Behavior* (Boulder, Colo.: Westview, 1997), pp. 11–45.
80. Reiter and Stam, *Democracies at War,* chap. 5.

place a significantly *lower* percentage of their populations under arms than do nondemocracies. For states at war the negative effect remains, though it is not statistically significant.[81]

A slightly different proposition than those above might be that democracies win wars not because they have higher levels of aggregate production, or that they have more soldiers under arms, but rather because democratic armies have more war materiel available to them on the battlefield than do autocratic armies. This is a potentially important distinction, as higher levels of corruption within autocratic societies may make the translation of industrial production into war goods less efficient. In other words, democracies and autocracies might have equal levels of industrial production, but the democracies may have more weapons available to their troops because of less corruption in their military-industrial complexes. We conducted statistical analysis using a data set of all large-scale interstate war battles from 1800–1982, exploring whether regime type is correlated with the competing militaries' numbers of tanks, artillery pieces, and air sorties. In the results, we found that higher levels of democracy mean *lower* levels of tanks, artillery pieces, and air sorties, and in the case of artillery pieces the relationship is statistically significant.[82] In short, we can confidently state that democratic armies do not enjoy significantly higher levels of war matériel on the battlefield than do autocratic armies.[83]

An alternative approach to estimating extraction effort, rather than looking at the proportion of society a state puts under arms or the number of weapons they place in the field, is to look directly at what proportion of the economy the state devotes to defense spending. One way to measure this is to look at the fraction of the nation's GDP that is devoted to military spending. Analysis of these data supports the proposition that democracies devote proportionally no more or less to their military efforts, in peace or in war.[84]

Thinking about this problem from the domestic political perspective, it is easy to understand the inability or unwillingness of democracies to extract more resources from their populations for war. Democratic states fund marginal increases in military expenditures through either higher taxes or deficit spending, both of which must ultimately lead to reductions in public consumption—the proverbial guns versus butter tradeoff. As Immanuel Kant

---

81. Ibid.
82. Ibid.
83. This result parallels the finding that there was not consistent evidence that among great powers since 1860 democracies built more capital ships than did nondemocracies. Sean Bolks and Richard J. Stoll, "The Arms Acquisition Process: The Effect of Internal and External Constraints on Arms Race Dynamics," *Journal of Conflict Resolution*, Vol. 44, No. 5 (October 2000), pp. 580–603.
84. Reiter and Stam, *Democracies at War*, chap. 5.

observed, the public prefers not to spend on military ventures at the expense of individual consumption, and the constraints imposed by democratic political institutions help impose limits on the abilities of democratic elites to raise levels of military spending. During the Vietnam War, for example, President Lyndon Johnson did not call up the reserves as part of the summer 1965 escalation of the war in part because he feared the economic costs of a reserve call-up would threaten his Great Society programs. Nor did he push for tax increases to fund the war effort. Three years later, he chose not to escalate the war following the Tet offensive, in part due to fears that such an action would threaten the U.S. economy, vulnerable because of the concurrent international gold crisis. A change in party in the White House in 1969 did not free up economic resources for the war. Budgetary pressures led to scaling back operations in Vietnam, including reducing the number of strategic and tactical air sorties flown, withdrawing two tactical fighter squadrons, and implementing naval reductions.[85] During the months preceding the 1991 Gulf War, President Bush defused Congressional grumbling over the financial cost of the coming war by securing substantial financial support for the war effort from other countries.[86] Similarly, in World War II the principal autocratic belligerents, Germany, Japan, Italy, and the Soviet Union, extracted far more from their societies (and imposed far greater suffering on their peoples) than did the principal democratic belligerents, the United States and Britain.[87] And, in

---

85. Berman, *Planning a Tragedy*, pp. 122–127; Robert S. McNamara, *In Retrospect: The Tragedy and Lessons of Vietnam* (New York: Times Books, 1995), p. 198; George C. Herring, *America's Longest War: The United States and Vietnam, 1950–1975*, 3d ed. (New York: McGraw-Hill, 1996), pp. 222–223; Lyndon Baines Johnson, *The Vantage Point: Perspectives of the Presidency, 1963–1969* (New York: Holt, Rinehart, and Winston, 1971), pp. 406–407; and Lewis Sorley, *A Better War: The Unexamined Victories and Final Tragedy of America's Last Years in Vietnam* (New York: Harcourt Brace, 1999), pp. 124–125.
86. Bush and Scowcroft, *A World Transformed*, pp. 359–360.
87. Jerome B. Cohen, *Japan's Economy in War and Reconstruction* (Minneapolis: University of Minnesota Press, 1949), especially p. 353; Richard B. Frank, *Downfall: The End of the Imperial Japanese Empire* (New York: Random House, 1999), p. 351; Akira Hara, "Japan: Guns Before Rice," in Mark Harrison, ed., *The Economics of World War II: Six Great Powers in International Comparison* (Cambridge: Cambridge University Press, 1999), p. 256; Vera Zamagni, "Italy: How to Lose the War and Win the Peace," in *The Economics of World War II*, pp. 190–191; Richard Overy, *Why the Allies Won* (London: Jonathan Cape, 1995); John Barber and Mark Harrison, *The Soviet Home Front, 1941–1945: A Social and Economic History of the USSR in World War II* (London: Longman, 1991); Alan S. Milward, *The German Economy at War* (London: Athlone, 1965), pp. 10–11; Lothar Burchardt, "The Impact of the War Economy on the Civilian Population of Germany during the First and Second World Wars," in Wilhelm Deist, ed., *The German Military in the Age of Total War* (Warwickshire, U.K.: Berg, 1985), pp. 62–64; Werner Abelshauser, "Germany: Guns, Butter, and Economic Miracles," in Harrison, *The Economics of World War II*, pp. 152–153; Richard J. Overy, *War and Economy in the Third Reich* (Oxford: Clarendon, 1994), pp. 283–285, 312; and Mark Harrison, "Resource Mobili-

World War I, autocratic Germany demanded more from its society than did its democratic adversaries.[88]

The evidence does not indicate that democracies win wars because they have greater military-industrial capability, or extract more from their economies during wartime. One remaining possibility is that regime type may affect how efficiently a belligerent manages its economy during wartime. Examining a handful of cases, we discovered a curvilinear effect between regime type and states' management of their wartime economies. Democratic and highly repressive states are more effective at wartime economic management than oligarchic regimes such as World War II Japan.[89]

Democracy and War Outcomes in the Twenty-first Century

Both selection effects theory and the warfighting conjecture that democracies have higher military effectiveness help explain why democracies win their wars. More generally, selection effects theory is consistent with a broad range of observed empirical patterns: (1) democracies are more likely to win the wars they initiate; (2) democracies are more likely to win the crises they initiate; (3) democracies fight shorter wars; (4) democracies suffer fewer military and civilian casualties in their wars; (5) democracies become more likely to initiate wars as their prospects for success increase; (6) as war endures, democratic public support for war declines, and chances for democratic victory drop; and (7) democracies sometimes initiate genocidal wars of empire, and sometimes use covert force against other democracies.

We wrote most of our 2002 book before the al-Qaida attacks of September 11, 2001, never imagining that in the first decade of the twenty-first century the United States would initiate two major wars. With the benefit of hindsight, how well do our arguments explain U.S. behavior and experiences in the Iraq and Afghanistan Wars?

These wars cast some favorable light on our principal arguments. The

---

zation for World War II: The U.S.A., U.K., U.S.S.R., and Germany, 1938–1945," *Economic History Review*, Vol. 41, No. 2 (May 1988), p. 184.

88. Millett and Murray, eds., *Military Effectiveness*, Vol. 1; and Burchardt, "The Impact of the War Economy," p. 43. On Britain, see Sir Ernest Llewellyn Woodward, *Great Britain and the War of 1914–1918* (London: Methuen, 1967), pp. 498–513; Sidney Pollard, *The Development of the British Economy, 1914–1990*, 4th ed. (London: Edward Arnold, 1992), p. 19; and Niall Ferguson, *The Pity of War: Explaining World War I* (New York: Basic Books, 1999), p. 276. On the United States, see James A. Huston, *The Sinews of War: Army Logistics, 1775–1953* (Washington, D.C.: United States Army, 1966), p. 328.

89. Reiter and Stam, *Democracies at War*, chap. 5.

United States and its democratic allies clearly won the conventional phases of both these wars. Both wars enjoyed popular consent of a majority of American voters when the Bush administration initiated them. The U.S. government sought in both cases to fight wars of short duration and limited cost. U.S. troops fought well, and Iraqi soldiers in particular surrendered in droves.

That being said, there were two areas of experience which were unexplained by our theory, and deserve further study. First, while the marketplace of ideas successfully predicted that the conventional phase of the wars would be straightforward and successful, it did fail to inform U.S. policymakers in other areas. President George W. Bush and many U.S. citizens and policymakers greatly overestimated the connections between Saddam Hussein and terrorist groups as well as the extent of Iraqi weapons of mass destruction programs. The American marketplace of ideas, if it had operated as the theory predicted, should have exposed the weakness of the evidence supporting these claims. Further, the Bush administration failed to enter both the Iraq war and the war in Afghanistan with a sufficiently well informed understanding of what the postconventional war phase would be like, and what policies would be necessary to maximize the chances for peaceful, democratic transitions. The theory would predict that a depoliticized governmental bureaucracy as well as an open debate in society should have better informed the president as to the risks of subsequent civil war and the costs that type of war creates for third parties.

Second, though the United States did quite well in the conventional phases of both wars, the same cannot be said about the insurgent civil war phases which followed the conventional phases. In Iraq, the United States eventually helped achieve something like an acceptable outcome, though after an unexpectedly costly and lengthy insurgency phase. The outcome in Afghanistan is at this writing (summer 2010) yet unknown. The puzzle is that while democracies seem to be quite effective at fighting conventional wars, they do not seem to enjoy such advantages fighting wars of insurgency. Our theory may offer one solution to this puzzle. Any war dragging on for years and imposing escalating costs in friendly casualties and fiscal expenditures will inevitably suffer declining public support, eventually forcing elected leaders to grasp for any politically palatable way out, even if it means forgoing clear victory. At this writing, it appears that just the sort of negotiated settlement our research suggest becomes likely after years of fighting may be what lies in store for President Obama as he makes decisions about the impending U.S. exit from Afghanistan.

Democracies' ability to defend themselves and moreover to win the wars they initiate augurs wells for the ongoing democratic experiment. Realists' fears that democracy is a luxury states cannot afford during times of military crises are overblown. Peaceful communities of democracies need not fear their destruction by authoritarian foes as many critics of democracy have asserted. Rather, it is now clear, from a security perspective, a world populated with democracies would be both a more secure as well as a safer place, something democracies' critics could not have imagined before the research of the past decade laid clear democracy's fourth virtue.

Powerful Pacifists

David A. Lake

Democratic States and War

$\mathbf{N}$o less likely to fight wars in general, democracies are significantly less likely to fight each other. The relative pacifism of democracies remains a puzzle. Although it has attracted significant attention, most of the recent work has been empirical[1] or philosophical.[2] No *theory* presently exists that can account for this striking empirical regularity.

Equally important, but far less widely recognized, is the propensity of democracies to win the wars that they do fight. In view of the drawbacks commonly associated with the democratic conduct of foreign policy[3] this finding poses even more of a conundrum. If democratic decisionmaking is often slow, inept, naive, and prone to stalemate, how and why do democracies typically triumph over their faster, more professional, sophisticated, and decisive autocratic brethren? In the ultimate contest of national strength and will, why are democracies more powerful?

I offer one possible explanation for this syndrome of powerful pacifism drawn from a larger theory of grand strategy.[4] I argue that autocratic states, which typically earn rents at the expense of their societies, will possess an imperialist bias and tend to be more expansionist and, in turn, more war-prone.

This article was originally published in the *American Political Science Review*, Vol. 86, No. 1 (March 1992), pp. 24–37. © Cambridge University Press. Reprinted by permission from Cambridge University Press. Earlier versions of this paper were presented at the annual meetings of the International Studies Association, Vancouver, 1991 and London, 1989, and the American Political Science Association, Washington, 1988. I would like to thank the participants in these meetings and especially Richard Anderson, Steven Brams, Bruce Bueno de Mesquita, Juliann Emmons, Jeff Frieden, Joanne Gowa, Robert Keohane, Wendy K. Lake, David Latzko, Joel Migdal, Mancur Olson, Steve Postrell, Eric Rasmussen, Ronald Rogowski, Bruce Russett, Gary Schwartz, Kenneth Waltz, and David Wilkinson for helpful comments. I am also indebted to Stephen Ansolabehere and Michael Harrington for their assistance with the data analysis and Michael Wallerstein for his help with the graphics. The generous financial support of the Academic Senate of the University of California, Los Angeles, the Center for International and Strategic Affairs, and the University of California Institute on Global Conflict and Cooperation is gratefully acknowledged.

1. Zeev Maoz and Nasrin Abdolali, "Regime Types and International Conflict, 1816–1976," *Journal of Conflict Resolution*, Vol. 33, No. 1 (March 1989), pp. 3–35; and Melvin Small and J. David Singer, "The War-Proneness of Democratic Regimes, 1816–1965," *Jerusalem Journal of International Relations*, Vol. 1, No. 4 (Summer 1976), pp. 57–69.
2. Michael W. Doyle, "Kant, Liberal Legacies, and Foreign Affairs," *Philosophy and Public Affairs*, Vol. 12, No. 3 (Summer 1983), pp. 205–235.
3. For an extreme view, see Michael J. Crozier, Samuel P. Huntington, and Joji Watanuki, *The Crisis of Democracy* (New York: New York University Press, 1975). Bruce Russett argues that cyclical majorities may be the greatest problem facing the democratic formulation of foreign policy. Russett, *Controlling the Sword: The Democratic Governance of National Security* (Cambridge, Mass.: Harvard University Press, 1990), pp. 115–117.
4. See David A. Lake, "Superpower Strategies: The State and the Production of Security," unpublished manuscript, University of California, Los Angeles, 1991.

To the extent that democratic states tend to be more constrained by their societies from earning rents, wage their own wars of expansion under more restricted conditions, possess greater incentives to intervene in the domestic political affairs of autocracies, and become objects of autocratic expansion, there should be no significant overall difference in frequency of war involvement. In the absence of this imperialist bias, however, democracies should be relatively pacific in their relations with each other.

Moreover, democratic states, because they are constrained from earning rents, will tend to create fewer economic distortions, possess greater national wealth, and devote greater absolute resources to national security. They will also tend to enjoy greater societal support for their policies. To the extent that states balance threats rather than power, democracies will also tend to form overwhelming countercoalitions against expansionist autocracies. Together, these three propositions imply that democratic states should be more likely to win wars.

A Theory of the Rent-Seeking State and Foreign Policy

Stimulated by the pioneering work of Lane, the microeconomic theory of the state conceives of the state as a profit-maximizing firm that trades services for revenues.[5] I define *profit* here as the difference between revenues acquired by the state and its real costs of producing services and collecting revenue. Real costs are determined, in turn, by the fair market value of the resources consumed by the state in the production and collection processes.[6] Thus, following normal usage, profit includes both what is commonly called *normal or*

---

5. See Frederic C. Lane, *Profits from Power: Readings in Protection Rent and Violence-controlling Enterprises* (Albany: State University of New York Press, 1979). This theory has also been referred to as the neoclassical theory of the state (see Douglas C. North, *Structure and Change in Economic History* [New York: Norton, 1981]) and the predatory theory (see Margaret Levi, "The Predatory Theory of Rule," in Michael Heckter, ed., *The Microfoundations of Macrosociology* [Philadelphia: Temple University Press, 1983]; and Margaret Levi, *Of Rule and Revenue* [Berkeley: University of California Press, 1988]). Both of these appellations have misleading normative implications. I prefer the more neutral term used here. See also Richard D. Auster and Morris Silver, *The State as Firm: Economic Forces in Political Development* (Boston: Martinus Nijhoff, 1979). Within this approach there is some disagreement about the objective that states pursue. North and Levi, for instance, both assume that states maximize revenue, while I assume that states seek profits. For an elaboration of this debate and defense of the profit-maximizing assumption, see David A. Lake, "The State and the Production of International Security: A Microeconomic Theory of Grand Strategy," paper presented at the annual meeting of the American Political Science Association, San Francisco, California, 1990.
6. In producing protection and collecting revenues, the state consumes resources—everything from paper clips and stationary to stealth bombers and nuclear weapons. States also consume labor. As noted, the real cost of these resources is determined by their fair market value or (in cases where the state is a dominant consumer) the price the state could obtain if it exercised its monopsonist power—a factor much more important for weapons than, say, for office supplies.

economic profit, defined by the opportunity cost of the factors of production or their return in their next-best occupation, and *supernormal profits or rents,* returns greater than what is necessary to sustain the factors of production in their present use.[7]

Foremost among the services provided by the state is protection. For purposes of the following analysis, protection is further restricted to mean only defense from external threats. While I focus on this single "industry," the argument can easily be generalized to other state services; and rents earned by the state on any service in any area under its control are sufficient to create an imperialist bias in its grand strategy.

Many competing definitions of the state exist. In this theory, the state, a service-producing firm, is defined by two functional attributes. Since the protection industry typically enjoys economies of scale over an extensive geographic area, the state forms a natural but local monopoly.[8] As a result, one and only one state will exist in any area at any given time. Similarly, as protection from foreign threats forms a local public good, whose suboptimal provision is otherwise ensured by the large number of citizens involved, the state will supply this service only if it is granted, or is able to obtain, some coercive ability over its society. These two aspects of the protection service, local monopoly and coercive supply, are consonant with Weber's classic definition of the state as a "compulsory organization with a territorial basis" that monopolizes the legitimate use of force.[9]

THE DEMAND AND SUPPLY OF PROTECTION

All individuals possess a positive demand for protection; and as price declines, they increase the quantity they demand. Accordingly, the demand curve for protection slopes downward and to the right, as in Figure 1.[10] The precise slope is a function of societal preferences (which are considered to be

---

7. See Gordon Tullock, "Rent Seeking as a Negative Sum Game," in James Buchanan, Robert D. Tollison, and Tullock, eds., *Toward a Theory of the Rent-seeking Society* (College Station: Texas A&M University Press, 1980).
8. Lane, *Profits from Power,* p. 23.
9. Max Weber, *Economy and Society,* 2 vols., ed. Guenther Roth and Claus Wittich (Berkeley: University of California Press, 1978), pp. 1, 58.
10. The public nature of the good protection complicates the definition of price and demand. By price, I mean the rate at which the state extracts wealth (for practical purposes, the tax rate [t] multiplied by the inverse of the probability (p) of successful evasion (free riding); that is, price = t (1 − p). By *positive demand* I mean that individuals prefer some level of protection at some positive cost to themselves (i.e., price) to no protection at zero cost. Strictly speaking, the demand for public goods is summarized in an evaluation schedule, which differs from a traditional demand curve only in that individual preferences are added vertically, rather than horizontally, in order to reflect

Figure 1. The Supply and Demand for Protection under Monopoly Provision

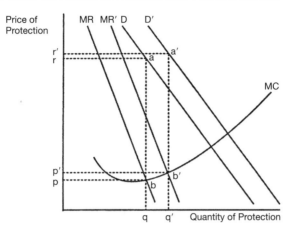

exogenous) and the availability of substitutes (which, given the nature of protection, tend to be few), suggesting that demand is relatively inelastic and the curve correspondingly steep. Although theoretically possible, it is unlikely that the demand for protection is perfectly inelastic (i.e., that the demand curve is perpendicular to the horizontal axis), since this implies that society is willing to pay virtually any price for even small amounts of the service. To the extent that society places any value on goods and services other than protection, or, if there is any trade-off between guns and butter, the demand curve must possess at least a slight negative slope. It is equally unlikely that the demand for protection can be saturated. Historical experience suggests that even high levels of defense spending do not create feelings of total security. To the extent that the security dilemma holds at all,[11] the demand for protection can be sated not at some high level of military spending but, rather, only in a world where everyone else is completely disarmed.[12] In short, under most feasible conditions, the demand curve for protection is sensitive to price.

nonrivalry in consumption. Following common practice, I nonetheless refer to this evaluation schedule as a demand curve. On the treatment of public goods within a Marshallian economic framework, see the seminal work by James M. Buchanan, *The Demand and Supply of Public Goods* (Chicago: Rand McNally, 1968). My approach differs from that of Buchanan (who adopts an essentially pluralist approach to politics) by including a theory of the state as a monopoly provider of public goods. This has important implications for the point of equilibrium production. See n. 13.

11. Robert Jervis, "Cooperation under the Security Dilemma," *World Politics,* Vol. 30, No. 2 (January 1978), pp. 167–214.

12. Plotting the probability of success in armed conflict against the quantity of resources mobi-

The level of protection demanded by society, in turn, is primarily a function of the level of external threat. The greater the external threat, ceteris paribus, the higher the demand for protection will be. This is depicted in Figure 1 as a shift outward in the demand curve ($D' > D$).

As in any monopoly, the state enjoys some measure of market power and can, within the limits set by the demand schedule, control the quantity of the good (protection) supplied; and as with any profit-maximizing firm, the state will set its output at the level that equates marginal cost with marginal revenue and charge what the market will bear.[13] In Figure 1, price p represents the normal profit level; but since in all monopolies the profit-maximizing level of production lies below the demand curve, the final price charged to consumers will be higher and may be as high as r. The difference between p and the price charged, say r, defines the rent or supernormal profit earned by the state—represented graphically by the rectangle p-r-a-b.[14]

Unlike other monopolists, however, states can also act opportunistically

---

lized for war, Richard M. Emerson argues that protection forms an S-shaped curve and that once the probability of success reaches 1.0, the country has entered a "zone of redundancy" in which it has "overmobilized." See Emerson, "Charismatic Kingship: A Study of State-Formation and Authority in Baltistan," *Politics and Society*, Vol. 12, No. 4 (December 1983), pp. 413–444. This clearly implies that the demand for protection can be sated. As I have argued however, the security dilemma (and the action-reaction syndrome implied therein) is likely to prevent the probability of success from reaching its upper bounds. Emerson further argues that charismatic leaders are most likely to arise when the probability of success is in its middling range and that once a state enters the zone of redundancy, it and its leadership will decay. Interestingly, Emerson's thesis is contradicted by his own evidence. The ancient states of Balti (the focus of his essay) began to decline in 1846 when they were absorbed by the maharaja of Kashmir and, behind him, the British Government of India (ibid., p. 438). This absorption, of course, implies that the Balti's probability of military success against these superior powers was significantly less than 1.0.
13. In pure competition, the firm also produces where marginal cost (MC) equals marginal revenue (MR). Facing a horizontal demand curve, however, marginal revenue equals price and, in turn, demand (D); and the competitive firm, in theory, produces where marginal cost equals demand, that is, where MR = D = MC = MR. In Buchanan's essentially pluralist view, this is also the Pareto optimal equilibrium for the production of public goods and the point against which all distortions must be measured (see Buchanan, *The Demand and Supply of Public Goods*). Adding a concept of the state as a monopoly provider of public goods significantly alters this equilibrium, however. With a downward-sloping monopoly firm demand curve, supplying protection where marginal cost equals demand would result in a significantly larger quantity; but all units to the right of q in Figure 1 would be produced at a loss (MC > MR). What is efficient for society is inefficient for the state, and vice versa. I shall argue that society can control, in part, the quantity of monopoly rents earned by the state; likewise, society can control in part the level of production, with the resulting equilibrium arising somewhere between MR = MC (the stat's preferred level) and MC = D (society's preferred level). For expositional clarity, I shall assume henceforth that production occurs where MR = MC and that political conflict occurs primarily over the level of monopoly rents earned by the state. Relaxing this assumption, however, only reinforces my arguments on democracy and the syndrome of powerful pacifism.
14. At the minimum, the protection service must generate normal profits for state officials, defined as the reservation wage; if the state fails to earn at least normal profits, the individuals who comprise it will leave and assume other employment.

against their own societies by artificially increasing the demand for their services through extortion or racketeering. Extortion occurs when states magnify, exaggerate, or "oversell" foreign threats to society, whether by supplying incomplete information or engaging in outright deception.[15] States conduct protection rackets by actively creating foreign threats, from which they then protect society.[16] In both cases, a state effectively shifts the demand curve outward ($D' > D$) and thereby earns greater rents ($p'-r'-a'-b' > p-r-a-b$).

Two important points follow from this analysis. First, the level of protection supplied by the monopoly state will always be less than that produced under conditions more closely approximating "pure competition."[17] While society would benefit from higher levels of protection, given prevailing costs of production, the profit-maximizing or optimal strategy for the state is to restrict supply—whether it can successfully capture the potential rents or not. Insecurity is an inherent feature of life under the modern state.

Second, the state faces strong incentives to seek rents at the expense of society. In other words, the state can benefit itself by charging consumers the monopoly price for protection (r in Figure 1) and by artificially inflating demand through extortion or racketeering.

SOCIETAL CONSTRAINTS ON STATE RENT SEEKING

Consumers clearly prefer to purchase protection at the lowest sustainable price (p in Figure 1). The state, on the other hand, clearly prefers to sell protection at the highest possible price, which is determined by the slope of the demand curve and represented by r. The actual price—and the level of rents extracted from society—is determined by how well individual citizens can control or regulate the rent-seeking behavior of the state. Society's ability to control the state, in turn, is influenced by the costs of three separate activities: monitoring state behavior, voice, and exit.[18]

In order to control the state, individuals must first monitor its performance and acquire information on the strategies it is pursuing, its real costs of protection production, the level of foreign threat, and the like. Monitoring the state is

---

15. See Edward Ames and Richard T. Rapp, "The Birth and Death of Taxes: A Hypothesis," *Journal of Economic History*, Vol. 37, No. 1 (March 1977), pp. 161–178; and Theodore J. Lowi, "Making Democracy Safe for the World," in James N. Rosenau, ed., *Domestic Sources of Foreign Policy* (New York: Free Press, 1967).

16. See Charles Tilly, "War Making and State Making as Organized Crime," in Peter B. Evans, Dietrich Rueschemeyer, and Theda Skocpol, eds., *Bringing the State Back In* (New York: Cambridge University Press, 1985).

17. This conclusion follows even without the assumption advanced in n. 13 as long as society faces positive costs for monitoring state behavior and engaging in exit and voice.

18. See Albert O. Hirschman, *Exit, Voice, and Loyalty: Responses to Decline in Firms, Organizations, and States* (Cambridge, Mass.: Harvard University Press, 1970).

analogous to principal-agent problems in publicly held corporations, where the stockholders (the principals) seek to ensure that the managers (the agents) work hard and faithfully in their interests.[19] The "problem" arises because no single stockholder typically has any incentive to invest in, or acquire information on, the manager's true performance—information, of course, it is assumed that the managers themselves possess. Collective action problems also stymie any group investment in information. The higher the costs of acquiring information, the less control the stockholders can exert over the firm and the more the managers can shirk or adopt policies that benefit themselves at the expense of their principals. The same is true of the state-society problem: no single citizen has any incentive to invest in information; and, because of the free-rider problem, collective investment in information occurs only at suboptimal levels.[20] The higher the costs of acquiring information regarding state performance, the greater latitude state officials possess to engage in rent-seeking behavior.

Once performance and the level of state rents have been assessed, individuals have two instruments through which to alter or change state behavior: exit and voice. Through exit, individuals move and deplete both the resource base of the state, raising its real costs, and its market for protection, lowering the price it can charge. Discipline is imposed upon the state, in other words, by reducing its profits, thereby punishing it for undesirable behavior.

As Hirschman notes, exit is the quintessential economic strategy: if a consumer ceases to like a product, he or she simply stops purchasing it.[21] Consumers of state-provided services do not have quite the same freedom. Because protection is a public good supplied by a local monopoly, individuals cannot choose to consume varying amounts or qualities; indeed, because it is coercively supplied, individuals cannot choose whether or not to consume any protection at all. Nonetheless, political exit can occur, although it takes different forms and usually entails a higher cost than exit from private goods con-

---

19. On principal-agent problems, see Eugene F. Fama, "Agency Problems and the Theory of the Firm," *Journal of Political Economy*, Vol. 88, No. 2 (April 1980), pp. 288–307; Eugene F. Fama and Michael C. Jensen, "Separation of Ownership and Control," *Journal of Law and Economics*, Vol. 26, No. 2 (June 1983), pp. 301–325; Eugene F. Fama and Michael C. Jensen, "Agency Problems and Residual Claims," *Journal of Law and Economics*, Vol. 26, No. 2 (June 1983), pp. 327–349; Michael C. Jensen and William H. Meckling, "Theory of the Firm: Managerial Behavior, Agency Cost, and Ownership Structure," *Journal of Financial Economics*, Vol. 3, No. 4 (October 1976), pp. 305–360; and Stephen A. Ross, "The Economic Theory of Agency: The Principal's Problem," *American Economic Review*, Vol. 63, No. 2 (May 1973), pp. 134–139.
20. A free press is an exception to this rule. News organizations specialize in collecting information (especially on state activities) and selling it to the public.
21. Hirschman, *Exit, Voice, and Loyalty*, p. 15.

sumption. Individuals can choose to migrate or "vote with their feet" (or assets), moving from high-rent to low-rent areas.[22] Freedom of emigration is often one of the first rights obtained in the process of democratization (as was recently witnessed in Eastern Europe), suggesting that the right to exit and democracy will tend to be conjoined, with the former serving as the ultimate guarantor against failures of the latter. Similarly, territories can secede or threaten to secede. If large enough, they can try to form an independent state or, failing that, they can try to merge with other secessionist or democratic territories into a larger union. While possible, exit is costly to the individuals or territories that choose it. The higher the cost of exit, the greater the ability of the state to earn rents.

Voice, or political participation, disciplines the state by separating or threatening to separate state officials from their offices: the citizens stay, but the composition of the state changes. Voice can take many forms, from voting, to campaign contributions, to mass unrest, to active rebellion.

At the individual level, the costs of political participation are unevenly distributed across society. This occurs for a host of idiosyncratic reasons. For instance, not all countries have universal suffrage, the opportunity cost of a campaign contribution is significantly less for a multimillionaire than for a welfare family, and the military can overthrow a leader with greater ease than can unarmed civilians.

At the aggregate or national level, the costs of political participation vary by regime type. For instance, in most democracies, where elections are the primary focus of political participation for the majority of citizens, it is relatively costless to vote and exercise voice.[23] At the other extreme, autocratic states typically suppress political dissent; and voting, if it occurs at all, is ineffective in removing officials from power. In these countries, to replace, or effectively threaten to replace, a ruler requires either mass unrest or some form of armed

---

22. See Charles M. Tiebout, "A Pure Theory of Local Expenditures," *Journal of Political Economy*, Vol. 64, No. 5 (October 1956), pp. 416–424. This is analogous to North's concept of external competition, where other states exist to act as alternative suppliers of protection (see North, *Structure and Change in Economic History*). On asset mobility, see Robert H. Bates and Da-hsiang Donald Lien, "A Note on Taxation, Development, and Representative Government," *Politics and Society*, Vol. 14, No. 1 (March 1985), pp. 53–70. The phrase *low rent* refers not to the overall tax rate but only to the level of supernormal profits earned by the state on the services it provides. Different states are likely to provide different mixes of services with corresponding tax levies. Following Tiebout, emigrants are likely to select the polity that offers the mix of services they want at the price they are willing to pay. The analysis here focuses only on state rents, which distort the implications of the now-standard Tiebout analysis.

23. Although the direct costs for voting are small to nil, there are certainly opportunity costs for activities forgone and possibly some indirect costs (e.g., transportation to the polling station).

rebellion—activities that carry considerably higher individual costs, including the possibility of death. In these polities, voice is very costly to the average citizen and, as a result, is seldom exercised. It follows that the higher the costs of political participation, on average, the greater the state's ability to earn rents will be.[24] The relatively low cost of political participation in democracies constrains the state's rent-seeking ability, whereas the relatively high cost of political participation in autocracies frees the state to earn rents.

In determining its level of feasible rents, the state will act as a discriminating monopolist, charging citizens in general and each separate citizen in particular a price for protection positively related to the costs of monitoring, exit, and political participation. Indeed, the state will charge up to—but in equilibrium not more than—the price at which individuals would be tempted either to exit or to engage in the lowest-cost form of political participation available to them that would effectively remove current state officials from power.

EXPANSION AND THE RENT-SEEKING STATE

In practice, there are always positive costs of monitoring state behavior and exercising voice and exit. As a result, all states possess some ability to earn rents, although the ability will be larger in autocracies than in democracies for the reasons just surveyed. To the extent that a state can earn rents, state and societal interests will diverge and the state will be biased toward an expansionary foreign policy.[25] This relationship is continuous. The higher the costs to society of controlling the state, the greater will be the rent-seeking ability of the state, the more the interests of state and society will diverge, and the more expansionist the state will become.[26] This imperialist bias arises for three reasons.

First, expansion may increase the state's rent-seeking ability by reducing the

---

24. This is not to deny that some groups within autocratic polities can participate at relatively low cost and thus have significant influence over the state. In these circumstances, it is expected that the powerful social groups would share in the rents earned by the state. The argument here hinges on the average cost for society as a whole. See also n. 25.
25. Expansionary policy is to be taken in opposition to unilateralism and cooperation. See Lake, "The State and the Production of International Security"; and Lake, "Superpower Strategies." This analysis contrasts with other explanations of so-called overexpansion or with a country's expansion beyond the point where the marginal costs equal the marginal gains to society. Rather than being driven by perverse incentives within the international system, cognitive bias, dominant social groups, or log rolls between concentrated social interests (see Jack Snyder, *Myths of Empire: Domestic Politics and International Ambition* [Ithaca, N.Y.: Cornell University Press, 1991]), the explanation developed here focuses on the monopoly structure of the protection industry and the costs to society of regulating state behavior.
26. This argument applies only for rents earned and retained by the state. If the state is merely a conduit for redistributing wealth between social groups, no imperialist bias will emerge. Thus, I differ crucially from much of the extant literature on rent seeking, which tends to focus on the ac-

benefits of exit. The net benefit of exit to any individual (and thus that individual's incentive to engage in this action) is determined by both the *push* of high rents at home and the *pull* of lower rents abroad. When all states extract equally high rents, there is no incentive to move. If through expansion a state can eliminate or engulf a low-rent competitor, it increases its own ability to earn rents. This suggests that low-rent states will often be objects of expansion for rent-seeking autocracies.

Second, a state may also expand so as to provoke others into threatening its own society. Both extortion and racketeering rest upon persuading citizens that foreign threats are larger and more real than they are or otherwise would be. If successful, the state convinces consumers to increase their demand for protection and, in turn, earns more rents. Even with incomplete and costly information, however, citizens may eventually discern the true level of threat and lower their demand. Through expansion short of universal empire, the state lends credibility to extortion and supports racketeering, thereby strengthening its ability to earn rents at the expense of society.

Third, and most important, the larger the state's rent-seeking ability, the higher the total revenue earned by the state. The more revenue (ceteris paribus), the larger the optimal size of the political unit. These relationships are depicted in Figure 2.

For all states, an optimal size exists defined by the costs of collecting revenue and producing protection and the revenues earned by providing this service to society. Each additional unit of territory acquired by the state produces additional revenue: the state becomes the new local monopoly supplier of protection, and it taxes consumers in that region accordingly.

On the other hand, the costs of governing rise with the size of the political unit, placing an effective cap on the size of nation-states. These resource costs occur primarily in the form of transactions costs of revenue

---

tions of *social* groups. See Mancur Olson, *The Rise and Decline of Nations: Economic Growth, Stagflation, and Social Rigidities* (New Haven, Conn.: Yale University Press, 1982). While social rent seeking will reduce national wealth as well, I have yet to see a convincing argument that democracies are more prone to social rent seeking than autocracies. Interestingly, even though Olson adopts a pluralist conception of the state, he argues that his theory is supported by evidence from both developed democracies and nondemocratic, non-Westernized polities (ibid., pp. 146–180). Although I do not find it persuasive, Robert B. Ekelund and Robert D. Tollison do argue that democracies will actually experience less societal rent seeking than autocracies. See Ekelund and Tollison, *Mercantilism as a Rent-seeking Society: Economic Regulation in Historical Perspective* (College Station: Texas A&M University Press, 1981). My intuition is that democratic and autocratic societies extract similar levels of rents but distribute them over greater and smaller sets of groups, respectively. I contend here, however, that rents earned by democratic *states* are significantly less than the rents extracted by autocratic *states*.

Figure 2. The Optimal Size of Political Units

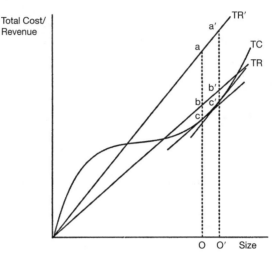

collection.[27] Over some limited range, the state may enjoy increasing returns to scale in revenue collection; but soon, the addition of more territory begins to strain the administrative abilities of the state, leading to diminishing returns.

When combined with the costs of producing protection, the state's total cost curve typically resembles that in Figure 2. Economies of scale in protection, in conjunction with initial increasing returns in revenue collection, suggest that the slope of the total cost-curve declines for some substantial distance, flattens as the marginal costs of revenue collection begin to rise more rapidly, and eventually increases as the costs of revenue collection accelerate and the economies of scale in protection are exhausted.

If revenues increase monotonically, there is a single optimal size of the political unit where marginal revenue equals the marginal costs of collection and production. Geometrically, this occurs where a line tangent to the cost curve is equal to the slope of the revenue line, as at size *O*. At *O*, the economic profit to the state is measured by the line segment bc. No profit-maximizing state has any incentive to expand beyond the point where marginal cost is equated with marginal revenue.

Rents earned by the state, however, cause the total revenue line to rotate counterclockwise from the origin (*TR'* > *TR*). Although the total cost curve

27. See Levi, *Of Rule and Revenue;* and North, *Structure and Change in Economic History.*

may also increase as state rent seeking stimulates higher transactions costs of revenue collection, the curve must rise at a lagging rate. Higher state rents do not increase the costs of producing protection from external threats per se, and this is likely to be the major component of the cost curve. And even if the transactions costs of revenue collection increase, because important social groups demand compensation or public unrest must be suppressed, collective action problems thwart a fully countervailing societal response. Assuming for expositional clarity that total costs remain constant, state rent seeking raises total revenue and expands the optimal size of the political unit from O to O'. Intuitively, with rents, each unit of territory produces greater revenue for the state. The greater the revenue, the greater the equilibrium costs that can be borne to capture that revenue. Thus, a state with an increased rent-seeking capacity has an incentive to expand until marginal revenue and marginal cost are once again equalized at a new, larger size.

While this expansion (from O to O') benefits the state, it harms society. With TR' representing the revenue of a successful rent-seeking state, the line segment $b'c'$ will be the state's economic profit and $a'b'$ its rent, with $a'b'$ being redistributed away from society to state officials. Not only is society exploited by state rent seeking, but it is doubly hurt by the additional expansion rent seeking induces. In the absence of expansion, the state's rents would be ab; with expansion, these rents increase to $a'b'$. Simple observation suffices to show that $a'b'$ will always be longer than ab.

Both the original citizens or consumers of state-produced services and the individuals newly incorporated into the territorial unit pay the new higher, expansion-induced price for protection. In other words, greater rents are extracted from both the original and augmented populations, although the exact rate at which these groups will be taxed is determined by their relative costs of controlling the state. The previously foreign population does not bear the burden of the higher rents alone; all consumers in that particular state face a higher price for protection. Imperialism is not simply a means for extracting wealth from "foreign" territories: it is a tool used by the state for exploiting its own society as well.

While this argument has been developed only in terms of a single state service (protection), any rents earned on any service provided by the state increase the optimal size of the political unit. Even if rents are earned from providing, say, the physical infrastructure, the net benefits to the state of providing that good still increase and provide an incentive for further expansion. Fully generalized to all areas of state service, the state optimizes where the marginal revenue (including rents) equals the marginal costs of providing all services and collecting all revenues. Given that the demand for protection is likely to be more inelastic than the demand for other state services, however, the highest rents are likely to be earned in this industry. Nor is there any apparent reason why societal constraints on the state should differ significantly across areas of service.[28] Hence, it is appropriate to focus on the production of protection. Nonetheless, it is important to note that in a fully generalized theoretical framework, any rents earned by the state are sufficient to generate an imperialist bias.

It is virtually impossible to measure state rents directly. Perhaps in past centuries, when the public fisc and the private purse of the ruler were one and the same, it might have been possible to observe the rents earned by the state by measuring the comparative opulence of the royal court. If the general argument developed here is correct, however, then surely one of the primary tasks of modern state budgeting techniques is to obscure the difference between normal and supernormal state profits. It also follows that in the countries where they are most easily observed, rents will be relatively low—even zero, at the extreme.[29] Given these measurement difficulties, no direct test of the theory is possible. Rather, it can be assessed—and the presence of state rents revealed—only by examining the theory's behavioral implications. Accordingly, I shall examine several hypotheses derived from the theory. In doing so, I also provide an explanation for one of the longest-standing puzzles of international relations.

The Propensity for War

In a recent review of the literature, Levy writes that (1) "the evidence shows that the proportional frequency of war involvement of democratic states has not been greater than that for nondemocratic states"; (2) "democracies may be slightly less likely than nondemocratic states to initiate wars, but the evidence is not yet conclusive on this question"; and (3) "although democracies have fought wars as frequently as have nondemocratic states, they almost never fight each other. . . . *This absence of war between democratic states comes as*

---

28. Monitoring costs will be higher, however, where the state can persuade society that "national security" considerations requirea higher degree of secrecy.
29. Just as the absence of supernormal profits by entrepreneurs in perfectly competitive economic markets does not vitiate the assumption of profit maximization and, in fact, follows from this assumption, the absence of rents does not undermine the assumption of state profit maximization. Rather, the receipt of only normal profits in ordinary economic enterprises and the state reflects the presence of acute competition and the constraints of society, respectively.

close as anything we have to an empirical law in international relations."[30] In Russett's view, this final result "is one of the strongest nontrivial or nontautological generalizations that can be made about international relations."[31]

For illustration, all interstate wars from 1816 to 1988 involving democratic states are listed in the Appendix.[32] Of these 30 conflicts, only 2 involve democratic states fighting each other: World War II, where Finland fought alongside the Axis (an easily explained exception), and the Spanish-American War of 1898.

This pattern of pacifism only among democratic states is inconsistent with most prevailing theories of international politics.[33] Realism, which focuses on the *universal* effects of international anarchy, the security dilemma, and the balance of power, cannot account for the relative pacifism of democracies. This is even more true of its contemporary variant, neorealism, which explicitly abstracts from the domestic characteristics of nation-states. Given the correlation between democratic political structures and capitalism, Marxist-Leninist theories predict a higher incidence of war between democracies. Conversely, liberal economic theory, which highlights the pacifying effects of commerce, predicts a lower *overall* incidence of democratic war involvement, not just a lower probability of war among democracies.

The most persuasive account of the relative pacifism of democracies was

---

30. See Jack S. Levy, "The Causes of War: A Review of Theories and Evidence," in Philip E. Tetlock, Jo L. Husbands, Robert Jervis, Paul C. Stern, and Charles Tilly, eds., *Behavior, Society, and Nuclear War*, Vol. 1 (New York: Oxford University Press, 1989), p. 270 (emphasis added).
31. Russett, *Controlling the Sword*, p. 123. For studies supporting these conclusions, see Steve Chan, "Mirror, Mirror on the Wall . . . : Are the Freer Countries More Pacific?" *Journal of Conflict Resolution*, Vol. 28, No. 4 (December 1984), pp. 617–648; Carol R. Ember, Melvin Ember, and Bruce Russett, "Peace between Participatory Polities: A Cross-Cultural Test of the 'Democracies Rarely Fight Each Other' Hypothesis," *World Politics*, Vol. 44, No. 4 (July 1992), pp. 573–599; Maoz and Abdolali, "Regime Types and International Conflict"; R.J. Rummel, "Libertarianism and International Violence," *Journal of Conflict Resolution*, Vol. 27, No. 1 (March 1983), pp. 27–71; Small and Singer, "The War-Proneness of Democratic Regimes"; and Erich Weede, "Democracy and War Involvement," *Journal of Conflict Resolution*, Vol. 28, No. 4 (December 1984), pp. 649–664.
32. Wars and participants are from Melvin Small and J. David Singer, *Resort to Arms: International and Civil Wars, 1816–1980* (Beverly Hills, Calif.: Sage, 1982), updated through 1988 from Singer, "Peace in the Global System." Regime type is from Ted Robert Gurr, ed., *Polity II: Political Structures and Regime Change, 1800–1986*, Inter-University Consortium for Political and Social Research, No. 9263 (Ann Arbor: University of Michigan, 1990). Democracy (DEMOC, variable 6.2) is an additive 11-point scale (0–10) based on the competitiveness of political participation, the openness and competitiveness of executive recruitment, and constraints on the chief executive. Countries that score 6 or more on this scale are classified here as democracies; countries that score 5 or less, as autocracies. There are relatively few countries that score in the middle of this range. Although countries that score 5 on this scale are not, in my view, normally considered to be democratic, lowering the threshold a point or two would not significantly alter this list of wars or the analysis.
33. Doyle, "Kant, Liberal Legacies, and Foreign Affairs," pp. 218–225; and Levy, "The Causes of War," p. 270.

first put forth by Immanuel Kant in 1795, when there were less than a handful of democracies in existence, and has been recently summarized and extended by Doyle.[34] There are three steps in Kant's argument. He first posits that republican forms of government, the inevitable result of political evolution, will replace monarchical (or autocratic) caprice with populist caution. Kant "argues that once the aggressive interests of absolutist monarchies are tamed and once the habit of respect for individual rights is engrained by republican government, wars would appear as the disaster to the people's welfare than he and other liberals thought them to be."[35]

While republican rule guarantees caution, international law guarantees mutual respect among liberal states—a step that separates Kant from other liberal thinkers:

As republics emerge . . . and as culture progresses, an understanding of the legitimate rights of all citizens and of all republics comes into play; and this, now that caution characterizes policy, sets up the *moral foundations for the liberal peace*. . . . In short, domestically just republics, which rest on consent, presume foreign republics to be also consensual, just, and therefore deserving of accommodation.[36]

Finally, drawing upon liberal economic theory, Kant concludes that cosmopolitan law, especially the "spirit of commerce," provides a material incentive for states to promote peace and avert war—a third step that reinforces the basis for mutual respect.

Doyle extends Kant by identifying and examining the moral imperative that liberalism creates for democracies in their relations with nondemocratic states.[37] "If the legitimacy of state action rests on the fact that it respects and effectively represents morally autonomous individuals," he argues, "then states

---

34. Doyle, "Kant, Liberal Legacies, and Foreign Affairs"; Michael W. Doyle, "Kant, Liberal Legacies, and Foreign Affairs, Part 2," *Philosophy and Public Affairs*, Vol. 12, No. 4 (Autumn 1983), pp. 323–353; and Michael W. Doyle, "Liberalism and World Politics," *American Political Science Review*, Vol. 80, No. 4 (December 1986), pp. 1151–1169. See also Russett, *Controlling the Sword*, pp. 124–132.
35. Doyle, "Kant, Liberal Legacies, and Foreign Affairs," p. 229.
36. Ibid., p. 230 (emphasis added). In a personal communication, Michael Doyle has suggested that it is this mutual republican recognition that separates modern democracies from the ancient Greek democracies, which appear not to have been substantially more pacific than their autocratic contemporaries. See Bruce Russett and William Antholis, "Democracies Rarely Fight Each Other? Evidence from the Peloponnesian War," typescript, Yale University, 1991. My explanation would focus on the relatively narrow franchise in these so-called democracies.
37. See Doyle, "Kant, Liberal Legacies, and Foreign Affairs, Part 2." Doyle distinguishes between strong and weak autocrats, suggesting that democracies will not war with the former (a point consistent with my analysis below) but will fail to take advantage of mutually advantageous agreements (such as arms control) or opportunities to exploit such states. I question the extent to which

that coerce their citizens or foreign residents lack moral legitimacy."[38] It follows, then, that "the liberal dictum in favor of nonintervention does not hold. Respecting a nonliberal state's state rights to noninterference requires ignoring the violations of rights they inflict on their own populations."[39] Thus, the moral foundations of the liberal peace are absent in relations with autocratic states, allowing war to be used as an instrument of statecraft and, at the extreme, necessitating active intervention by liberal states in the internal affairs of autocracies.

Kant's theory and Doyle's extension constitute a normative philosophy for the conduct of foreign affairs but not a positive theory of international relations. This philosophy rests, fundamentally, upon a moral imperative of *restraint*, in which democratic states must forgo potentially welfare-improving actions that would be damaging to their liberal brethren, and *necessary action*, in which a democracy might actually reduce its own material welfare in order to overthrow an autocracy and free a repressed people.[40] While it forms a sophisticated and coherent worldview, liberalism nonetheless contains a curious combination of motivations—a combination that can be reconciled only if we assume that individuals are in fact essentially moral actors willing to forgo material gains for normative ends. But if we accept this assumption, then the exploitative behavior of autocratic rulers is theoretically unmotivated: if individuals are moral and will act morally given the opportunity, why should autocrats, who are unconstrained by society, act immorally? And if power corrupts the ruler, why should collectivities be immune—especially in their relations with other states? Kant's suggestion that autocratic rulers might "resolve on war as a pleasure party for the most trivial reasons" does not resolve this contradiction.[41] Kant's liberalism ultimately falls short of being a positive *theory* of democratic pacifism because it lacks what today we would call a fully developed and consistent set of micro motives.[42]

The theory I have summarized above offers an alternative explanation for

this "stylized fact" holds, as democracies often do ally with one autocrat against another even in times of peace. The United States, for instance, played the "China card" in the 1970s. Nonetheless, to the extent that this tendency does exist, it cannot be reconciled with the theory summarized here.
38. Ibid., p. 325.
39. Ibid., p. 330.
40. Although only implied in their writings, this emphasis on forgoing material gains and incurring material losses is essential for both Kant and Doyle; in its absence, democratic pacifism is not moral but is economically efficient—a position that both authors claim to supersede.
41. Quoted in Doyle, "Kant, Liberal Legacies, and Foreign Affairs," p. 229.
42. Several scholars have recently attempted to formalize a model of foreign policy decision-making that focuses on the greater domestic constraints democratic societies exert on their leaders.

the relative pacifism of democratic states that nonetheless builds upon, and subsumes, many of Kant's essential insights. First, as I have demonstrated, democratic states will tend to be less expansionist than autocratic states. The larger the rent-earning ability of the state, the greater the optimal size of the territorial unit, and the greater the incentives for the state to try to reach this optimal size. To the extent that war is a necessary byproduct of expansion, it follows (ceteris paribus) that autocratic states will be more war-prone. Where, for Kant, republican institutions restrain the capriciousness of the ruler, democracy, in this approach, constrains the ability of the state to extract monopoly rents at society's expense.

Second, the theory suggests that democracies will often be the object of expansion by autocratic states. Democracies pose two threats to the rent-seeking ability of autocracies. As I have noted, democracies, by their very existence, serve as magnets that pull individuals out of autocratic polities; in the absence of a low-rent democratic haven, exit is less likely to occur. By eliminating democracies, autocratic states can reduce the gains from, and incentives for, emigration. In addition, through their political openness and richer informational environments, democratic states reduce the costs to autocratic societies of monitoring state behavior; by observing democracies, citizens in autocracies are more likely to become aware of the magnitude and consequences of state rent seeking.[43] Again, by eliminating democracies abroad, autocratic states solidify and reinforce their rent-earning abilities. For both of these reasons, democracies, perhaps unwittingly, may become targets of the expansionary activity of autocracies—thus expanding their overall war involvement despite their own pacific nature.[44]

These "neo-Kantians," however, suffer from the same problem as their intellectual progenitor. T. Clifton Morgan and Sally Howard Campbell base their analysis on "two key assumptions: (1) that a democratic political structure imposes constraints on the decision-making process by restricting the key decision makers' freedom of choice; and (2) that with regard to decisions for war, these constraints push toward peace (i.e., heads of state would be more likely to opt for war than those they govern)." See Morgan and Campbell, "Domestic Structure, Decisional Constraints, and War: So Why Kant Democracies Fight?" *Journal of Conflict Resolution*, Vol. 35, No. 2 (June 1991), pp. 187–211, at pp. 189–190. Bruce Bueno de Mequita and David Lalman find empirical support for Morgan and Campbell's first assertion but do not examine the second, more fundamental proposition. Like Kant, these authors do not deduce motivations or explain why leaders are more war-prone and the populace more pacific. While it might be understood as an alternative explanation of democratic pacifism, I see this work as complementary to the analysis offered here. See Bueno de Mesquita and Lalman, *War and Reason: A Confrontation between Domestic and International Imperatives* (New Haven, Conn.: Yale University Press, 1992).

43. This suggests why "insulation" is an essential element of the totalitarian model. See Carl Friedrich, Michael Curtis, and Barbara Benjamin, *Totalitarianism in Perspective: Three Views* (New York: Praeger, 1969).

44. This argument also implies that autocracies should initiate wars with democracies. Given the complexities of military strategy and the often-considerable gains from surprise attack and pre-

Finally, democratic states may engage in expansion and even intervene in other countries, as Doyle claims, but only under restricted conditions. Democracies will expand *only* when the initial costs of conquest and ongoing costs of rule are less than the discounted present value of future economic profits. Under these conditions, expansion is socially optimal regardless of regime type. Nineteenth-century European imperialism may provide examples of such socially efficient expansion; in many areas of the periphery the cost of war against more "primitive" peoples was relatively low and the potential gain from the erection of a "modern" state with its greater extractive capacity disproportionately large. With the rise of "European-style" states in the periphery during the twentieth century, however, the cost of imperialist expansion has substantially increased, apparently foreclosing this option.

By the same principle, democracies may also preemptively intervene in the domestic affairs of an autocracy to construct democratic political structures as long as the costs of the intervention are less than the expected costs of a war stimulated by state rent seeking. In this view, the proactive policy of democracies rests not on a moral imperative but on a rational calculus of preemption.

Together, these propositions imply that democracies are, on average, no more or less war prone than other states. On the other hand, democracies are less likely to fight each other, for only in this area is the absence of an imperialist bias manifest. Indeed, the almost complete lack of war between democracies suggests just how important state rent seeking may be as a source of international conflict.

The Propensity for Victory

Democracies are not only less likely to wage war with each other, they are also significantly more likely to win the wars they fight against autocracies. Liberalism offers a possible explanation for this correlation: democracies tend to turn international conflicts into ideological crusades and demand total victory.[45] Yet, as I have noted, the micro motives for democratic crusading are, at

---

emption, however, who actually fires the first shot or crosses the border (the basis of most standardized codings of war "initiation") is a poor indicator of who actually provoked, or "caused," the war. In addition, since history is typically written by the victors (and, as I shall show, democracies win a disproportionate number of wars), even more subjective estimates of war initiation may well be biased. As a result, I do not place much faith in patterns of war initiation as a test of theory presented here. Following the coding rules used in Table 1, however, the correlation (gamma) between democracy and war initiation is correctly signed but very weak ($-.04$) and insignificant. To the extent that this result is valid, it must count against the theory.

45. Contradicting the "crusading" hypothesis, Russett finds that the surge in public support for the president following an aggressive foreign policy action (often referred to as the "rally-round-

best, unclear. Moreover, within this framework, ideological fervor need not translate into victory; for the fear that the winner will transform the loser's social structure may only spur the failing side onto greater efforts. The drive for victory need not yield the desired result.

The theory summarized in Part I suggests a second explanation. To the extent that democratic states earn fewer rents, it follows that they tend (1) to create fewer economic distortions, possess greater national wealth, and devote more resources to security; (2) to enjoy greater societal support for their policies and therefore a greater extractive capacity; and (3) to form overwhelming countercoalitions against expansionist autocracies. I shall develop each of these propositions in turn.

State rents, like the rents earned by private actors, distort patterns of production and consumption, divert resources into directly unproductive activity, create social deadweight losses, and thereby reduce the total product of goods and services within an economy. While the successful rent earner is better off, the economy as a whole suffers. At the very least, growth rates lag behind their potential. North has examined the effects of state rent seeking on property rights and, in turn, growth; but the argument is more general.[46] Any state rents, including monopoly rents earned through the exchange of state-produced services such as protection, distort the economy and, over time, lower national wealth relative to its potential.[47]

---

the-flag" phenomenon) is typically short-lived, with a half-life of no more than a month or two. See Russett, *Controlling the Sword,* p. 94.

46. See North, *Structure and Change in Economic History.* See especially Mancur Olson, "Autocracy, Democracy, and Prosperity," paper presented at the annual meeting of the American Political Science Association, Washington, D.C., 1991. This extends to the state Olson's earlier work on social coalitions. Echoing North and the analysis here, he finds that democracies should experience lower state rents and higher growth.

47. This is a long-run, dynamic argument; it does not imply that all democracies are wealthier than all autocracies. The entire literature on the microeconomic theory of the state directly contradicts that on the so-called developmental state. The latter approach asserts that strong, autonomous states have to deter social rent seeking and stimulate growth under conditions of late industrialization. See, for example, Peter Evans, *Dependent Development: The Alliance of Multinational, State, and Local Capital in Brazil* (Princeton, N.J.: Princeton University Press, 1979). As I suggest in n. 25, I do not expect there to be a systematic difference between the *level* of rents extracted by social groups in democracies and autocracies. Moreover, the literature on the developmental state is generally mute on the motivations of state officials: Why do political leaders choose the "public" interest over their own private interests? The microeconomic theory of the state suggests that strong, autonomous states will typically exploit, rather than develop, their societies except where they face what North has termed severe *external constraints* (analogous here to very low costs of exit). See North, *Structure and Change in Economic History.* Interestingly, most, but not all, of the developmental states have experienced clear threats from other states (e.g., South and North Korea, Taiwan, and China). In such polities, especially those producing far within their production possibility frontiers, this analysis predicts that states will opt for increased normal profits from growth over increased rents from redistribution. In the long run, the theory suggests that these de-

The wealthier the country (ceteris paribus), the more absolute resources it devotes to producing security. For any given set of costs and preferences, wealthier countries produce and consume greater amounts of all goods and services, including protection; under normal conditions, an increase in wealth affects only the level, not the mix, of goods in the economy. It follows, then, that wealthier countries produce and consume greater amounts of security (i.e., provide greater absolute resources to the state for producing protection). In contests where sheer resources matter, the cumulative effects of lower state rents may prove decisive.

Democratic states should also possess greater societal support for their policies, suggesting that they will enjoy a greater extractive capacity for any given level of national wealth. All societies make trade-offs between consuming what are mostly private goods and services and consuming state-produced protection; while both constitute current consumption, the latter serves as insurance that societies will be able to enjoy their present and future holdings of wealth. During hostilities, when the external threat to national wealth, territorial integrity, and the present form of rule is most acute, individuals and, in turn, society will tend to purchase greater quantities of protection (pay higher taxes), thereby transferring greater resources to the state for the waging of war.

In view of their exploitative nature, however, autocratic states pose a greater threat than do nonautocratic states to the current and future ability of democratic societies to produce and consume wealth; as a result, democracies should demand greater protection against these threats and contribute proportionately more to ensure victory in war. Within the theory developed here, individuals are indifferent between being ruled by two equally democratic regimes; each provides similar levels of protection close to the normal profit price (p in Figure 1). As a result, there is no incentive for citizens of one democracy to purchase protection against another equally democratic state unless there is some uncertainty about the likely behavior of the democratic conqueror. Autocratic states, on the other hand, provide protection only at a higher price, with the difference between the normal profit and higher prices (p and r in Figure 1) being captured by the state as rents. In this case, citizens of democracies do possess incentives to purchase protection against this threat to their current and future wealth; indeed, the greater the expected rent seeking from an autocratic conqueror, the more protection citizens will demand, and pay for, from their present democratic state.

---

velopmental autocracies will either ossify and once again become a drag on development (as in the former Soviet Union) or liberalize (as in South Korea).

Conversely, autocratic societies may actually benefit from defeat—if the victorious democratic states remake the autocracies in their own images. As a result, societies in autocratic polities should be willing to contribute proportionately less than democratic societies. While autocratic states will seek to offset this tendency by vilifying the enemy (instilling fear that defeat will mean national destruction, rape, and slavery), a fifth column remains a real possibility. Autocracies may also increase the degree of coercion used to extract resources; but this implies higher transactions costs for revenue collection, which will further disadvantage them relative to democracies. At the very least, the lack of societal support places a real constraint on their extractive capabilities. Lamborn's study of the great powers during the late nineteenth century supports this expectation, as does the tendency of states to expand the franchise or otherwise liberalize politically during or immediately after major wars.[48]

Finally, to the extent that states balance threats, rather than power (as Walt has argued and as is consistent with the logic I have developed), democratic states should form overwhelming countercoalitions against autocratic states.[49] Not only are autocracies more likely to seek territorial expansion, they are more likely to target democracies (to reduce exit options). In addition, autocratic expansion poses a greater threat to democracies because of the larger rents the state is likely to extract if successful. The greater the threat (ceteris paribus), the greater the balancing reaction by other states.

Given its imperialist bias and likely behavior if successful, the threat posed by an autocracy is proportionately greater than the sum of its aggregate resources, the traditional measure of national power. It follows that the countercoalition that forms against any autocracy should be disproportionately large or overwhelming. If autocracies have greater incentives to expand, democracies have greater incentives to resist. As a result, this coalition should also be disproportionately composed of democratic states.[50] The overlarge "demo-

---

48. See Alan C. Lamborn, "Power and Politics of Extraction," *International Studies Quarterly*, Vol. 27, No. 2 (June 1983), pp. 125–146; and Alan C. Lamborn, *The Price of Power: Risk and Foreign Policy in Britatin, France, and Germany* (Boston: Unwin Hyman, 1991). For the domestic consequences of war, see Arthur A. Stein, *The Nation at War* (Baltimore, Md.: Johns Hopkins University Press, 1978). On political reform in Germany during World War I, see Gerald D. Feldman, *Army, Industry, and Labor in Germany, 1914–1918* (Princeton, N.J.: Princeton University Press, 1966). On the Soviet Union during World War II, see William O. McCagg Jr., *Stalin Embattled, 1943–1948* (Detroit: Wayne State University Press, 1978), especially pp. 18–23.
49. See Stephen M. Walt, *The Origins of Alliances* (Ithaca, N.Y.: Cornell University Press, 1987).
50. Randolph M. Siverson and Juliann Emmons find that democracies have a greater propensity to ally with one another than with other states. See Siverson and Emmons, "Birds of a Feather: Democratic Political Systems and Alliance Choices in the Twentieth Century," *Journal of Conflict Resolution*, Vol. 35, No. 2 (June 1991), pp. 285–306.

cratic" coalition should *deter* autocratic expansion (by raising the costs of conquest) and, if deterrence fails, be more likely to win. Combined with the greater wealth and extractive capabilities of the individual states, this suggests that the democratic coalition should be virtually invincible. The overlarge coalitions formed during World Wars I and II and, more strikingly, the Cold War (often understood as anomalies in realist theories of international relations)[51] bear out this expectation.

EVIDENCE

Many of the concepts central to the theory summarized here are difficult to operationalize or lie beyond current data-gathering techniques. Extractive capability, for example, is unmeasurable: military spending as a proportion of gross national product (a commonly used indicator) may be distorted and inflated by state rents. Nonetheless, the propositions I have developed suggest that democratic states should tend to win wars—a derived hypothesis that provides an indirect test of the theory.

The historical record is striking. Of the 30 wars listed in the Appendix, 3 (the Korean, Israeli-Egyptian, and Israeli-Syrian Wars) must be excluded from this analysis for want of a clear victor. I also exclude the Spanish-American War, fought between two democracies, but include World War II despite Finland's exceptional position. Of the 26 wars fought since 1816 between democracies and autocracies, the former have won 21 (81 percent) and lost 5 (19 percent). In other words, democratic states, either singly or in combination with other states, have won four times as many wars as autocratic states. Excluding the First and Second Balkan Wars, where Greece was the sole democracy on the winning side, does not appreciably change these results: the democracies still win 19 (79 percent) of 24 wars.

Scoring each participant individually yields a strong and significant correlation between democratic victory and autocratic defeat. Table 1 breaks down all 121 participants in the 26 wars according to regime type and outcome. This construction biases the results against the hypothesis by coding as winners the not-inconsequential number of autocracies who fought as members of victorious democratic coalitions. Nonetheless, the degree of association is strong, indicating that even with this bias democratic states are significantly more likely to win—and autocratic states more likely to lose—than the converse.

Rather than relying upon a simple dichotomy, it is possible to examine the

51. Kenneth Waltz, *Theory of International Politics* (Reading, Mass.: Addison-Wesley, 1979).

Table 1. Regime Type and Victory in War (Individual Participants)

Success in War	Nature of Regime		
	Autocratic	Democratic	Total
Loser	42	9	51
Winner	32	38	70
Total	74	47	121

Note: All wars in the Appendix are included except the Spanish-American War of 1898 and wars in which no clear winner emerged. Gamma = .694; chi squared = 16.673; df = 1; p = .000046.

average degree of democracy in the sets of winners and losers of these 26 wars. Using an 11-point scale of democracy (0–10), the mean of the 70 winners is 5.60 and the mean of the 51 losers is 2.55.[52] The probability that these figures would emerge by chance is less than 0.001 (t = 4.43; df = 119).

Finally, the relationship between democracy and victory is quite robust. Table 2 presents a logit analysis performed with the 121 war participants as the units of observation. Along with democracy, the analysis included military personnel, a measure of military strength, iron and steel production, a proxy for industrial capacity, and a dummy variable indicating whether or not the country initiated the war.[53] Common sense and, for war initiation, previous research,[54] suggest that all of these relationships should be positively related to victory.

Democracy is consistently positive and significant, offering strong support for the argument developed here. Military personnel and iron and steel pro-

52. See n. 31 above. As noted, at least one participant in each war is democratic (defined by a minimum score of 6 on an 11-point scale). The mean score of 5.60 suggests that the "average" winner in these 26 wars was autocratic, although there were more democracies (38) than autocracies (32) in the population (see Table 1).
53. Data on military personnel and iron and steel production are from the July 1990 Correlates of War, National Materials Capabilities data set (J. David Singer and Melvin Small, principal investigators). An attempt was made to include data on military expenditures and energy usage from the same data set, but there were too many missing values for the early nineteenth century. Data on war initiation is from Small and Singer, "The War-proneness of Democratic Regimes." Unfortunately, this set covers only the period through 1965 and does not include several wars later added in Small and Singer, *Resort to Arms;* this accounts for the reduction in the number of observations in Table 2, equation 2. The proviso stated in n. 42 holds here, as well. I also assessed the possible presence of interaction effects among these measures by multiplying democracy with military personnel, iron and steel production, and war initiation. In every case, these interaction terms were insignificant. These results are not reported here.
54. See Bruce Bueno de Mesquita, *The War Trap* (New Haven, Conn.: Yale University Press, 1981), p. 22.

Table 2. Logit Analysis of Victory in War

Variable	Equation 1	Equation 2
Constant	.4214	−.7527*
	(.4461)	(.3346)
Democracy (0–10)	.1933*	.2524***
	(.0763)	(.0597)
Military personnel (millions)	.3701	.4968
	(.5325)	(.4188)
Iron and steel production		
(millions of tons)	.0024	−.0334
	(.0542)	(.0201)
Initiator[a]	−1.3982**	—
	(.5126)	
Log likelihood	−48.05	−71.63
n	87	121
Percent correctly predicted	72.41	67.77

Note: Standard errors are in parentheses. Dependent variable is war outcome (loser = 0;
 winner = 1).
[a]0 = *no;* 1 = *yes.*
*$p < .02$.
**$p < .01$.
***$p < .001$.

duction, on the other hand, are insignificant at standard levels in both equations, suggesting that in this set of wars neither military nor economic strength is associated with victory. When included, war initiation is significant but in the wrong direction; surprisingly, in wars between democracies and autocracies, noninitiators win more frequently.

The failure of these alternative explanations to predict victory or loss correctly does not imply that common sense or previous research is wrong. Rather, it highlights the exceptional nature of war between democracies and autocracies. In these conflicts, military strength, industrial capacity, and the ability to choose to wage war appear to be far less important determinants of victory than governmental form.

Conclusion

Regime type does matter in international politics. Democracies are less likely to fight wars with each other. They are also more likely to prevail in wars with autocratic states. This syndrome of powerful pacifism accords, in part, with "Kantian" liberalism; but because of inconsistent behavioral assumptions, this normative frame cannot be said to constitute a positive theory of international relations.

I have offered an alternative explanation drawn from the microeconomic theory of the state. Specifically, state rent-seeking creates an imperialist bias in a country's foreign policy. This bias is smallest in democracies, where the costs to society of controlling the state are relatively low, and greatest in autocracies, where the costs are higher. As a result, autocracies will be more expansionist and, in turn, war prone. To the extent that democracies do wage occasional wars of expansion, intervene in the domestic affairs of autocracies, and are targets of autocratic expansion, there should be no significant overall difference in their frequency of war involvement. Only in their relations with each other does the relative pacifism of democracies appear. In addition, democracies (constrained by their societies from earning rents) create fewer economic distortions and possess greater national wealth, enjoy greater societal support for their policies, and tend to form overwhelming countercoalitions against expansionist autocracies. Thus, democracies will be more likely to win wars.

If democracies are powerful pacifists, why do autocracies persist within the international system? It follows from the arguments I have outlined that democracy is an evolutionarily superior and stable form of rule. If so, why has democracy not displaced autocracy?

On the one hand, democracy *has* expanded. The number of democratic countries has grown from a mere handful in the eighteenth century to over 60 in the early 1980s.[55] The recent transformations in Eastern Europe suggest the promise of further liberalizations elsewhere, although in my opinion the tide could easily be reversed. But given the relative infrequency of war, there is no reason to presume that political change will be rapid; the evolutionary pace of the global system may well be glacial.

On the other hand, democracies only *tend* to win wars; historically, for every four they win, they lose one. Nor is there any consistent trend in favor of the democracies: autocracies were most successful in the 1960s, winning half of the wars in which they were engaged, but entirely unsuccessful in the 1970s and 1980s (see Appendix). Any evolution toward greater democracy is likely to be characterized by fits and starts. Four steps forward, one step back.

---

55. On the diffusion of democracy, see Harvey Starr, "Democratic Dominoes: Diffusion Approaches to the Spread of Democracy in the International System," *Journal of Conflict Resolution*, Vol. 35, No. 2 (June 1991), pp. 356–381.

Appendix: All Wars Involving Democratic States, 1816–1988

In the following list, boldface in the righthand column indicate democratic states. Bulgaria, Romania, Italy/Sardinia, and France which fought on both sides of World War II are listed here on their initial sides.

War	Participants (Winners First)
1. Mexican–American 1846–48	**United States** vs. Mexico
2. Roman Republic 1849	The two Sicilies, **France,** and Austria–Hungary vs. the Papal States
3. Crimean 1853–56	**United Kingdom,** Italy/Sardinia, France, and Turkey/ Ottoman Empire vs. USSR/Russia
4. Anglo–Persian 1856–57	**United Kingdom** vs. Iran/Persia
5. Sino–French 1884–85	**France** vs. China
6. Greco–Turkish 1897	Turkey/Ottoman Empire vs. **Greece**
7. Spanish—American 1898	**United States** vs. **Spain**
8. Boxer Rebellion 1900	Japan, **United Kingdom,** USSR/Russia, **France** and **United States** vs. China
9. Spanish—Moroccan 1909–10	**Spain** vs. Morocco
10. First Balkan 1912–13	**Greece,** Yugoslavia/Serbia, and Bulgaria vs. Turkey/ Ottoman Empire
11. Second Balkan 1913	Turkey Ottoman Empire, **Greece,** Yugoslavia/Serbia, and Romania vs. Bulgaria
12. First World War 1914–18	Japan, Belgium, **United States,** Yugoslavia/Serbia, **United Kingdom, Portugal,** Romania, **France, Greece,** Italy/Sardinia, and USSR/Russia vs. Turkey/Ottoman Empire, Austria–Hungary, Germany/Prussia, and Bulgaria
13. Hungarian–Allies 1919	Romania and **Czechoslovakia** vs. Hungary
14. Russo–Polish 1919–20	**Poland** vs. USSR/Russia
15. Russo–Finnish 1939–40	USSR/Russia vs. **Finland**
16. Second World War 1939–45	**United Kingdom, Australia, Canada,** Ethiopia, Poland, **United States,** USSR/Russia, **Belgium,** Brazil, China, Yugoslavia/Serbia, **Netherlands, New Zealand, France, South Africa,** Greece, **Norway,** and Mongolia, vs. Japan, Italy/Sardinia, Germany/Prussia, Romania, Bulgaria, Hungary, and **Finland**
17. Palestine 1948	**Israel** vs. Jordan, Iraq, Syria, Egypt/UAR, and Lebanon
18. Korean 1950–53 (no clear victor)	**Greece, United Kingdom, Canada,** Thailand, **Belgium, Turkey/Ottoman Empire, United States, Australia, Netherlands,** Ethiopia, Columbia, **Philippines, France,** and South Korea vs. North Korea and China

(Continued)

War	Participants (Winners First)
19. Sinai 1956	**United Kingdom, France, and Israel** vs. Egypt/UAR
20. Sino–Indian 1962	China vs. **India**
21. Second Kashmir 1965	Pakistan vs. **India**
22. Vietnamese 1965–75	North Vietnam vs. Thailand, South Vietnam, **United States,** Kampuchea/Cambodia, **Republic of Korea, Australia,** and **Philippines**
23. Six Day 1967	**Israel** vs. Egypt/UAR, Jordan, and Syria
24. Football 1969	**El Salvador** vs. Honduras
25. Israeli–Egyptian 1969–70 (no clear victor)	**Israel** vs. Egypt/UAR
26. Bangladesh 1971	**India** vs. Pakistan
27. Yom Kippur 1973	**Israel** vs. Egypt/UAR, Iraq, Syria, Jordan, and Saudi Arabia
28. Turko–Cypriot 1974	**Turkey/Ottoman Empire** vs. Cyprus
29. Falklands 1982	**United Kingdom** vs. Argentina
30. Israeli–Syrian (Lebanon) 1982 (no clear victor)	**Israel** vs. Syria

Part II:
The Democratic Victory Debate

Making Military Might: Why Do Democracies Succeed?

Risa A. Brooks

A Review Essay

Dan Reiter and Allan C. Stam III
Democracies at War. Princeton, N.J.:
Princeton University Press, 2002.

Throughout history, states have exhibited a puzzling degree of variation in their capacities to create military power. Some consistently excel at warfare. Germany in the late nineteenth and early twentieth centuries, for example, has long been singled out for its military's exceptional tactical proficiency, especially in the battles of both world wars. Others—such as Italy in those same conflicts—often perform poorly, demonstrating endemic weaknesses in their abilities to generate military force.[1] Yet a third category of states exhibits curious variation in military effectiveness over time. Most analysts, for example, count Egypt's performance in the 1967 Six Day War against Israel among the most dismal of contemporary military history. They also tend to agree, however, that in the 1973 October War Egypt was able to dramatically improve its effectiveness, manifest most vividly in the military's textbook crossing of the Suez Canal.

Why are some states, at some times, better able to translate their basic material and human strengths into military power? Sociologists and military historians have long been interested in this question, often focusing on the human relations among soldiers and quality of leaders that make some militaries exceptional. More recently, political scientists have evinced growing interest in exploring the sources of states' military effectiveness. In the 1990s Stephen Rosen, for example, analyzed how colonial India's caste system, and states' social structures more broadly, influence the capacity to mobilize and utilize the

Risa Brooks is Assistant Professor of Political Science at Marquette University. This article is adapted from Risa A. Brooks, "Making Military Might: Why Do States Fail and Succeed? A Review Essay," *International Security,* Vol. 28, No. 2 (Fall 2003), pp. 149–191.

The author wishes to thank Stephen Biddle, Michael Desch, Elizabeth Kier, David Lake, and Elizabeth Stanley-Mitchell for helpful comments and reactions. She wishes also to thank Lee Seymour for research assistance.

1. On Italy see, for example, "Italian Military Efficiency—A Debate," *Journal of Strategic Studies,* Vol. 5, No. 2 (June 1982), pp. 248–278. On overviews of Egypt's performance in the wars, see Trevor N. Dupuy, *Elusive Victory: The Arab Israeli Wars, 1947–1974* (New York: Harper and Row, 1978); and Chaim Herzog, *The Arab-Israeli Wars: War and Peace in the Middle East from the War of Independence through Lebanon* (New York: Random House, 1982). Germany is discussed later in this essay.

armed forces in war.[2] Stephen Biddle and Robert Zirkle explored how differences in Iraqi and North Vietnamese civil-military relations affected assimilation of new military technologies.[3] Others wrote on the effects of culture and technology on effectiveness, and a more recent generation of scholars has investigated the effects of ideology and nationalism on fighting motivation.[4] This complements a more established literature on military change, innovation, and politico-military doctrine, which at times addresses (tangentially) issues of military effectiveness.[5] The common focus in all this scholarship is on how social, political, and cultural factors shape the capacity to use material or other resources to create military power.

This attention to military effectiveness by political scientists is arguably long overdue. The study of military effectiveness is central to the discipline's academic and practical concerns. Most fundamental are its implications for measuring and conceptualizing state power—a (if not the) basic variable in the study of international relations since its inception. Traditionally, assessments of power have been heavily materialist and crude in conception.[6] They often ig-

2. Stephen Peter Rosen, *Societies and Military Power: India and Its Armies* (Ithaca, N.Y.: Cornell University Press, 1996).
3. Stephen Biddle and Robert Zirkle, "Technology, Civil-Military Relations, and Warfare in the Developing World: Conventional Proliferation and Military Effectiveness in Developing States," *Journal of Strategic Studies*, Vol. 19, No. 2 (June 1996), pp. 171–212.
4. See Christopher S. Parker, "New Weapons for Old Problems: Conventional Proliferation and Military Effectiveness in Developing States," *International Security*, Vol. 23, No. 4 (Spring 1999), pp. 119–147. Kenneth M. Pollack, "The Influence of Arab Culture on Arab Military Effectiveness," Ph.D. dissertation, Massachusetts Institute of Technology, 1996; Norville de Atkine, "Why Arab Armies Lose Wars," *MERIA*, Vol. 4, No. 1 (March 2000), http://meria.idc.ac.il/journal/2000/issue1/jv4n1a2.html#Author; Tania M. Chacho, "Why Did They Fight? American Airborne Units in the Second World War," paper presented at the annual meeting of the American Political Science Association, San Francisco, California, August 30–September 2, 2001; and Jasen Castillo, "The Will to Fight: Explaining a Nation's Determination in War," paper prepared for delivery at the annual meeting of the International Studies Association, Chicago, Illinois, February 21, 2001.
5. Works on doctrine and military innovation include Barry R. Posen, *The Sources of Military Doctrine: France, Britain, and Germany between the World Wars* (Ithaca, N.Y.: Cornell University Press, 1984); Jack Snyder, "Civil-Military Relations and the Cult of the Offensive, 1914 and 1984," in Steven E. Miller, ed., *Military Strategy and the Origins of the First World War* (Princeton, N.J.: Princeton University Press, 1991), pp. 20–58; Stephen Van Evera, "The Cult of the Offensive and the Origins of the First World War," in ibid., pp. 59–108; Elizabeth Kier, *Imagining War: French and British Military Doctrine between the World Wars* (Princeton, N.J.: Princeton University Press, 1997); Jeffrey W. Legro, *Cooperation under Fire: Anglo-German Restraint during World War II* (Ithaca, N.Y.: Cornell University Press, 1995); Kimberly Marten Zisk, *Engaging the Enemy: Organization Theory and Soviet Military Innovation, 1955–1991* (Princeton, N.J.: Princeton University Press, 1993); Deborah D. Avant, *Political Institutions and Military Change: Lessons from Peripheral Wars* (Ithaca, N.Y.: Cornell University Press, 1994); Emily O. Goldman and Richard B. Andres, "Systemic Effects of Military Innovation and Diffusion," *Security Studies*, Vol. 8, No. 4 (September 1998), pp. 67–85; and Theo Farrell and Terry Terriff, eds., *The Sources of Military Change: Culture, Politics, Technology* (Boulder, Colo.: Lynne Rienner, 2002).
6. On this point, see Stephen Biddle, *Military Power: Explaining Victory and Defeat in Modern Battle*

nore the complex social, organizational, and political processes that affect how states and militaries prepare for and employ force in war. Yet those factors may be as important as any indicator of industrial capacity or population size in determining how (well) force is likely to be used in any given war, and therefore the actual balance of power among states in the international arena.

Equally important, military effectiveness has major ramifications for practical debates about defense planning and net assessment. Analysts working on the applied arm of combat analysis—and especially in military-related operations research—often compare states' material resources (such as weapons' firepower and technological attributes) in predicting how states will fare in war, with much less appreciation of the diverse, intangible factors that may affect their capacities to use these resources effectively. If the social scientists discussed above are right, however, this heavily materialist approach may be fundamentally flawed. A major correction in how policy analysts measure military power may be in order.

Given the academic and practical issues at stake, the appearance of a recent work on military effectiveness provides a welcome contribution to debate about the origins of states' military power. Dan Reiter and Allan Stam's monograph, *Democracies at War*, asks critical questions about the nature of states' advantages in warfare and posits a provocative thesis about why states perform better and worse on the battlefield.[7] Because of the importance of the issues the book addresses, it is likely to be widely read and cited by policy analysts and academics alike.

For these reasons, the book merits careful evaluation before its claims are accepted as conventional wisdom. This review essay attempts such an assessment. Its conclusions about the quality of *Democracies at War*, are, unfortunately, less than positive. The book exhibits some significant strengths: it is by serious and accomplished scholars, addresses important questions, presents its arguments in a clear and intelligent way, and attempts rigorously to evaluate its claims. The authors should be especially commended for their efforts to an-

---

(Princeton, N.J.: Princeton University Press, forthcoming), chap. 2. For a recent materialist, wealth-centered approach to measuring "power," see John J. Mearsheimer, *The Tragedy of Great Power Politics* (New York: W.W. Norton, 2001), pp. 55–82. Despite emphasizing wealth, Mearsheimer does note that variations in states' "efficiency" and in the decisions they make about what forces to procure affect their military might. For an approach that explicitly incorporates social, political, and other factors into calculations of power, see Ashley J. Tellis, Janice Bially, Christopher Layne, Melissa McPherson, and Jerry M. Sollinger, *Measuring National Power in the Postindustrial Age* (Santa Monica, Calif.: RAND, 2000).

7. Dan Reiter and Allan C. Stam III, *Democracies at War* (Princeton, N.J.: Princeton University Press, 2002). Further references appear parenthetically in the text.

alyze how states' historical and political traditions affect their capacities to create military power. Reiter and Stam explore vital academic and practical issues in their efforts to identify the strengths and weaknesses of different regime types in war. Yet these strengths aside, close reading reveals major conceptual and methodological weaknesses in. The book, in the end, is not fully convincing in its claims about democracies and states' military effectiveness.[8] Scholars and policy analysts should read this study, but with a dose of skepticism.

This essay details the reasons for skepticism. I offer three sets of critical arguments. First, I analyze the overarching theses of the study and argue that Reiter and Stam do not provide wholly complete or original arguments. In the end, to the extent we accept that democracies are exceptionally good at warfare, the authors do not provide compelling explanations for why they are so. Second, I analyze in detail the logic of the book's main suppositions and analytical approaches and, again, find both lacking. Reiter and Stam's key hypotheses, in turn, are based on questionable assumptions and a crude conceptualization of the relationships among key variables. Third, I evaluate the book's empirical tests and method of evaluation and conclude that Reiter and Stam poorly measure key variables and inadequately test key hypotheses. Combined, these conceptual and empirical flaws undermine these scholars' claims about the unique qualities of democracies in war.

The essay begins with an overview of the book's arguments and how they fit into the broader study of military effectiveness. It then presents the three sets of critiques noted above, concluding with a commentary on their implications for research on military effectiveness.

One final note to the reader is warranted before proceeding. Throughout this essay, the term "military effectiveness" primarily refers to the tactical level of warfare or tactical effectiveness. This emphasis is consistent with how Reiter and Stam use the concept. It is essential to remember, however, as Allan Millett, Williamson Murray, and Kenneth Watman's study nicely highlights, that militaries operate on multiple levels—strategic, operational, and tactical—and their effectiveness at one level does not necessarily translate into effective-

---

8. This essay complements Michael Desch's recent effort in which he critiques *Democracies at War* and other works that make the claim that democracies are especially likely to prevail in war. Michael C. Desch, "Democracy and Victory: Why Regime Type Hardly Matters," *International Security*, Vol. 27, No. 2 (Fall 2002), pp. 5–47. Because of the wealth of material he covers, Desch understandably could devote only a few pages to military proficiency arguments. Yet because *Democracies at War* will likely be seminal for work on military effectiveness, which involves a distinct set of research questions and scholarly community, I am giving special emphasis to Reiter and Stam's theses about military effectiveness.

ness on another; hence focusing only on the tactical level offers a somewhat truncated view of war.[9] More important, as this essay points out, these levels of military activity often interact and affect one another such that tactical activity can be heavily influenced by strategic- and operational-level activity and vice versa.

Democracies in Context

Analysts of military effectiveness are interested in how states actually create military power from their given human and material resources.[10] They often explore the social, political, cultural, organizational, and technological factors that shape military activity and influence a state's capacity to use its resources efficiently. Scholars are concerned, for example, with understanding why some states excel at training, motivating, and leading soldiers in combat; developing efficacious weapons systems and innovative doctrines; and integrating military strategies and operational plans with broader political objectives. They are also interested in why other states exhibit systematic weaknesses in these areas.

As such, the study of a state's military effectiveness and the analysis of its material strengths are complementary, but distinct, endeavors. A poorly endowed military can be highly effective if it uses its resources well. And a wealthy military can be extremely ineffective if it wastes them. Hence a military may procure a large arsenal of weapons and therefore rate high on traditional indices of power. But if those weapons are ill suited for the kinds of wars the state is likely to fight, if they are poorly integrated with doctrine, and if soldiers are insufficiently trained and therefore incapable of using and maintaining those weapons, the state would rank as ineffective in key areas of military activity. Its actual military power would be much lower than material indicators would suggest.

Sociologists and operations researchers working on military issues have long sought to understand the nature of these relationships between military resources and effectiveness, and each tackles a different dimension of the issue. Traditionally, military-related operations research, which generally dates to

---

9. Allan R. Millett, Williamson Murray, and Kenneth H. Watman, "The Effectiveness of Military Organizations," in Millett and Murray, eds., *Military Effectiveness*, Vol. 1: *The First World War* (Boston: Allen and Unwin, 1988), pp. 1–30.
10. For this definition, see Millett, Murray, and Watman, "The Effectiveness of Military Organizations," pp. 1–2.

British and U.S. efforts to model the effects of different technologies and systems on battle outcomes in World War II, has been primarily concerned with evaluating the quality, utilization, and assimilation of weapons systems by military organizations and how these affect a state's power in war.[11] The sociological approach also dates to World War II, and specifically, to influential studies on the sources of motivation of soldiers in combat. Much of this literature is framed around a central debate about the degree to which the interpersonal bonds among individuals ("unit cohesion") or ideology and morale motivates them in combat.[12] Also important, in the 1980s, the military historians Allan Millett, Williamson Murray, and Kenneth Watman offered an approach that emphasized (sometimes implicitly) the influence of organizational factors on military effectiveness.[13] They analyzed military activity across its four dimensions (political, strategic, operational, and tactical), arguing that military effectiveness depends on the internal consistency, sensitivity to resource and other constraints, and integration across and within each level of activity.[14]

---

11. War gaming, more broadly, has been practiced for centuries and was formally adopted by the Germans as a training tool in the early nineteenth century. Military-related operations research is pervasive in defense policy circles. Analysts have a major professional society, numerous journals, and a growing of body of texts and degree programs. In its current incarnation, military-related operations research involves mathematical modeling and computer simulations of battle and war outcomes; the models draw heavily on technological and numerical indicators of military power, and seek to evaluate the relative strength of different military forces, both in an absolute sense and in particular battlefield contexts.
12. The unit cohesion school posits that the immediate experience of being in combat, and the relations among soldiers on the battlefield, are the primary forces driving individuals to fight. The original research in this area was done by the U.S. Army Research Branch during World War II and is published in Samuel A. Stouffer et al., *The American Soldier: Combat and Its Aftermath* (Princeton, N.J.: Princeton University Press, 1949). In a seminal study of Nazi soldiers, Morris Janowitz and Edward A. Shils came to a similar conclusion. See Shils and Janowitz, "Cohesion and Disintegration in the Wehrmacht in World War II," *Public Opinion Quarterly*, Vol. 12, No. 2 (Summer 1948), pp. 280–315. See also S.L.A. Marshall, *Men against Fire* (New York: William Morrow, 1964); and William Darryl Henderson, *Cohesion: The Human Element in Combat* (Washington, D.C.: National Defense University Press, 1985). A second, albeit less influential, strand of sociological literature emphasizes the effects of ideology on morale and motivation. In a widely noted study, for example, Omar Bartov challenged the conclusions of Janowitz and Shils. Bartov, *Hitler's Army: Soldiers, Nazis, and War in the Third Reich* (Oxford: Oxford University Press, 1991). Related are arguments about the effects of ideology and beliefs on the will to fight. On ideology see also Charles C. Moskos, *The American Enlisted Man: The Rank and File in Today's Military* (New York: Russell Sage, 1970). More broadly, see James Burk, "Morris Janowitz and the Origins of Sociological Research on Armed Forces and Society," *Armed Forces and Society*, Vol. 19, No. 2 (Winter 1993), pp. 167–186. Some political scientists also focus on these issues. See Chacho, "Why Did They Fight?"; and Castillo, "The Will to Fight."
13. Millett, Murray, and Watman, "The Effectiveness of Military Organizations."
14. Accordingly, one of the major contributions of their study is to highlight potential conflicts in

Although important, most research on military effectiveness, especially that in sociology and operations research, has been relatively limited in its level of analysis and analytical focus, exploring issues such as the microrelations among individuals and weapons utilization.[15] Relatively unexamined are potentially dozens of social, cultural, political, technological, and systemic factors—many within the research domain and expertise of political scientists—that might, through their effects on various dimensions of military activity, influence how states generate power. The book by Reiter and Stam, political scientists by training, thus represents an important step forward in their discipline's growing research on military effectiveness. The book adopts a novel approach in trying to isolate particular social and political attributes of states—democratic institutions—that render them exceptional in war, and it offers a provocative argument about what makes states better and worse at fighting wars.

DEMOCRACIES AT WAR

Reiter and Stam begin with a general observation about the effectiveness in war of a particular subgroup of states: *Democracies at War* sets out to explain why democracies appear disproportionately to win the wars they fight. To explore the issue, the authors adopt a novel research design. Unlike traditional political science texts, in which an author presents a general theory in the first chapters of the book and then uses the remainder to evaluate it, Reiter and Stam's book is organized around four relatively independent chapters. Each tests arguments drawn from a different body of conceptual literature in political science, in what the authors refer to as "perspectives" on why democracies win wars.

Reiter and Stam first evaluate what they term the "skeleton of democracy." Here they test the claim that because democratic institutions purportedly create accountability and responsive political leadership, democratic leaders are likely to initiate only winnable wars. Later, in a chapter entitled "The Family of Democracy," the authors evaluate the argument that democracies make better

---

effectiveness across different tiers of military activity—an issue implicit in many contemporary debates about the use of force. For example, conflict between tactical and political/strategic effectiveness is often implicit in debates about the military lessons of the 1999 Kosovo intervention. See, for example, Benjamin S. Lambeth, *NATO's Air War for Kosovo* (Santa Monica, Calif.: RAND, 2001), pp. 234–235.
15. Millett, Murray, and Watman's approach is much broader, but they focus on presenting a typology of ineffectiveness at different levels of military activity. They are not concerned with systematically hypothesizing about the causes of those various incarnations of ineffectiveness.

friends and allies than other states and are more likely to prevail in war as part of overwhelming military coalitions. Yet another chapter probes whether democratic victory stems from the "power of democracy" and a superior ability to mobilize economies, extract resources from populations, and build large militaries. After evaluating these arguments, Reiter and Stam then offer their conclusions. Democracies, they find, do not make better allies; nor are they richer and better at mobilizing their resources and therefore better prepared to win long wars. In contrast, Reiter and Stam do find that democratic leaders are more discriminating in choosing which wars to start—they tend to select wars they are favored to win. Yet, critically, while this might explain why democracies may avoid some challenging wars, it does not explain why they often prevail once they are in them.

Instead, the reason why democracies win the wars they fight, Reiter and Stam assert, has to do with their superior military effectiveness. Entitled the "spirit of democracy," this argument explains why, once democracies go to war, they are far more likely to prevail over their autocratic counterparts. Here the authors offer four subarguments. First, Reiter and Stam posit that because democracies are consent-based systems (i.e., their leaders are elected), soldiers should have higher morale and hence fight harder than those governed by authoritarian systems. Second, they hypothesize that because democracies have a liberal political culture that emphasizes individualism, their soldiers will demonstrate higher levels of initiative, and therefore tactical effectiveness, on the battlefield. Third, they argue that civil-military relations in democracies foster superior leadership because unlike in authoritarian regimes, officers in democracies are more likely to be promoted on the basis of skill and talent, as opposed to political loyalty and partisanship. The final hypothesis, which the authors do not test, is that soldiers in authoritarian militaries are more likely to surrender to democratic armies because they can expect to be well treated as prisoners of war, which weakens the strength and morale of autocratic armies.

Reiter and Stam evaluate these hypotheses with the frequently used POLITY data and the underutilized Historical Evaluation and Research Organization (HERO)/U.S. Army Concepts Analysis Agency data on battles.[16] To evaluate their hypotheses in chapter three, Reiter and Stam first use HERO data to test the effects of morale, initiative, and leadership on battle outcomes to see if

---

16. The HERO data were originally compiled by military historians who evaluated and coded dozens of aspects of different battles drawing from both primary and secondary historical sources. HERO is discussed on p. 70 of Reiter and Stam, *Democracies at War*.

higher levels correspond with a greater propensity to prevail in battles. Here they find that all three factors are important determinants of victory. Second, using HERO and POLITY data, they evaluate whether regime type affects a military organization's levels of morale, initiative, and leadership. Here they find that democracy tends to correspond with higher initiative and better leadership, although not necessarily with higher morale. From this two-step analysis, they thus infer that democracies win battles because they produce soldiers who exhibit superior leadership and initiative on the battlefield. For Reiter and Stam, good commanders and aggressive soldiers are the essence of the democratic advantage in war.

The Explanatory Vacuum

Reiter and Stam's volume is impressive in scope and objective. It is refreshing in the clarity of its arguments and in the authors' efforts to test them rigorously and creatively. The book is stimulating and interesting to read. This is an ambitious work.

Democracies at War, however, does not fully succeed in its ambitions. It suffers from several serious conceptual and methodological problems, which when combined significantly undermine the persuasiveness of the authors' claims.

DEMOCRACIES AT WAR: SUBSTITUTING FACTS FOR ARGUMENT

Democracies at War fails to present a complete and fully satisfying argument. At first glance the book seems weighty in analytical scope and insight. The authors draw on a diverse array of scholarship to generate hypotheses about democratic success in war. In chapter two, for example, they use institutional and bargaining theory to develop insights about how electoral constraints should influence leaders' selection of wars to fight. In chapter three, they draw from political philosophy to explain how liberal political culture promotes individualism and therefore initiative in soldiers. And in chapter four, the authors employ Immanuel Kant's protoconstructivist ideas in hypothesizing about the relationships among democracies and their propensities to ally.

This analytical eclecticism would be admirable if the objective of the book was to derive and test hypotheses from extant theories. Yet Reiter and Stam aspire to more than evaluating the insights of alternative theories: In chapter one, they propose to offer an original argument about the causes of democratic victory. Reiter and Stam's volume draws heavily from an extant literature in an effort to identify factors that could account for democratic success in war, but their book never offers a deductive argument for why some factors should be more powerful explanations than others; their deductive logic is not sufficient to distinguish among the alternative explanations and tell readers, on analytical grounds, which should best explain democratic success in war. Instead, Reiter and Stam test a diverse array of hypotheses in the book's chapters, find empirical support for three, and then offer these findings as an explanation of democratic victory. Consequently, the argument about why democracy is such a sui generis phenomenon reads like a cumulation of disparate hypotheses. There is no true analytical engine driving the testing machine.

Reiter and Stam would likely contest this observation. They might, for example, point to their claim in chapter one: "Our central argument is that democracies win wars because of the offshoots of public consent and leaders' accountability to the voters" (pp. 3–4). But this "central" argument appears largely to have been constructed to fit the hypotheses that their empirical analysis supports. Hence Reiter and Stam go on to report that these consensual systems confer two advantages. First, "being vulnerable to the will of the people restrains democratic leaders and helps prevent them from initiating foolhardy or risky wars." Second, "democracies' emphasis on individuals and their concomitant rights and privileges produces better leaders and soldiers more willing to take the initiative on the battlefield" (p. 4). In short, the central argument in chapter one is simply a description of their positive findings in chapters two and three.[17]

Even if we accept the post hoc nature of the argument, it provides a thin explanation for democratic success in war. Take the key propositions from chapter three. According to Reiter and Stam, once democracies enter wars, we should expect them to win battles because of superior leadership and more aggressive soldiering. These are important aspects of military organization and activity, but they are only a fraction of the factors that determine states' tactical effectiveness in war. Additional factors that Reiter and Stam might have examined include: weapons handling, information management/intelligence, logistics, training, and unit cohesion.

---

17. Perhaps the post hoc nature of the argument is best illustrated by Reiter and Stam's treatment of the morale variable. They spend pages explaining why democracies should foster superior morale—respect for the individual, legitimacy of the system, and so on. Yet when they find no significant relationship between morale and democracy, they do not interpret this as disconfirming evidence for their central argument about the merits of consensual systems. Instead they appear to formulate their argument to fit the hypotheses for which they find empirical support. Reiter and Stam, *Democracies at War*, pp. 60–64, 73–74.

Reiter and Stam, however, neglect all these aspects of military activity. That they do so is all the more puzzling given that some seem natural areas in which democracies might excel, and therefore integral to their broader goal of identifying the origins of democracies' advantages in war. For example, one area the authors fail to investigate is intelligence and information management. As Reiter and Stam themselves argue in *Democracies at War* and elsewhere, military leaders and their organizations may be less likely to misrepresent information in meritocratic democracies than in politicized autocracies and therefore be superior in gathering and relaying intelligence.[18] Despite the plausibility of this argument, Reiter and Stam offer no analytical reason why they do not investigate the role of intelligence, which incidentally appears in earlier versions of the argument,[19] or other military activities in their study. Of course, excluding these factors from the empirical analysis might have been necessary for practical reasons (e.g., because data were not available), but that is not a sufficient justification for excluding them from the conceptual argument. Reiter and Stam themselves admit as much by including in their study a hypothesis that posits that soldiers from autocracies will surrender relatively quickly to democratic armies, even though the authors lack data to test it.

Finally, Reiter and Stam do not give adequate attention to the most important factor determining any state's ultimate success in war: the quality of its strategic and operational plans and preparations.[20] As Millett, Murray, and Watman emphasize, the decisions that top political and military leaders make about when and how to prosecute war are vital to the success of any battle and war.[21] These include the number and type of campaigns that military leaders plan and prioritize; the quality of those war plans; and the weapons they pro-

---

18. Ibid., p. 23; and Dan Reiter and Allan C. Stam, "Democracy and Battlefield Military Effectiveness," *Journal of Conflict Resolution*, Vol. 42, No. 3 (June 1998), p. 266.
19. In a 1998 article, they hypothesize that meritocratic norms should enhance democracies' efficacy in intelligence gathering. They dismiss the argument, however, after finding that military intelligence is not a significant cause of battle success for states in war, regardless of regime type, which is in itself a bizarre and suspect finding. See Reiter and Stam, "Democracy and Battlefield Military Effectiveness," pp. 266, 271.
20. Note that in their initial tests in chapter two, where Reiter and Stam establish that democracies disproportionately prevail in war, they do include a control variable for strategy. But in the tests in chapter three, where the authors evaluate the effects of initiative and leadership on battlefield success, they do not control for broader operational/strategic considerations. Since the unit of analysis there is the battle, they might argue that controlling for strategy is less important than if the outcome analyzed was victory in war. But that neglects the effects that broader strategy has on preparations and planning for discrete battles—effectiveness at the strategic level affects effectiveness at the campaign and tactical levels, and vice versa. On the connections between strategic/operational/tactical effectiveness, see Millett, Murray, and Watman, "The Effectiveness of Military Organizations."
21. Ibid.

cure, the manpower they mobilize and train, the intelligence they gather, and the deployments they authorize in support of those plans. Absent sound strategic decisionmaking and operational planning, a military organization's tactical advantages may be rendered irrelevant to its ultimate success in war: Even the most effective forces run out of equipment and personnel and become exhausted if they are forced to fight in battles for which the state is not more broadly prepared. Germany's experience in both world wars provides a vivid illustration. That country lost both wars despite its remarkable tactical initiative and innovations largely because of poor strategic and operational decisions at the political-military apex.[22] In the end, tactical effectiveness may be only a second-order issue: Ultimately, how much does a military's tactical proficiency matter, if its strategy and operational plans doom it to defeat?[23]

In short, Reiter and Stam's argument about democratic success rests on a very narrow conception of what makes states effective in war. The authors do not give adequate attention to a variety of aspects of tactical activity that affect battlefield success, and more important, the broader strategic and operational plans and decisions that are essential to victory. Reiter and Stam's own data analysis suggests that they left a lot unexplained about the causes of democratic victory; the low r^2, as they acknowledge, reveals that neither initiative nor leadership explains the huge amount of the variance in the data (p. 79). Ultimately, Reiter and Stam may be right that better soldiering and leadership contribute to democratic effectiveness, and this is an important contribution to the debate. But the authors' claims are pitched much more strongly than this. They find that democracies are significantly more likely than autocratic states to win the wars they actually fight: According to Reiter and Stam, democracies prevail in 93 percent of those wars they initiate and in 63 percent of those in which they are targets of another's attack.[24] Are initiative and leadership really all that is at work here?

---

22. For example, in World War I, Germany—not the Allies—innovated breakthrough tactics to overcome trench warfare. Despite the innovation, the Germans were too exhausted logistically and otherwise to prosecute the war. See Holger Herwig, "The Dynamics of Necessity," in Millett and Murray, *Military Effectiveness*, Vol. 1, p. 102. On German tactical proficiency in World War I, see Niall Ferguson, *The Pity of War: Explaining World War I* (New York: Basic Books, 1999), pp. 308–310; and Michael Geyer, "German Strategy in the Age of Machine Warfare, 1914–1945," in Peter Paret, ed., *Makers of Modern Strategy: From Machiavelli to the Nuclear Age* (Princeton, N.J.: Princeton University Press, 1986), pp. 539–542. On World War II, see Martin van Creveld, *Fighting Power* (Westport, Conn.: Greenwood, 1974).

23. The inverse to the German case is also sometimes observed: Some states, such as Britain during World War I, prevail in wars despite unremarkable tactical performances.

24. This compares with 60 percent and 34 percent for dictatorships, respectively. Reiter and Stam, *Democracies at War*, p. 29.

Conceptual and Methodological Flaws

I argue above that Reiter and Stam do not provide fully developed and compelling explanations for democratic effectiveness in their book. But are the arguments and hypotheses they do offer otherwise sound? Do the links Reiter and Stam posit between democracy, initiative, and leadership make conceptual sense? In fact, their study suffers from conceptual and methodological weaknesses.

LOGICAL MISSTEPS IN *DEMOCRACIES AT WAR*
Reiter and Stam's study exhibits serious analytical shortcomings that stem from logical and empirical flaws in their arguments. First, the authors conflate culture with institutions, and second, they lump together autocratic states.

CONFLATING CULTURE AND INSTITUTIONS. The first analytical shortcoming is vividly illustrated in the book's argument about the effects of liberal political culture on democratic soldiers' initiative in combat. Essential to this argument is the assumption that democracy necessarily coincides with a liberal political culture that emphasizes the individual as the unit of value in society. Hence Reiter and Stam regularly make references such as: "Democratic, liberal culture emphasizes the importance of the individual. Indeed, the very essence of liberal democracy is the guarantee that the individual is minimally fettered." They refer to "the emphasis on individual prerogatives in democratic culture," and to the "the emphasis on the rights and prerogatives of the individual" (pp. 59–60). In short, Reiter and Stam not only equate democracy with liberal political culture; they also treat an individualistic political culture as a universal attribute of democratic states.

Such an assumption may seem plausible at first glance: Individuals often do enjoy greater civil and personal rights and privileges in democracies than they do in authoritarian regimes. But Reiter and Stam are asserting far more than this. When they talk about "liberal political culture," they are not interested only in the rule of law, constitutionalism, and elections, which might be associated with a universal set of norms involving civil liberties and personal rights; they are also referencing ingrained social conventions and worldviews associated with the value and role of the individual, and how this affects soldiers' actions in war. The logic of their argument suggests that within liberal societies, where the individual is elevated above collective identity, individuals are more willing to act independently and seize opportunities in battle. This is culture at its deepest—working at the societal and individual level—not just at the level of government and institutions.

Herein lies the problem with Reiter and Stam's assumption: Having elections and a constitution does not imply a liberal society in this deepest sense. Philosophically, popular rule (democracy) and a social order that favors individualism (liberalism) are distinct phenomena.[25] More practically, states can be procedurally democratic without being individualistic socially.[26] Some observers might question, for example, whether Japan and other Asian democracies qualify as having individualistic societies, such that the individual—versus the group—is the ultimate unit of value. Instead the conventional characterization of these societies' cultural norms is that they often reflect a high degree of comfort with social convention and group conformity. And if this is the case, following Reiter and Stam's logic, readers might then expect these states' soldiers to reflect those social values, and show low initiative on the battlefield, despite having elected their leaders through competitive elections.

Finally, even those states that do appear to meet Reiter and Stam's criteria for liberal democratic states may not have militaries that reflect the individualistic values they associate with them. Take, for example, the British military in the later nineteenth and early twentieth centuries. Victorian and Edwardian Britain was the epitome of a class-based society; it represented the antithesis of individualism in the way that Reiter and Stam characterize the concept in which individual ambition is unconstrained and initiative is encouraged and rewarded. Even more interesting, this class system has been blamed for exacerbating tactical inefficiencies within the British military. For example, during World War I the army chain of command on the western front was accused of being rigid and unresponsive, in part as a result of an overweening deference to authority and hierarchy among army personnel.[27] As Paul Kennedy characterizes the British army chain of command, "there was no real method for protesting against [the lack of communication] in favor of a full and open discussion among the commanders of alternative tactics, or for reassessment of

---

25. See, for example, discussion in Tarak Barkawi and Mark Laffey, *Democracy, Liberalism, and War: Rethinking the Democratic Peace Debate* (Boulder, Colo.: Lynne Rienner, 2001), pp. 13–14.
26. For recent commentary on the differences between liberalism and democracy, see Fareed Zakaria, "The Rise of Illiberal Democracy," *Foreign Affairs*, Vol. 76, No. 6 (November/December 1997). Zakaria goes so far as to suggest that not all democracies will respect civil liberties, let alone exhibit an individualistic culture in the way Reiter and Stam use the concept.
27. This, moreover, was long after the Cardwell reforms of the 1870s abolished the purchase of commissions and sought to create a more meritocractic officer corps and enlisted ranks. Prior to those reforms, it is especially difficult to argue that British military culture promoted anything like individuality and freedom of expression and action; rather it was mired in class politics and aristocratic conventions. For an overview of the reforms, see W.S. Hamer, *The British Army: Civil-Military Relations, 1885–1905* (Oxford: Oxford University Press, 1970).

basic assumptions."[28] In Kennedy's view, this conservative military culture hurt the development and innovation of new tactics that might have tested the stalemate on the western front.[29] Niall Ferguson makes a similar critique of Britain in World War I: "The entire culture of the British regular army militated against effective improvisation. The command structure was based on obedience to superiors and suspicion of subordinates. . . . Inhibitions existed at every level. Orders were issued at the top and fed down the line; there was little traffic in the other direction. . . . Partly for this reason, the proponents of a more technocratic approach made slow headway against the traditional believers in war as a moral rather than material contest. Too much emphasis was placed on morale, courage and discipline; not enough on firepower and tactics."[30] This description of Britain sounds little like Reiter and Stam's paradigmatic democratic army in which personal initiative flourishes and promotes aggressive innovation and exploitation of opportunities on the battlefield. And if Britain does not fit the model, then it is unclear that we should expect other democratic states that qualify as even less liberal to as well.

The British case underscores one last critical and arguably controversial assumption in Reiter and Stam's argument. The authors assume that a society's values will automatically permeate its military culture. As they put it, "Armies are microcosms of the societies they come from, and to a large extent, the military mirrors the qualities of the society from which it is drawn" (p. 65). There are huge counterveiling forces at work, however, that might prevent the diffusion of civilian values into military culture. The premise of Samuel Huntington's seminal *Soldier and the State,* for example, is that militaries are not creatures of their societies.[31] Militaries are hierarchical organizations; the nature of the job requires that individuals act by reflex and in conformity with es-

---

28. Paul Kennedy, "Britain in the First World War," in Allan R. Millett and Williamson Murray, *Military Effectiveness,* Vol. 1, p. 53.

29. See, for example, ibid., pp. 59–60; John Gooch, "The Armed Services," in Maurice W. Kirby, Mary B. Rose, and Stephen Constantine, eds., *The First World War in British History* (London: Edward Arnold, 1995), pp. 191–192; David Woodward, *Lloyd George and the Generals* (East Brunswick, N.J.: Associated University Presses, 1983), p. 76; and Ferguson, *The Pity of War,* pp. 292–294. Of course this is not the only view; on debate about the quality of Douglas Haig's leadership see Kennedy, "Britain in the First World War"; see also Ferguson, *The Pity of War,* pp. 302–305.

30. Ferguson, *The Pity of War,* pp. 306–308, goes on to provide examples of how this culture impeded innovation and initiative in World War I.

31. In fact, for Samuel P. Huntington the fundamental problem for U.S. civil-military relations was the incongruity between liberal American society and an inherently conservative military organization. Huntington, *The Soldier and the State: The Theory and Politics of Civil-Military Relations* (Cambridge, Mass.: Harvard University Press, 1957).

tablished conventions of behavior. Most military organizations deliberately socialize and train the "individualism" out of their troops. There are good reasons for this: Submission to authority is essential to maintaining both good discipline in the chain of command and the effectiveness of the fighting organization.

In summary, not all democracies are liberal in the way that Reiter and Stam characterize the concept. Nor is it self-evident that even in democratic states that qualify as liberal, their militaries' ethos is fundamentally shaped by those societal values. And if democracies do not all have liberal culture and not all militaries in liberal democracies reflect liberal values, then the causal argument about why democratic states should foster soldiers with more initiative breaks down.[32]

LUMPING AUTOCRACIES. A second weakness in *Democracies at War* lies in the generalizations the authors make about civil-military relations in autocratic states. Reiter and Stam argue that democracies should have superior leadership because they tend to promote individuals based on merit. Autocracies, in contrast, promote for political reasons, often at the expense of skill and talent, which leads to inferior leadership. The logical connection between partisanship and poor leadership in highly politicized civil-military settings seems tenable and has been argued by many scholars.[33] As noted earlier, promotions on the basis of partisanship are often one component in a broader effort to maintain political control over the military in places where leaders rely heavily on the armed forces' support to maintain office.

Reiter and Stam assume, however, that all authoritarian regimes use this model of political control. In reality, the degree of politicization and its impact on military effectiveness vary considerably within nondemocratic regimes. Not all authoritarian regimes are created equally. Biddle and Zirkle provide an excellent illustration of how they differ. They contrast civil-military relations in North Vietnam and in Iraq to evaluate how civil-military relations in each affected the assimilation of new technologies, and therefore the state's military effectiveness. Much of what they say about Iraq is consistent with Reiter and Stam's arguments. Hence they observe that promotions were governed by po-

---

32. Desch's finding that three countries—the United States, Britain, and Israel—account for 75 percent of Reiter and Stam's results on tests affirming democracies' propensities to win wars only reinforces the sense that something particular about these countries, and not liberalism in general, is driving the findings. Desch, "Democracy and Victory," pp. 16–17.
33. See my discussion above and especially note 21.

litical considerations and initiative was suppressed; in addition, there were multiple lines of command, substantial surveillance of military authorities, training that was politicized, and the exclusion of the military at the highest decisionmaking levels. When combined, these factors undermined Iraq's military effectiveness, as Reiter and Stam might predict.

North Vietnam during the 1960s, however, exhibited a markedly different pattern of civil-military relations compared with Iraq. According to Biddle and Zirkle, and contrary to Reiter and Stam's conjectures about leadership in autocratic regimes, merit was paramount in officer promotions; North Vietnamese military leaders were selected largely on the basis of skill and talent. Moreover, the military and political leadership exhibited a strong unity of command, initiative was not suppressed, and military training was notably apolitical. Consequently, relative to Iraq and by many objective standards, North Vietnamese soldiers exhibited substantial initiative and were highly proficient at using their weapons. Cases such as these suggest that there is significant variation in civil-military relations in authoritarian regimes, and that the predictions Reiter and Stam make about politicized selection criteria and poor leadership only logically fit certain cases. In short, Reiter and Stam make vast generalizations about civil-military relations in authoritarian regimes. Finer distinctions would have sharpened the argument considerably and revealed variations within and across authoritarian (and democratic) regimes critical to anticipating how well they will perform in war.

Empirical Failings

Reiter and Stam's best response to any of these critiques is to point to their empirical results. One might quibble with their arguments, but the proof is in their results. The problem is that in addition to conceptual shortcomings, *Democracies at War* has weaknesses that cast doubt on their empirical findings.

TESTING THE LOGIC OF TESTING IN *DEMOCRACIES AT WAR*

Reiter and Stam's study suffers from important empirical shortcomings. First, and most important, the empirical analysis in *Democracies at War* does not effectively test one of the book's key arguments: that because democracies have a liberal political culture, their soldiers will demonstrate higher levels of initiative on the battlefield. The indicators they use for their independent and dependent variables do not really correspond with the argument they are trying to test. Of course, many large-*n* studies must contend with having imperfect measures for concepts, given the difficulty of quantifying intangibles or gaining access to data. Yet the weaknesses in Reiter and Stam's empirical analysis are particularly significant. Two specific flaws in their operationalizations are evident.

The first is that they do not actually measure liberal political culture, even though the causal pathway they posit emphasizes the ultimate importance of culture on military effectiveness. Rather than measuring culture, their explanatory variables are measures of procedural democracy, drawn from the widely used POLITY data: "Democracy and autocracy scores are aggregations of the degree of openness of the system, the degree of participation, and the degree of competitiveness of candidate selection" (p. 40). In effect, these are indicators that relate to the competitiveness of elections—of democratic institutions and not of liberal political culture. None captures a society's philosophical underpinnings and how its citizens value individual versus group rights, or the degree to which they interact within hierarchical versus egalitarian social structures—elements that observers might expect to be associated with liberal societies. Consequently, culture effectively does little real work in their empirical analysis. As discussed above, Reiter and Stam instead rely on a problematic assumption that procedural democracy implies liberal culture (as they characterize it). Hence, given these tenuous links between culture and democratic institutions, and by not coding culture directly, their tests can say little about the effects of culture on initiative.

Second, and even more problematic, is Reiter and Stam's measurement of initiative. Recall that by "initiative" Reiter and Stam mean the willingness of individuals to take risks, to be assertive, and to seize opportunities in combat. As they put it, "Modern warfare requires the display of individual initiative on the battlefield to exploit best emergent but fleeting opportunities and to cope with unanticipated conditions. The exercise of individual initiative is especially important for the successful execution of mobile and maneuver-based strategies, as their very premise is allowing lower-ranked officers and troops the freedom to exploit the fluid battlefield" (p. 59). In effect, Reiter and Stam are interested in how culture affects the assertiveness of soldiers in the heat of combat.

To test these relationships, the authors rely on HERO data, which conveniently includes a measure of initiative. Yet herein lies the problem for Reiter and Stam. HERO coding rules are ambiguous as to the meaning of initiative. The most straightforward interpretation of the rule suggests that HERO's "initiative" measures something significantly different from Reiter and Stam's interpretation of the concept.[34]

HERO defines initiative in the following terms: "Initiative is 'an advantage gained by acting first, and thus forcing the opponent to respond to one's own plans and actions, instead of being able to follow one's own plans'" (quoted on p. 78). This coding definition suggests that the analysts compiling data for HERO were looking to see which side attacked first in a battle and "seized the initiative" in this traditional sense, and not passing judgment on whose soldiers had more aggressively and better exploited opportunities in the heat of combat. If so, this would be unsurprising because the unit of analysis in HERO is the "battle." It would make sense that these analysts would be focusing on an operational level—a campaign or theater level of warfare[35]—concept of initiative and measuring who attacked first in the battle. Yet Reiter and Stam's arguments are really about the nature of individual action and behavior in the tense moments of combat—about "exploit[ing] fleeting opportunities and cop[ing] with unanticipated conditions," as they put it (p. 65). Their argument is much more subtle than just looking to see who struck who first in battle.[36] As such, HERO's definition of initiative has, at best, an ambiguous, and at worst, a completely unrelated, relationship to the concept of individual initiative, in the spirit in which Reiter and Stam use it.[37]

These ambiguities, in turn, cast doubt on whether Reiter and Stam have really demonstrated that democratic soldiers exhibit greater initiative in combat. Rather than revealing anything about the intrinsic strengths of soldiers, their finding that democracy and initiative are related may reflect that because democracies win wars, they are also more likely to be on the operational offensive in those wars, initiating engagements and pursuing their adversaries. The list of wars included in the HERO data set appears to contain a large number of wars in which the more democratic side ultimately prevailed: the U.S. Civil War, the U.S.-Mexican War, the Boer War, World War I, and World War II, as

---

34. On problems with HERO data and ambiguity in coding, see Desch, "Democracy and Victory," pp. 39–42.
35. For this definition, see Millett, Murray, and Watman, "The Effectiveness of Military Organizations," p. 12.
36. It is possible that what HERO coders meant is consistent with the meaning that Reiter and Stam impute to the concept of initiative. But the ambiguity of the coding rules makes certainty impossible.
37. These problems are rendered even worse if one considers the significant measurement error that Desch identifies in his article, "Democracy and Victory," pp. 38–41. Hence it appears that not only does initiative fail to capture the concept Reiter and Stam are trying to test, but it may be that the data on initiative as originally defined in HERO are plagued by error.

well as four Arab-Israeli wars.[38] In addition, if these more democratic states are winning the great majority of wars they fight, as Reiter and Stam contend, one might also expect them to be attacking first in battles, at least in the latter stages of war when they may be pursuing their adversaries. Therefore democracies should have high initiative in the way that HERO appears to code the concept. Accordingly, Reiter and Stam's finding that democracies exhibit a high level of initiative may just be an artifact of these states' propensities for victory: Democracies' high initiative may be the by-product—rather than a cause—of their success at war.

CONFOUNDING VARIABLES. In addition to poorly operationalizing their variables, Reiter and Stam do not control for some potential alternative explanations for why some states' soldiers might exhibit higher initiative and superior leadership.[39] One obvious factor that they neglect is the effect of variation in states' human capital resources on the quality of their military activities. There are good reasons to think that human capital should matter for military effectiveness, and more specifically for initiative and leadership. A better-educated population should yield military personnel with higher literacy and technological skills; these men and women should then be more trainable, better able to assimilate sophisticated weaponry, more confident in battle, and therefore more ready to seize the initiative. Similarly, higher investment in both regular and military education should produce a better officer class and therefore superior leaders.[40]

Most important, for Reiter and Stam's analysis, democratic institutions and high levels of human capital appear to correlate positively: Democracies tend to spend more money on education, and by implication, produce better-educated populations.[41] Consequently, the methodological effect of including

---

38. See Reiter and Stam, *Democracies at War*, pp. 82–83. Without the relative democracy scores, it is impossible to confirm whether HERO contains a disproportionate number of wars in which the relatively more democratic side prevailed. But given the findings earlier in the book that show that democracies win wars disproportionately, one might logically expect that a large number of victorious democracies would appear in the HERO sample as well.

39. They do control for a variety of factors in their initial analysis using COW/POLITY data where they establish that democracies disproportionately win wars. However, in the empirical analysis in chapter three, in which they evaluate whether initiative and leadership influence success in war, and whether democracies excel in those areas, the only control variable included is home territory.

40. See, for example, "Anthony Pascal, "Are Third World Armies Third Rate? Human Capital and Organizational Impediments to Military Effectiveness," Working Paper P-6433 (Santa Monica, Calif.: RAND, January 1980).

41. See David A. Lake and Matthew A. Baum, "The Invisible Hand of Democracy: Political Control in the Provision of Public Services," *Comparative Political Studies*, Vol. 34, No. 6 (August 2001), pp. 597–621.

democracy in the empirical analysis, but not human capital, is to inflate the beneficial impact of democratic institutions on military effectiveness.[42] A recent study by Stephen Biddle and Stephen Long of military effectiveness suggests that this may be a problem in Reiter and Stam's analysis.[43] After including human capital (along with democracy, culture, and a variety of other control factors) in their empirical tests, Biddle and Long found that the sign for the democracy coefficient reversed: Instead of helping, having democratic institutions appears to degrade a state's effectiveness, when other factors are controlled for.[44]

Generally, this suggests that the benefits of procedural democracy may be an artifact of Reiter and Stam's neglect of alternative causes of effectiveness that co-vary with democracy (intuitively democracy is acting as a proxy for the true causal variable). Specifically, it suggests that democracy alone has no direct benefit on states' military effectiveness; democracy matters little (and may be harmful) if a state otherwise has low human capital resources. In other words, states with low human capital may perform poorly in war, even if they are democratic (e.g., India or the many newly democratizing and developing countries). Conversely, even those states without democracy can still perform well in war if they achieve higher levels of human capital (e.g., Belarus and Singapore).

MAKING RELATIVE DEMOCRACIES FROM ABSOLUTE AUTOCRACIES. All of these problems are compounded by yet other oddities in Reiter and Stam's empirical analysis, including how they code states' levels of democracy in their tests. Their approach for determining states' democracy scores is largely driven by the fact that their variables for morale, leadership, and initiative are measured in a relative manner in the HERO data: "That is, [HERO] does not assess an army's level of morale on a scale against all other armies at the time, but rather it compares the morale [and leadership and initiative] levels of two armies in a

---

42. Methodologically, these results suggest that Reiter and Stam may have an omitted variable bias problem; because human capital and democratic institutions likely co-vary, by including an indicator for the latter and not the former in the analyses, their results may overstate the real causal impact of democratic institutions. On such methodological issues, see Gary King, Robert Keohane, and Sidney Verba, *Designing Social Inquiry: Scientific Inference in Qualitative Research* (Princeton N.J.: Princeton University Press, 1994), pp. 168–176.
43. Stephen Biddle and Stephen Long, "Democratic Effectiveness? Reassessing the Claim That Democracies Are More Effective in Battle," paper prepared for delivery at the annual meeting of the American Political Science Association, Boston, Massachusetts, August 29–September 1, 2002.
44. Not only does the sign reverse, but the significance level of the democracy coefficient increases when the control variables are included. The overall fit of the model to the data also improves significantly. See ibid., p. 9.

particular engagement."[45] Thus, a state is not absolutely high in morale, but rather exhibits more or less morale than its counterpart in a battle. In turn, as Reiter and Stam go on to explain, because their independent variables are measured in comparative terms, "efficient hypothesis testing" requires that they convert regime type into a relative score (p. 77). Hence a state's level of democracy is measured against its adversary's to determine which side is more liberalized than the other.[46] Consequently, two states may in an absolute sense qualify as autocratic, but if one is slightly less repressive, then it ranks higher on the relative democracy score. Reiter and Stam's empirical analysis, then, effectively evaluates whether a state's slightly higher ranking on this democracy score corresponds with at least a slightly higher level of morale, leadership, or initiative relative to its counterpart in a particular battle.

Although perhaps necessary for methodological reasons, the relative democracy score is more than a little odd conceptually. First, it assumes that regime type is a continuous, rather than a discrete, variable such that the nature of governance in democratic and authoritarian regimes differs in degree rather than kind. Within comparative politics, there has been a large debate about this issue with arguably the most methodologically sophisticated study to date deciding against the view that regime type should be measured on a continuum from democracy to authoritarianism. Rather the study asserts that democracies are qualitatively distinct from even the most liberalized autocracies.[47]

Second, and more important, Reiter and Stam's methodology implies that the organization and culture of military and society are going to change continuously with subtle differences in regime type. Hence two states might both qualify as autocratic, but if one liberalizes slightly, this logic implies that societal culture will become incrementally more liberal; military organization will become somewhat more meritocratic; and soldiers' initiative and leadership will therefore concomitantly improve.[48] Consequently, the state will perform at

45. In particular, morale, leadership, and initiative are each measured on five-part index (−2, −1, 0, 1, 2), which reflects a judgment about which side (attacker or defender) had an advantage in the battle. Reiter and Stam, *Democracies at War*, pp. 76–78.
46. Here they produce a scale of −20 to 20. Hence an army that is significantly more democratic than its adversary might be assigned a score of 20, with one significantly less democratic achieving a −20.
47. See Mike Alvarez, Jose Antonio Cheibub, Fernando Limongi, and Adam Przeworski, "Classifying Political Regimes," *Studies in Comparative International Development*, Vol. 31, No. 2 (Summer 1996), pp. 21–22. As the authors conclude, "The analogy with the proverbial pregnancy is thus that while democracy can be more or less advanced, one cannot be half democratic: there is a natural zero point" (p. 21).
48. In other words, Reiter and Stam's empirical approach implies that the organization and struc-

least slightly better than its fellow dictatorship on the battlefield. But is this how the relationship between regime type and military activity works? It seems equally, if not more, plausible that more dramatic shifts in the internal logic and structure of a regime are necessary before one would observe changes in the nature of its society and military organization. If Bahrain, for example, were at war with Saudi Arabia, does it make sense to expect that the former's soldiers would exhibit more initiative because Bahrain's emir recently introduced highly circumscribed parliamentary elections, and Saudi Arabia has no elected legislature? Is it plausible to argue that the reason the German army during World War I demonstrated superior tactical proficiency relative to its autocratic counterparts was because Wilhemine Germany had a semi-democratic parliament and states such as Imperial Russia were closer to pure autocracies?[49]

HISTORICAL REALITY CHECK. Finally, perhaps Reiter and Stam's analysis might stand in better stead if the argument passed the (admittedly unscientific) test of historical common sense. Is there really good reason to believe that democracies are inherently superior at tactical initiative? There may well be, but a large amount of historical and anecdotal evidence belies such a relationship. Many nondemocratic states, for example, have had soldiers who exhibit enormous tactical initiative on the battlefield. For instance, in his lengthy examination of Libya's wars with Chad, Pollack details how Chadian soldiers outfought their counterparts, repeatedly demonstrating remarkable tactical ingenuity and leadership.[50] As noted above, Biddle and Zirkle discuss how during the Vietnam War, the North Vietnamese demonstrated substantial initiative in employing their air defense system.[51]

One has only to look to the conflicts listed in HERO such as the Zulu (1879) and the Sudan (1882, 1898) wars to find examples of nondemocratic regimes whose soldiers exhibited exceptional tactical initiative. For example, although the Zulus ultimately lost the war, most analysts agree that their soldiers exhib-

---

ture of military activity in a state varies continuously with differences in its regime type. More likely, for a state to stop using partisanship as the basis of military appointments and for its society to reflect truly liberal values would require a substantial and qualitative difference in regime type.

49. Reiter and Stam might respond to this critique by noting that, statistically, cases with large differences in regime type likely do most of the work, relative to the smaller effects of dyads with subtle regime type differences. But scholars' empirical analyses should always be grounded in coherent theories of the relationship among the variables tested. Hence even if less important empirically, the inclusion of cases with subtle differences is nonetheless problematic conceptually.

50. Pollack, *Arabs at War*, especially his assessment on p. 412.

51. Biddle and Zirkle, "Technology, Civil-Military Relations, and Warfare in the Developing World," pp. 193–194.

ited amazing initiative battling the British. When the war started, the British had been expecting the proverbial cakewalk. Instead they faced an army that fought aggressively and with substantial initiative, most notably at the battle at Islandwana, in which 20,000 Zulu forces attacked the British, and at Rorke's Drift when the Zulu commander showed so much "initiative" that he disobeyed orders from King Cetshwayo and seized the moment to launch a surprise attack.[52] Similarly, in the Anglo-Egyptian campaigns against the nascent Mahdist state in Sudan, the Mahdi are reputed to have shown significant initiative and leadership.

Indeed the army that is most often singled out as the historical exemplar of tactical effectiveness is not democratic: Germany from the wars of unification in the mid-nineteenth century through World War II is almost universally acclaimed for its devastating tactical proficiency.[53] Many scholars have puzzled over the German phenomenon. The renowned sociologists Morris Janowitz and Edward Shils, for example, were so intrigued by the sources of German effectiveness in World War II that they undertook extensive interviews and analysis of German prisoners of war after the war to determine its sources. This study subsequently spawned a half-century debate about the sources of human motivation in war. Martin van Creveld, for example, later wrote a book investigating the Germans' capacity to generate what he saw as amazing levels of "fighting power" during World War II.[54] Even more interesting, while lauding the German army, van Creveld offered a scathing critique of U.S. forces, arguing that American solders' initiative had been suppressed by an overly managerial style of administration, "an addiction to information," and "less than mediocre" tactical leadership.[55] In a study of war outcomes Trevor Dupuy, who oversaw the collection of the HERO data that Reiter and Stam utilize, himself acknowledged the tactical inferiority of U.S. and British forces

---

52. On the Zulu war, see J.J. Guy, "A Note on Firearms in the Zulu Kingdom with Special References to the Anglo-Zulu War, 1879," *Journal of African History*, Vol. 12, No. 4 (1971), pp. 557–570; and John Labland, *Kingdom in Crisis* (Manchester, U.K.: Manchester University Press, 1992). This example raises some interesting points. First, it suggests that initiative is a double-edged sword: What looks like initiative in a winning battle such as Islandwana looks like a breakdown in discipline in a battle like Rorke's Drift. Second, it illustrates the importance of viewing military effectiveness not only at the tactical level, but also in context of larger strategic/operational objectives. The attack at Rorke's Drift is significant because it compromised the larger political objectives of the Zulus not to appear to be the aggressors in the conflict. In so doing, the Zulus hoped to maintain international sympathy for their cause.
53. For a similar observation, see Desch, "Democracy and Victory," p. 41.
54. Van Creveld, *Fighting Power*.
55. Ibid., pp. 33, 166–168.

when compared with the Germans in World War II.[56] After controlling for material factors, he found that the outcome of a battle (victory or defeat) between German, American, and British forces could be predicted only if it was assumed that, as van Creveld puts it: "Man for man, unit for unit the Germans were twenty to thirty percent more effective than the British and American forces facing them."[57] In a footnote, Reiter and Stam do acknowledge the German phenomenon, but dismiss it as an aberration, "an isolated case," rather than an indication that the connections between regime type and tactical effectiveness they posit merit further consideration (p. 214, n. 17). In sum, a historical reality check suggests that autocratic states often seem capable of producing innovative, risk-acceptant soldiers who seize opportunities on the battlefield and that democracies do not automatically excel in these areas.

Conclusion

Democracies at War is a noteworthy book because it addresses a truly important question: What explains why some states are better than others at creating militaries that are effective at warfare? It raises provocative and interesting questions about the efficacy of democracies in war. The book, however, exhibits a variety of conceptual and empirical shortcomings. Combined, these weaknesses seriously undercut the book's persuasiveness.

The volume does not offer a satisfying argument for the patterns of military effectiveness it purportedly identifies. Reiter and Stam string together the positive results of their empirical tests, offering them up as an explanation for democratic effectiveness, but never provide a coherent, deductive argument about why democracies excel at war. In the end, Reiter and Stam never convincingly explain why the reader should expect democracies to perform so competently in war.

The arguments these authors do attempt to articulate are also flawed. Reiter and Stam mistakenly conflate liberalism and individualism with procedural democracy. They make broad generalizations about the civil-military relations of autocratic states, ignoring important variation in the mechanisms of political

---

56. Trevor N. Dupuy, *Numbers, Predictions, and War: Using History to Evaluate Combat Factors and Predict the Outcome of Battles* (New York: Bobbs Merrill, 1979); and Trevor N. Dupuy, *A Genius for War: The German Army and General Staff, 1807–1945* (London: Prentice Hall, 1977), pp. 234–235.
57. Van Creveld, *Fighting Power*, p. 5.

control and their implications for the quality of military leadership in these states.

Democracies at War exhibits important weaknesses in how the authors evaluate their various hypotheses. Inadequacies in the way Reiter and Stam measure concepts such as liberal culture and initiative mean that the authors do not effectively test what they assert is a central finding of the book: that democratic soldiers exhibit higher individual initiative on the battlefield. The authors also fail to control for alternative explanations and therefore obscure the relationship between democracy and military effectiveness.

In short, *Democracies at War* exhibits some serious flaws in its arguments, logic, and methodology. Yet, notwithstanding its imperfections, this study does merit reading by both academics and policy analysts.

Especially important, the book positions the study of military effectiveness center stage on the scholarly agenda. The publication of this book is likely to focus significant, and well deserved, attention on how states create and employ military force. Reiter and Stam point to several particular factors that might affect how states organize, prepare, and execute plans for war. But, equally important, they open the door to new avenues of inquiry about how other forces might influence states' military effectiveness. For example, while focusing on regime type, Reiter and Stam raise natural questions about variation within the democratic/authoritarian categories. For instance, do some democracies' institutional structures encourage military organizations to be responsive and innovative in formulating doctrine and procuring new weapons, while others create a status quo bias? Do, in fact, civil-military relations vary in predictable ways within authoritarian regimes such that some regimes, at some times, are better able to produce professional, skilled officers and personnel than others? *Democracies at War* also raises interesting questions about how societal culture (ingrained social norms and values) and military organizational culture (universal bureaucratic prerogatives or variations in the cultures of different organizations) might influence military activity in a state, and consequently its capacity to generate force from its human and material resources. In short, by drawing attention to factors such as regime type, political culture, and civil-military relations, Reiter and Stam are paving the way for the emergence of a research program that examines the implications of a diverse array of phenomena for military effectiveness.

Such a research program, moreover, could potentially spur new thinking about how scholars conceptualize the sources of states' military power. Almost universally, political scientists have in the past used very rough indictors of military power—gross national product, indications of industrial capacity (e.g., steel production), expenditure on the military, and expenditure per solider—to discriminate among states. These analyses therefore rely on very crude, heavily materialist indicators of power. Yet, as Biddle demonstrates, these gross indicators have only weak predictive power in gauging victory in war. As he puts it, "An enormous scholarly edifice thus rests on very shaky foundations" in its assumptions about the nature of military power.[58]

In contrast, studies such as Reiter and Stam's suggest that the immaterial and less easily quantifiable aspects of military organization and activity are essential sources of state power. In the end, materialist indicators of power could prove to be highly misleading: We may well discover that there is a tremendous disconnect between a state's access to technological, financial, and human resources and the social, political, and other intangible factors that allow it to translate those resources into military power in war against another state. Simply put, given the same basic resources, not all states may be equally capable of generating military power from them.

Such findings, in turn, could have practical implications for debates about U.S. defense spending and priorities. They might, for example, inspire at least a partial reorientation in how the United States utilizes its defense dollars: toward greater expenditure on training, doctrine, and enhancement of organizational efficiency and integration and away from a preoccupation with the procurement of expensive, high-technology weapons systems. In so doing, research on military effectiveness represents an important antidote to the trends of the day, which increasingly privilege technological, over human and organizational, solutions to enhancing U.S. military capabilities.

Research on military effectiveness could also yield new insight into practical efforts to assess states' relative capabilities. As the United States embarks on an era in which military action may be contemplated less against stable and predictable states, and more against states and nonstate actors whose political and social structures are complex and in flux, policy analysts will likely need to develop new ways of incorporating nontraditional measures of power into capabilities assessments. Failure to do so could heighten the risk for the United States of miscalculating its adversaries', its allies', and even its own capabilities in a future war or military engagement. Scholarship on military effectiveness thus represents an important first step toward developing the methodologies essential to measuring capabilities in the twenty-first century.

---

58. Biddle, *Military Power,* table 2.1 and chap. 2.

Democracy and Victory | *Michael C. Desch*

Why Regime Type Hardly Matters

Whether democracies are more or less likely to win wars has long been a contentious issue. The Greek general Thucydides' chronicle of the defeat of democratic Athens in its twenty-four-year struggle with authoritarian Sparta in *The Peloponnesian War*, particularly his account of the Sicilian debacle, remains the classic indictment of the inability of democracies to prepare for and fight wars.[1] Indeed, for most of Western history, pessimism dominated thinking about democracy and war. "Democratic defeatists," from the French aristocrat Alexis de Tocqueville to mid-twentieth-century realists such as E.H. Carr, George Kennan, and Walter Lippmann, believed that democracy was a decided liability in preparing for and fighting wars. Particularly during the Cold War, the pessimistic perspective on the fighting power of democracies was dominant.[2] Even leaders of the free world, such as John F. Kennedy, believed that when democracy "competes with a system of government . . . built primarily for war, it is at a disadvantage."[3] Despite the end of the Cold War, a few Cassandras remain concerned that democracies are unprepared to meet the next major military threat from authoritarian states such as China or international terrorist organizations such as al-Qaeda.[4]

Not everyone shared this pessimism, however. The Greek historian Herodotus argued that democracy increased military effectiveness: "As long as the

Michael C. Desch is Professor and Associate Director of the Patterson School of Diplomacy and International Commerce at the University of Kentucky.

For their extremely helpful comments on earlier drafts of this article I thank Eugene Gholz, Douglass Gibler, Hein Goemans, Samuel Huntington, Stuart Kaufman, Edward Mansfield, Daniel Markey, John Mueller, John Odell, Robert Pape, Dan Reiter, Helena Truszczynska, Steven Voss, Stephen Walt, participants in seminars at the Program for International Security Policy at the University of Chicago and the Belfer Center for Science and International Affairs at the John F. Kennedy School of Government, Harvard University, the anonymous reviewers for *International Security*, and especially John Mearsheimer. I received generous financial support from the Smith Richardson Foundation and indispensable research assistance from Glenn Rudolph and John Hajner.

1. Thucydides, *The Peloponnesian War*, Bk. 7, trans. Rex Warner (Middlesex, U.K.: Penguin, 1954).
2. See, for example, Jean-François Revel, *How Democracies Perish* (Garden City, N.Y.: Doubleday, 1984), p. 3.
3. Quoted in Melvin Small, *Democracy and Diplomacy: The Impact of Domestic Politics on U.S. Foreign Policy, 1789–1994* (Baltimore, Md.: Johns Hopkins University Press, 1996), p. xiii.
4. Donald Kagan and Frederick W. Kagan, *While America Sleeps: Self-Delusion, Military Weakness, and the Threat to Peace Today* (New York: St. Martin's, 2000), pp. viii, 307; and Robert Kagan and

International Security, Vol. 27, No. 2 (Fall 2002), pp. 5–47
© 2002 by the President and Fellows of Harvard College and the Massachusetts Institute of Technology.

Athenians were ruled by tyrants they were no better warriors than their neighbors, but once they got rid of the tyranny they became best of all by a long shot."[5] With the democratic West's victory in the Cold War, a renewed optimism about the military prowess of democratic states has taken root. "Democratic triumphalists" note that an examination of major wars since 1815 reveals that the more democratic states have been on the winning side in the overwhelming majority of cases.[6] "There is something about democratic regimes," two triumphalists suggest, "that makes it easier for them to generate military power and achieve victory in the arena of war."[7]

Democratic triumphalists offer different explanations for why this should be the case, and sometimes they dissent from each other's arguments; taken as a whole, however, they suggest two reasons why democracies tend to win wars.[8] Some argue that democracies are better at picking the wars they get into, starting only those they know they can win. This is the "selection effects" argument. Others maintain that once at war, democracies fight more effectively: They have bigger economies, form stronger alliances, make better decisions, have higher levels of public support, or can count on greater effort from their soldiers. This is the "military effectiveness" argument.

The aim of this article is to question this sanguine view about democracy and military victory. I make three arguments. First, an examination of the historical data and methodological approach does not strongly support the triumphalists' claim that, all other things being equal, democracies are more likely to win in war.

Second, the logic that underpins the triumphalists' case is unpersuasive. Specifically, there is no reason to believe, nor is there much evidence to suggest, that leaders of democracies are more careful in selecting their wars than their authoritarian counterparts. The same charges can be made against the military effectiveness argument.

William Kristol, "Getting Serious," *Weekly Standard*, November 19, 2001, http://www. weeklystandard.com/Content/Public/Articles/000/000/000/518hrpmo.asp.
5. Quoted in Donald Kagan, *Pericles of Athens and the Birth of Athenian Democracy* (New York: Free Press, 1991), p. 16. See also Victor Davis Hanson, *The Soul of Battle: From Ancient Times to the Present Day, How Three Great Liberators Vanquished Tyranny* (New York: Free Press, 1999).
6. David A. Lake, "Powerful Pacifists: Democratic States and War," *American Political Science Review*, Vol. 86, No. 1 (March 1992), pp. 24–37. See also Dan Reiter and Allan C. Stam III, *Democracies at War* (Princeton, N.J.: Princeton University Press, 2002).
7. Dan Reiter and Allan C. Stam III, "Democracy and Battlefield Military Effectiveness," *Journal of Conflict Resolution*, Vol. 42, No. 3 (June 1998), p. 259. For similar sentiments, see Aaron L. Friedberg, *In the Shadow of the Garrison State: America's Anti-Statism and Its Cold War Grand Strategy* (Princeton, N.J.: Princeton University Press, 2000), p. 340; and Hanson, *The Soul of Battle*, p. 4.
8. William Reed and David H. Clark, "War Initiation and War Winners," *Journal of Conflict Resolution*, Vol. 44, No. 3 (June 2000), pp. 378–395.

Third, explanations other than those based on regime type more plausibly explain how states perform in war. Some of these explanations are well known. For example, an advantage in military power is often a reliable indicator of which side is likely to win a war.[9] The nature of the conflict can also influence military outcomes. In particular, the opposing sides in a war often have asymmetrical interests, which sometimes produce a paradoxical outcome where the weaker state defeats its more powerful adversary.[10] Moreover, states that imitate the military organization and doctrines of the leading states in the international system are likely to prevail in war.[11] Nationalism has also proven to be a potent source of increased military effectiveness in democracies (e.g., revolutionary France, 1789–94) and in autocracies (e.g., Prussia and Spain, 1807–15).[12] Other explanations are less well known. It is possible, for instance, that the correlation between democracy and victory is spurious: Certain factors that make it more likely that a state will be democratic also increase the likelihood that it will win most of its wars.[13] Finally, whether a regime is consolidated or not could determine its performance in war.

My case against the triumphalists should not be read as support for the pessimists' claim that democracies are especially inept at fighting wars, and therefore likely to be defeated by rival authoritarian states. Rather, it supports the

---

9. Ivan Arreguín-Toft, "How the Weak Win Wars: A Theory of Asymmetric Conflict," *International Security*, Vol. 26, No. 1 (Summer 2001), p. 97, finds that in interstate wars between 1800 and 1998, the stronger actor won nearly 71 percent of the time. See also John J. Mearsheimer, "Assessing the Conventional Balance: The 3:1 Rule and Its Critics," *International Security*, Vol. 13, No. 4 (Spring 1989), pp. 54–89; and John J. Mearsheimer, *The Tragedy of Great Power Politics* (New York: W.W. Norton, 2001).
10. Andrew M. Mack, "Why Big Nations Lose Small Wars," *World Politics*, Vol. 27, No. 2 (January 1975), pp. 175–200; and Arreguín-Toft, "How the Weak Win Wars," pp. 93–128.
11. Kenneth N. Waltz, *Theory of International Politics* (Reading, Mass.: Addison-Wesley, 1979), pp. 76–77, 127–128. See also João Resende-Santos, "Anarchy and the Emulation of Military Systems: Military Organization and Technology in South America, 1870–1914," *Security Studies*, Vol. 5, No. 3 (Spring 1996), pp. 193–260.
12. Aside from Carl Maria von Clausewitz, *On War*, ed. Anatol Rapoport (Middlesex, U.K.: Penguin, 1968), pp. 384–385, the best general discussions of military consequences of increasing nationalism are Peter Paret, "Nationalism and the Sense of Military Obligation," *Military Affairs*, Vol. 34, No. 1 (February 1970), pp. 2–6; Robert R. Palmer, "Frederick the Great, Guibert, Bülow: From Dynastic to National War," in Peter Paret, ed., *Makers of Modern Strategy: From Machiavelli to the Nuclear Age* (Princeton, N.J.: Princeton University Press, 1986), pp. 91–122; Barry R. Posen, "Nationalism, the Mass Army, and Military Power," *International Security*, Vol. 18, No. 2 (Fall 1993), pp. 80–124; and Stephen Van Evera, "Hypotheses on Nationalism and War," *International Security*, Vol. 18, No. 4 (Spring 1994), p. 30.
13. John Mueller makes a similar argument about the spurious relationship between democracy and peace in "Is War Still Becoming Obsolete?" paper prepared for the annual meeting of the American Political Science Association, Washington, D.C., August 1991, pp. 50–52. See also John Mueller, *Quiet Cataclysm: Reflections on the Recent Transformation of World Politics* (Reading, Mass.: Addison-Wesley, 1995).

view that, on balance, democracies share no particular advantages or disadvantages in selecting and waging wars. In other words, regime type hardly matters for explaining who wins and loses wars.

The remainder of the article is laid out as follows. The triumphalists' case is presented in the next section. In the following section, I critique the data and approach that undergird the triumphalists' claim that in war democracies are more likely to be victorious. The logic and evidence that underpin the triumphalists' case—selection effects and military effectiveness—are analyzed in the next two sections. Throughout the article I use, among other cases, Israel since 1948 to illustrate the problems with these arguments. Israel is a big winner in the triumphalists' data sets and so should be an easy test for their claim. The standard view is that Israel, a small, embattled democracy, has won its wars despite overwhelming odds for many of the reasons that triumphalists suggest.[14] If their theories do not in fact explain these victories, there are even more grounds for discounting them.[15] The article concludes with a brief discussion of the implications of my findings for scholarly debates on the relationship between democracy and war. It also offers some policy recommendations on how to think about the sources of military effectiveness.

The Triumphalists' Case

The foundation of the triumphalists' claim that democracies are more likely to win wars is based on two studies that employ different sets of cases selected from the same databases. In a 1992 study, David Lake looked at every war since 1815 listed in the Correlates of War (COW) data set and selected those involving states with a democracy score of 6 or higher based on the widely used POLITY democracy index.[16] This criterion makes sense because states with such scores exhibit the characteristics that we expect of democracies.[17] Using

---

14. Reiter and Stam, *Democracies at War*, p. 65.
15. This is essentially what Stephen Van Evera calls a "hoop test": that is, if a theory is correct it should easily pass this test; if it does not, there are grounds to doubt the theory. See Van Evera, *Guide to Methods for Students of Political Science* (Ithaca, N.Y.: Cornell University Press, 1997), p. 31.
16. Lake, "Powerful Pacifists." The COW data set refers to J. David Singer and Melvin Small, *Correlates of War Project: International and Civil War Data, 1816–1992*, No. 9905 (Ann Arbor, Mich.: Inter-University Consortium for Political and Social Research [ICPSR], 1994). This data set contains information about war participation, outcomes, various indices of military power, and war initiators. For the most recent version of the POLITY democracy index, see Keith Jaggers and Ted Robert Gurr, *Polity III: Regime-Type and Political Authority, 1800–1994*, No. 6695 (Ann Arbor, Mich.: ICPSR, 1996). From that data, most analysts calculate a composite democracy score (-10 to 10) by subtracting their AUTOC from DEMOC scores.
17. On a 21-point scale from -10 to 10, 6 is the generally accepted cutoff point for democracy.

Lake's method, I have determined that in the most current versions of the COW and POLITY data sets, there have been 31 wars involving democracies, 3 of which are excluded because they were draws (Korean War, 1969 War of Attrition, and 1982 Lebanon War). Democracies won 23 of the remaining 28 wars, or 82 percent (see Table 1).[18]

In a more recent study, Dan Reiter and Allan Stam examined most of the wars since 1815 in the COW data set to determine how often, controlling for other factors, the more democratic state prevailed over the less democratic state. Like Lake, Reiter and Stam used the POLITY democracy index to measure the level of democracy in the warring states. Utilizing that criterion and the most current versions of the COW and POLITY III data sets, I counted 75 wars, 24 of which were excluded because (1) data are missing on the level of democracy for all participants, (2) the wars involved states with the same democracy score, (3) the war ended in a draw, or (4) the conflict was ongoing. The more democratic state won 36 of the remaining 51 wars, or 71 percent (see Table 2).[19]

In sum, the historical record appears to support the triumphalists' claim that whether one looks at wars involving states with democracy scores greater than 6 or expands the universe to consider all wars in which more democratic states battled less democratic ones, there is a strong correlation between democracy and victory.

Do Democracies Really Win Wars More Often?

To determine whether regime type really explains a state's military performance, it is to necessary to look more closely at both the data and the approach that lead triumphalists' to conclude that democracies are more likely to win their wars.

Karen Rasler and William R. Thompson, "Predator Initiators and Changing Landscapes for Warfare," *Journal of Conflict Resolution,* Vol. 43, No. 4 (August 1999), p. 7.
18. Lake, "Powerful Pacifists," pp. 24–37. Because there was often more than one state on each side in these wars, Lake actually has an *N* of 121. The Gulf War, which occurred subsequent to the publication of Lake's article, should also count according to his criteria.
19. The 1994 version of the COW data set, including these wars, has an *N* of 269. Reiter and Stam, *Democracies at War,* pp. 52–57, employ many (but not all) of these wars, leaving them with an *N* of 197. Specifically, they do not include the following wars: First Schleswig-Holstein (1848–49), Spanish-Chilean (1865–66), Sino-French (1884–85), Franco-Thai (1893), Central American (1906), Lithuanian-Polish (1919–20), Franco-Turkey (1919–22), Sino-Japanese (1937–41), Chankufeng (1938), Franco-Thai (1940–41), Korean (1950–53), Second Kashmir (1965), Football War (1969), Sino-Vietnamese (1979), Iran-Iraq (1980–88), Gulf War (1990–91), and Azeri-Armenian War (1992–98). They also disaggregate World War I, World War II, the Vietnam War, and the Yom Kippur War into distinct phases.

Table 1. Outcomes of COW Wars (democracy score > 6).

Pessimists	Triumphalists	Not Counted
	Mexican-American (1848)	
	Roman Republic (1849)	
	Crimean (1853–56)	
	Anglo-Persian (1856–57)	
	Sino-French (1884–85)	
Greco-Turkish (1897)		
	Spanish-American (1898)	
	Boxer Rebellion (1900)	
	Spanish-Moroccan (1909–10)	
	First Balkan (1912–13)	
	Second Balkan (1913)	
	World War I (1914–18)	
	Hungarian-Allies (1919)	
	Russo-Polish (1919–20)	
Russo-Finnish (1939–40)		
	World War II (1939–45)	
	Palestine (1948)	
		Korea (1950–53)
	Sinai (1956)	
Sino-Indian (1962)		
Second Kashmiri (1965)		
Vietnam (1965–75)		
	Six-Day (1967)	
	Football (1969)	
		War of Attrition (1969–70)
	Bangladesh (1971)	
	Yom Kippur (1973)	
	Turko-Cypriot (1974)	
	Falklands (1982)	
		Lebanon (1982)
	Gulf War (1990–91)	
Total 5	23	3

DATA

There are at least six problems with the data that the triumphalists use to support their claim that democracies excel at winning wars. First, conflicts are misaggregated in a number of cases. Misaggregation could—and sometimes does—bias the results in favor of democracy.[20] Second, there are cases of democracies winning wars as members of mixed alliances where the nondemocracy accounted for the majority of the winning alliance's military

---

20. Other scholars have recognized this problem too. See, for example, D. Scott Bennett and Allan C. Stam III, "The Declining Advantages of Democracy: A Combined Model of War Outcomes and Duration," *Journal of Conflict Resolution*, Vol. 42, No. 3 (June 1996), p. 246; and Reiter and Stam, *Democracies at War*, p. 39.

Table 2. Outcomes of COW Wars (winner democracy > loser democracy).

Pessimists	Triumphalists	Not Counted
	Franco-Spanish (1823)	
		Russo-Turkish (1828–29)
	Mexican-American (1848)	
		Austro-Sardinian (1848–49)
First Schleswig-Holstein (1848–49)		
		Roman Republic (1849)
La Plata (1851–52)		
	Crimean (1853–56)	
	Anglo-Persian (1856–57)	
		Italian Unification (1859)
		Spanish-Moroccan (1859–60)
		Italo-Roman (1860)
	Italian-Sicilian (1860–61)	
	Franco-Mexican (1862–67)	
Ecuador-Columbia (1863)		
Second Schleswig-Holstein (1864)		
	Lopez (1864–70)	
	Spanish-Chilean (1865–66)	
		Seven Weeks (1866)
		Franco-Prussian (1870–71
		Russo-Turkish (1877–78)
	Pacific (1879–83)	
	Sino-French (1884–85)	
Central America (1885)		
	Franco-Thai (1893)	
	Sino-Japanese (1894–95)	
Greco-Turkish (1897)		
	Spanish-American (1898)	
	Boxer Rebellion (1900)	
	Russo-Japanese (1904–05)	
Central America (1906)		
		Central America (1907)
	Spanish-Moroccan (1909–10)	
		Italian-Turkey (1911–12)
	First Balkan (1912–13)	
	Second Balkan (1913)	
	World War I (1914–18)	
	Hungarian-Allies (1919)	
	Russo-Polish (1919–20)	
	Lithuanian-Polish (1919–20)	
		Greco-Turkey (1919–22)
		Franco-Turkey (1919–22)
Sino-Soviet War (1929)		
	Manchuria (1931–33)	
Chaco (1932–35)		
Italo-Ethiopian (1935–36)		
	Sino-Japanese (1937–41)	
	Chankufeng (1938)	

Table 2. (continued)		
Pessimists	Triumphalists	Not Counted
Nomohan (1939)		
Russo-Finnish (1939–40)		
	World War II (1939–45)	
	Franco-Thai (1940–41)	
	Palestine (1948)	
		Korea (1950–53)
	Sinai (1956)	
		Russo-Hungarian (1956)
Sino-Indian (1962)		
Second Kashmir (1965)		
Vietnam (1965–75)		
	Six-Day (1967)	
	Football (1969)	
		War of Attrition (1969–70)
	Bangladesh (1971)	
	Yom Kippur (1973)	
	Turko-Cypriot (1974)	
		Vietnamese-Cambodian (1975–79)
		Ethiopian-Somali (1977–78)
		Uganda-Tanzania (1978–79)
		Sino-Vietnamese (1979)
		Iran-Iraq (1980–88)
	Falklands (1982)	
		Lebanon (1982)
		Sino-Vietnamese (1985–87)
	Gulf War (1990–91)	
		Azeri-Armenian (1992–98)
Total 15	36	24

strength.[21] A "mixed alliance" is one in which the democratic participant accounts for less than 50 percent of the power potential in two out of three power categories, such as iron and steel production, number of troops, and total population. Third, in some cases a democracy was much more powerful than its adversary and used that advantage to overwhelm its rival. A "gross mismatch" is a conflict in which one side has a better than 2:1 advantage in two out of three power indices. Such gross mismatches should be considered only if the triumphalists' can prove that regime type caused the imbalance of

---

21. Bennett and Stam, "The Declining Advantages of Democracy," p. 248, n. 20, also identified this problem. It is not clear, however, given the large number of missing data points and the fact that capabilities may not measure real contribution to the war effort, that their solution of gauging each participant's role in the alliance based on their individual capabilities solves the problem of who contributed what in a mixed alliance.

power.[22] Fourth, in several cases the triumphalists' coding is questionable and, when corrected, weakens their case. Fifth, there are cases in which the belligerents' interests in the outcome of the conflict are so asymmetrical that it is impossible to ascribe the outcome to regime type and not to the balance of interests. Sixth, many of the cases involve states that cannot really be considered democratic and therefore are not strong tests of the triumphalists' theories.

A number of the cases in the COW data set are not fair tests of whether regime type affects the likelihood of a state winning its wars. A fair test of a theory involves identifying crucial cases that clearly rule out alternative explanations.[23] For example, in Lake's data set, World War II is treated as a single war involving the same belligerents from 1939 to 1945 in which the democracies prevailed. This characterization is misleading, however, because the war comprised at least three distinct conflicts involving different actors and different scenarios: the Battle of France (May–June 1940), the European War (June 1941–May 1945), and the Pacific War (December 1941–August 1945). Treating World War II as single war overstates the effectiveness of the democracies and misses the real reasons why they were on the winning side.

In the spring of 1940, Nazi Germany went to war against Britain, Belgium, France, and the Netherlands. Early in the war, the Germans, who were about as powerful (0.8:1 in iron and steel production, 0.9:1 in military manpower, and 0.8:1 in population) as their democratic adversaries, nonetheless defeated them decisively, thus contradicting the triumphalists' expectations.[24]

In the ensuing war in Europe, a mixed alliance including Britain, the Soviet Union, and the United States defeated an alliance of fascist states led by Nazi Germany and Italy. Although the democracies—Britain and the United States—were on the winning side, this case does not strongly support the triumphalists' claim for two reasons.[25] First, the Soviet Union—not Britain and

---

22. Because Reiter and Stam, *Democracies at War*, p. 58, reject other triumphalist arguments that democracies win wars because of a preponderance of power—either their own or their allies—they ought to be particularly eager to find cases of democracies being relatively evenly balanced with nondemocracies.

23. On the importance of "crucial cases" for devising "fair tests" for comparative theory testing, see Arthur Stinchcombe, *Constructing Social Theories* (New York: Harcourt, Brace, and World, 1968), pp. 24–28.

24. Classic accounts include William L. Shirer, *The Collapse of the Third Republic: An Inquiry into the Fall of France in 1940* (New York: Da Capo, 1994); Eugen Weber, *The Hollow Years: France in the 1930s* (New York: W.W. Norton, 1994); and Alistair Horne, *To Lose a Battle: France 1940* (New York: Penguin, 1988).

25. In *Democracies at War*, Reiter and Stam, who do disaggregate the war in Europe, separately credit the United States and Britain with defeating Nazi Germany. Their data set also credits demo-

the United States—was principally responsible for defeating Nazi Germany. Most historians agree that the war in Europe was settled mainly on the eastern front.[26] Indeed, roughly 85 percent of the *Wehrmacht* was deployed along that front for most of the war; not surprisingly, about 75 percent of German casualties were suffered there.[27] Second, this case is a gross mismatch: The Allies had a 3.8:1 advantage in iron and steel, a 1.7:1 advantage in military manpower, and a 2.47:1 advantage in population over the Axis.

In the Pacific War the United States, with support from Australia, Britain, China, and New Zealand, inflicted a decisive defeat on Japan in 1945. Although the democracies were on the winning side in this conflict, Japan lost because it was far less powerful than its rivals. Although the military manpower balance was roughly even, the Allies had a 13:1 advantage in iron and steel production and a 10:1 advantage in population.

Several Arab-Israeli cases also illustrate the problems with miscodings in the triumphalists' data set. Reiter and Stam, for example, code the 1969–70 Israeli-Egyptian War of Attrition and the 1982 Lebanon war as victories for democratic Israel. Most analysts, however, including the original compilers of the COW data set, regard them as draws. As Ezer Weizman concluded, "It is no more than foolishness to claim that we won the War of Attrition. On the contrary, for all their casualties it was the Egyptians who got the best of it."[28] Even a few miscodings can bias the triumphalists' findings about the propensity of democracies to win their wars.

Other Arab-Israeli cases illustrate how asymmetric interests might be a better determinant of military success. Israel did well in conventional wars in which its survival was at stake (e.g., 1948 and 1967). In contrast, Israel fought poorly in unconventional wars where its survival was not on the line (e.g., Lebanon in 1982 and the first Palestinian *intifada* [uprising] in 1987).[29] This is not surprising because, as Martin Gilbert notes, the 1982 Lebanon war "was

cratic Israel with not one but two victories in the 1973 Yom Kippur War by dividing it into two wars: Israel versus Egypt and Israel versus Syria. This coding tilts the scale in favor of democracies, although it is balanced by their counting as separate victories Germany's defeats of Belgium, the Netherlands, Denmark, and France.
26. Alan Clark, *Barbarrossa: The Russian-German Conflict of 1941–45* (New York: Quill, 1965); Richard Overy, *Why the Allies Won* (New York: W.W. Norton, 1995), pp. 63–100; and Richard Overy, *Russia's War: A History of the Soviet War Effort, 1941–45* (New York: Penguin, 1997).
27. W. Victor Madej, "Effectiveness and Cohesion of the German Ground Forces in World War II," *Journal of Political and Military Sociology*, Vol. 6, No. 2 (Fall 1978), pp. 233–248.
28. Quoted in Martin van Creveld, *The Sword and the Olive: A Critical History of the Israeli Defense Force* (New York: PublicAffairs, 1998), p. 215. On Lebanon, see Ze'ev Schiff and Ehud Ya'ari, *Israel's Lebanon War* (New York: Simon and Schuster, 1984).
29. Quoted in van Creveld, *The Sword and the Olive*, p. 296.

the first war in Israel's history for which there was no national consensus. Many Israelis regarded it as a war of aggression."[30] The abysmal performance of the Israel Defense Forces (IDF), and indeed the Israeli government as a whole, was even more marked in Israel's efforts to suppress the first *intifada*. As Martin van Creveld wrote: "Never known for its discipline, the IDF's traditional strengths—originating in the *Yishuv*'s prestate military organizations—had been initiative and aggressiveness in defeating Arab armies in short, sharp wars. Now those very qualities started turning against it in a prolonged conflict that demanded patience, professionalism, and restraint."[31] The late Israeli Prime Minister Yitzak Rabin concurred: "It is far easier to resolve classic military problems. . . . It is far more difficult to contend with 1.3 million Palestinians living in the Territories, who do not want our rule, and who are employing systematic violence without weapons."[32]

Of the 75 wars since 1815 listed in the most recent version of the COW data set, 54 are clearly unfair tests. This leaves 21 cases of fair fights. Of these, the more democratic state won 12 times, and the less democratic state won 9 times (see Appendix).[33] This approach of looking at wars involving states that are relatively more democratic increases the number of relevant cases; however, it also results in the inclusion of many cases of wars between states where at least one of the belligerents does not score a 6 or above on the democracy scale—for example, the Pacific War (1879–83), the Sino-Japanese War (1894–95), the Russo-Japanese War (1904–05), the Manchurian War (1931–33), the Sino-Japanese War (1937–41), and Changfukeng (1938). There were 31 wars involving states that were clearly democratic; however, 22 of these involve misaggregations, mixed alliances, gross mismatches, or asymmetric interests. Thus, of the remaining 9 cases, 3 support the pessimists and 6 support the triumphalists.[34]

In both cases, democracies do better than their rivals. They seem to do better in wars involving one clearly democratic state (democracies win in 67 percent

---

30. Martin Gilbert, *Israel: A History* (New York: William Morrow, 1998), p. 504.
31. Van Creveld, *The Sword and the Olive*, p. 344.
32. Quoted in Gilbert, *Israel*, p. 526.
33. This adds 12 cases to those listed in note 34. Five support the defeatists: Ecuadorian-Colombian (1863), Second Schleswig-Holstein (1864), Central American (1906), Sino-Soviet (1929), and Chaco (1932–35); and seven support the triumphalists: Pacific (1879–83), Central American (1885), Sino-Japanese (1894–95), Russo-Japanese (1904–05), Manchurian (1931–33), Sino-Japanese (1937–41), and Chankufeng (1938).
34. These are the first part of World War I (1914–17), the Battle of France (May–June 1940), and the Sino-Indian War (1962), which seemingly support the defeatists; and the Russo-Polish War (1919–20), the Israeli War for Independence (1948), the Six-Day War (1967), the Football War (1969), the Yom Kippur War (1973), and the Falklands War (1982), which appear to support the triumphalists.

of the 9 cases) as opposed to all wars (democracies win in 57 percent of the 21 cases). Yet, based on these findings, it is difficult to have confidence in the proposition that democracy is the reason states are more likely to win their wars.

Table 3 illustrates the potential impact of misaggregations, mixed alliances, gross mismatches, asymmetric interests, and miscodings on the triumphalists' findings. Model 1 is a simple probit model using Reiter and Stam's data estimating the effects of the level of democracy (without any control variables) on the likelihood of a state winning a war. Not surprisingly, the model supports their argument that a democracy is more likely than a nondemocracy to achieve victory. Model 2 shows what happens when the misaggregations in World War II (crediting Britain and the United States with defeating Nazi Germany) and the Yom Kippur War (crediting Israel with two victories) are corrected, the miscodings are eliminated (Israel should be credited with draws rather than victories in the 1969–70 War of Attrition and the 1982 Lebanon war), and the focus is exclusively on cases that are fair tests of the triumphalists' theories. With these changes, the democracy variable is no longer significant.

APPROACH

Some might argue that a better approach would be to keep the unfair tests and control statistically for other factors that may account for why democracies win wars more often than nondemocracies. The major advantage of this approach, proponents argue, is that it offers a large number of cases that make advanced statistical analysis possible. Yet even if one accepts the validity of all the historical cases and tries to control for competing explanations, there are still reasons to question the triumphalists' claim that democracy is the key to military victory.

First, Lake as well Reiter and Stam employ approaches that utilize "pooled data" consisting of a number of countries, some of which are involved in multiple wars, to generate each data point. A central assumption of statistical analysis is that each data point is independent (the outcome of one war is not affected by the outcome of previous ones), homogeneous (the wars are roughly comparable), and exchangeable (if a democracy can beat one nondemocracy, it should be able to defeat all similar nondemocracies). Reiter and Stam, for example, have an N of 197, but this actually consists of only 66 countries, a small number of which are looked at repeatedly. Among the most democratic states in their data set (scores of 9 or 10 on the democracy index), three—Britain, Israel, and the United States—comprise approximately 56 percent of the cases.

Table 3. Probit Results (win/lose).

Variables	Model 1	Model 2 (fair fights > 6)
Constant	0.1410283	−0.3440138
	(0.097201)	(0.227655)
Democracy		0.0364302
	0.0359429**	
	(0.0137452)	(0.0313352)
Pseudo R^2	0.0248	0.0332
LL	−133.04446	−21.342535
N	197	34

NOTE: Data are available at http://www.yale.edu/plsc151a/. I used the variables politics and wl.
* ≤ 0.05 (all tests two-tailed)
** ≤ 0.01
*** ≤ 0.001
[Robust standard errors]

Of the most democratic states that won wars, these three countries account for 75 percent of the results. Given that three states play such a large role in the triumphalists' findings, it makes sense to ask whether there are particular circumstances in each case, or variables not contained in the triumphalists' models, that explain their propensity for winning particular wars. This is the potential problem of "fixed unobserved effects" that a recent article suggests affects much large-N research in international relations.[35] Some scholars argue that this problem can be solved simply by reporting robust standard errors.[36] The optimal solution to the fixed effects problem, however, is to collect more and better data that would make it possible to control directly for the unobserved variables that might be unique to each case.[37] This is by no means an easy task. Unobserved variable bias would not be much of a problem if it were easy to identify and measure those variables. Therefore, another way to address the problem is through in-depth process tracing in obviously related cases to ascertain whether factors unique to those cases can explain the outcome.

35. Donald P. Green, Soo Yeon Kim, and David H. Yoon, "Dirty Pool," *International Organization*, Vol. 55, No. 2 (Spring 2001), pp. 441–468.
36. John R. Oneal and Bruce M. Russett, "Clear and Clean: The Fixed Effects of the Liberal Peace," *International Organization*, Vol. 55, No. 2 (Spring 2001), p. 471.
37. Gary King, "Proper Nouns and Methodological Propriety: Pooling Dyads in International Relations Data," *International Organization*, Vol. 55, No. 2 (Spring 2001), pp. 497–507.

Second, although there is a correlation between democracy and victory, correlation does not mean causation.[38] To establish causation, the most likely alternative explanations need to be ruled out. There are, however, alternative explanations that the triumphalists cannot rule out by controlling for them statistically. For example, a large body of scholarship argues that democracy takes root and flourishes as the result of a distinct set of preconditions, including high levels of aggregate wealth, equitable wealth distribution, free markets, high levels of social development, a strong feudal aristocracy, a strong bourgeoisie/middle class, high levels of literacy and education, a liberal political culture (e.g., toleration, compromise, and respect for the law and individual rights), Protestantism, strong intermediary organizations, capable political institutions, low levels of domestic political violence, moderate politics, occupation by a democratic state, geographical security (water, mountains, etc.), strong allies, and weak adversaries.[39]

Some of these preconditions for democracy confer decided military advantages as well.[40] For example, wealthy, highly developed, well-educated, strongly institutionalized states that are geographically secure and have strong allies and weak adversaries are also more likely to win wars. Rather than democracy explaining this outcome, it is possible that certain preconditions of democracy produce both a democratic political system and an impressive record of military success. If this argument is correct, then the correlation between democracy and military victory is spurious: The preconditions, not democracy per se, account for both.

If the preconditions argument is correct, there should be little variation in the military effectiveness of states over time, especially pre- and post-democracy, but significant variation across cases with different preconditions. Some democracies, such as the United States and Israel, were founded on democratic principles, so they are not useful for assessing the preconditions argument. Two other democracies—Britain and France—have long predemocratic

38. Important cautions about overreliance on correlational findings include Jack S. Levy, "Domestic Politics and War," *Journal of Interdisciplinary History,* Vol. 18, No. 4 (Spring 1988), p. 669; and David Dessler, "Beyond Correlations: Toward a Causal Theory of War," *International Studies Quarterly,* Vol. 35, No. 3 (September 1991), pp. 337–355.

39. Samuel P. Huntington, *The Third Wave: Democratization in the Late Twentieth Century* (Norman: University of Oklahoma Press, 1991), pp. 37–38.

40. Brian M. Downing, *The Military Revolution and Political Change: Origins of Democracy and Autocracy in Early Modern Europe* (Princeton, N.J.: Princeton University Press, 1992), pp. 78–79. For the classic discussion of how a benign security environment is more conducive to democracy, see Otto Hintze's treatment of Great Britain in "Military Organization and the Organization of the State," in Felix Gilbert, ed., *The Historical Essays of Otto Hintze* (New York: Oxford University Press, 1975), pp. 178–215.

histories. They have also had strikingly different records of military success since 1648. Britain has fought about 43 wars since the end of the Thirty Years' War, winning 35 (81 percent) of them. Britain's record in the COW data set is slightly better: It has fought 9 wars and won 8 (89 percent).[41] The preconditions argument would attribute these results to the fact that Britain is a wealthy, geographically secure state with many allies, allowing it to win wars with little domestic mobilization. Conversely, France has few of the preconditions necessary for democracy and military success, and thus has been both an inconsistent democracy and a less successful belligerent. France has fought 31 wars since 1648 and won 18 of them (58 percent). In the COW data set, it fought 16 wars, winning only 9 (56 percent).

Another possible explanation for how a state performs in war is whether its government is consolidated. The mean democracy score for Lake's winners is 0.59, which is well below the democracy range.[42] The average democracy score for winners in Reiter and Stam's data set is even lower: −1.41. The distribution of winners in all wars since 1815 by democracy score shows that this remarkably low average is due to the large numbers of highly authoritarian states that won their wars too (see Figure 1). This leads Reiter and Stam to propose that the effect of the level of democracy is curvilinear (i.e., the most democratic and most autocratic states win, but those in the middle tend to lose).[43] This pattern, however, is also compatible with an argument that ascribes victory not to the level of democracy but to whether a regime has been politically consolidated, as one would expect with highly democratic and authoritarian states. The mixed regimes in between high democracy and high autocracy, which are referred to as "anocracies," may perform poorly in war because they are unconsolidated, transitional regimes.[44] The primary reason for characterizing anocracies as transitional regimes is that they do not stay at this level as long as regimes do when they are either in the democracy or autocracy range.[45]

---

41. British and French military track records since 1648 were calculated from R. Ernest Dupuy and Trevor N. Dupuy, *The Harper Encyclopedia of Military History: From 3500 B.C. to the Present*, 4th ed. (New York: HarperCollins, 1993).
42. Lake, "Powerful Pacifists," p. 31, n. 31.
43. Reiter and Stam, *Democracies at War*, pp. 25, 129.
44. This logic parallels Edward D. Mansfield and Jack Snyder, "Democratization and the Danger of War," *International Security*, Vol. 20, No. 1 (Summer 1995), p. 35, who suggest that an alternative explanation for their finding about the increased likelihood of international conflict in democratizing states is that states undergoing any sort of political change are more likely to engage in war.
45. For evidence that anocracies are short-lived, see Håvard Hegre, Tanja Elligson, Scott Gates, and Nils Peter Gleditsch, "Toward a Democratic Civil Peace? Democracy, Political Change, and Civil War, 1816–1992," *American Political Science Review*, Vol. 95, No. 1 (March 2001), p. 34.

Figure 1. The Distribution of Winners By Democracy Score.

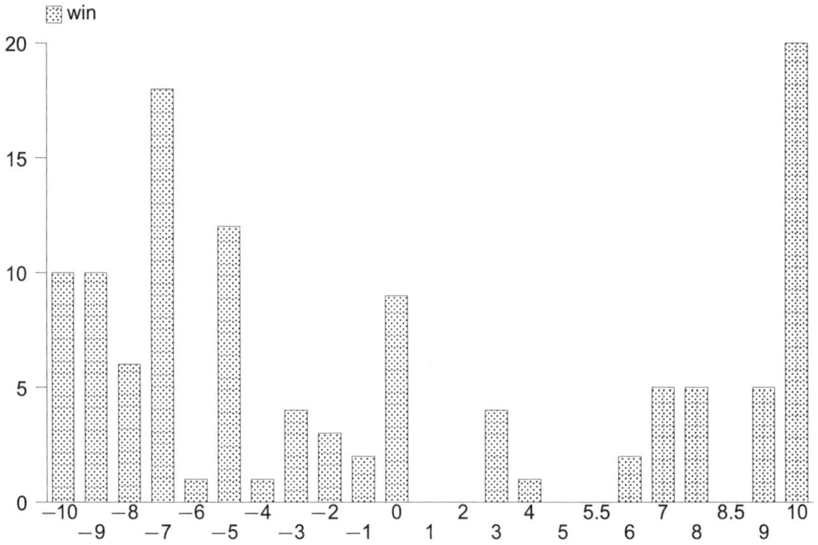

SOURCES: J. David Singer and Melvin Small, *Correlates of War Project: International and Civil War Data, 1816–1992,* No. 9905 (Ann Arbor, Mich.: Inter-University Consortium for Political and Social Research [ICPSR], 1994); and Keith Jaggers and Ted Robert Gurr, *Polity III: Regime-Type and Political Authority, 1800–1994,* No. 6695 (Ann Arbor, Mich.: ICPSR, 1996).

In sum, the historical data do not strongly support the triumphalists' claim that democracies are more likely to win wars than nondemocracies. In particular, many of the cases they employ are not fair tests of their claim and therefore cannot be used to support (or refute) it. Nor does the triumphalists' approach effectively rule out two alternative factors that may explain why states win wars: (1) the existence of common preconditions for democracy and victory and (2) the degree of regime consolidation. In the following two sections, I assess the causal mechanisms that the triumphalists use to explain why, in their view, democracies are more likely than other types of regimes to win their wars.

Selection Effects

According to the selection effects argument, democracies win wars because they start them only if they have a high probability of being victorious. The reason for this caution is that democratic leaders must run for office, and voters will punish those who initiate unsuccessful wars. Authoritarian leaders, on the other hand, are rarely held accountable by their populations, and thus can more easily weather a losing war.[46]

Lake as well as Reiter and Stam use statistical methods that aim to show that whether or not democracies start wars matters tremendously for the outcome. Their data show that even controlling for power and other factors, democracies are more likely to win the wars they initiate; triumphalists interpret this as support for the selection effects argument (see Table 4).[47]

Despite this apparent support for the triumphalists' case, there are three reasons for skepticism. First, victory in war is a complex and overdetermined phenomenon in which many factors play a role. The key question is: Which factors play the biggest roles? As Table 5 makes clear, a calculation of the "marginal effects" for each variable in Table 4 shows that democracy has one of the smallest effects of any variable. Marginal effects are derivatives of the probability that the dependent variable will equal 1 (in this case that the state wins) with

---

46. Levy, "Domestic Politics and War," pp. 658–659; Bruce Bueno de Mesquita and Randolph M. Siverson, "War and the Survival of Political Leaders: A Comparative Study of Regime Types and Political Accountability," *American Political Science Review*, Vol. 89, No. 4 (December 1995), pp. 841–855; Kenneth A. Schultz, "Domestic Opposition and Signaling in International Crises," *American Political Science Review*, Vol. 92, No. 4 (December 1998), p. 830; Bennett and Stam, "The Declining Advantages of Democracy," pp. 346, 365; Dan Reiter and Allan C. Stam III, "Democracy, War Initiation, and Victory," *American Political Science Review*, Vol. 92, No. 2 (June 1998), p. 378; and Bruce Bueno de Mesquita, Randolph M. Siverson, and Alistair Smith, "Policy Failure and Political Survival: The Contribution of Political Institutions," *Journal of Conflict Resolution*, Vol. 43, No. 2 (April 1999), pp. 147–161.
For the claim that initiators are more likely to win wars, see Kevin Wang and James Lee Ray, "Beginners and Winners: The Fate of Initiators of Interstate Wars Involving Great Powers since 1495," *International Studies Quarterly*, Vol. 38, No. 1 (March 1994), pp. 139–154.
47. Lake, "Powerful Pacificists," uses a logit model to measure the impact of two independent variables—democracy and initiation—on the dependent variable, which is the likelihood of winning or losing a war. Based on that, for example, going from a democracy score of 5 to 10 (e.g., from Syria in 1948 to the United States in 1941) more than doubles the likelihood of victory. Logit makes it possible to calculate the odds likelihood ratio by applying anti-logs to both sides of the basic equation $logit(\pi) = \alpha + \beta X$,
which yields the odds likelihood ratio from the formula

$$\frac{\pi}{1 - \pi} = e^{\alpha + \beta x} = e^{\alpha}(e^{\beta})^{x}.$$

This reveals the effect of a one-unit increase in the democracy score on the likelihood of victory.
Reiter and Stam, *Democracy at War*, p. 45 (Table 2.2), particularly model 4 (which best captures the argument that democracies are better able to pick winning wars), employ more sophisticated probit models (including more control variables and a broader spectrum of cases). Unlike Lake, who measures the interaction effect between democracy and war initiation by including both variables in the same equation, Reiter and Stam assess selection effects by including a number of interaction terms between democracy and war initiation in their equations along with various control variables.

Table 4. Probit Results (win/lose).

Variables	R&S Model 4
democracy*initiation	0.0675943*
	(0.0298018)
democracy*target	0.0639582*
	(0.0275639)
initiation	0.9142049**
	(0.3422103)
capabilities	3.726842***
	(0.5249923)
allies capabilities	4.721843***
	(0.6837011)
quality ratio	0.0522075
	(0.0329194)
terrain	−10.93261***
	(2.937978)
strategy*terrain	3.560021***
	(0.9689448)
strategy1	7.235081*
	(2.886022)
strategy2	3.478767
	(1.993146)
strategy3	3.35718*
	(1.428867)
strategy4	3.069146*
	(1.252304)
Constant	−5.517191**
	(1.698374)
Pseudo R^2	0.5244
LL	−64.886064
N	197

NOTE: I used the following variables: polini, poltarg, init, wl, concap, qualrat, capasst, terrain, strat1, strat2, strat3, strat4, and staterr. These variables are discussed at length in Dan Reiter and Allan C. Stam III, *Democracies at War* (Princeton, N.J.: Princeton University Press, 2002), pp. 40–44. I also estimated this model using only the fair-fight cases and found no selection effects for democracies.
* ≤ 0.05 (all tests two-tailed)
** ≤ 0.01
*** ≤ 0.001
[Robust standard errors]

Table 5. Marginal Effects of Variables in Probit.

Variable	dy/dx
democracy*initiation	0.0267582
democracy*target	0.0253188
initiation*	0.3469761
capabilities	1.475326
allies capabilities	1.869212
quality ratio	0.0206671
terrain	−4.327838
strategy*terrain	1.409287
strategy1*	0.6914264
strategy2*	0.5623581
strategy3*	0.851552
strategy4*	0.5051578

*dy/dx is for discrete change of dummy variable from 0 to 1.

respect to each independent variable by itself. The marginal effects calculation measures the sensitivity of that probability of winning to changes in the values of various independent variables. The higher the absolute value of the marginal effect of an independent variable (i.e., the larger the value of dy/dx), the more sensitive the probability of the dependent variable equaling 1 is to changes in each independent variable, and thus the greater the effect of that independent variable. In other words, the marginal effects calculation measures how much a state's chance of winning changes because of variations in the independent variables. The interaction between democracy and initiation has one of the smallest effects (0.0267582), whereas terrain (−4.327838) and power—both the state's (1.475326) and its allies' (1.869212)—and the interaction between strategy and terrain (1.409287) have the largest effects on who wins.[48]

Second, there is little reason to think that caution about starting a war should be unique to democratic leaders. In fact, even some triumphalists concede that leaders of every kind of regime incur significant costs from starting a losing war, and thus they are apt to be careful about blundering into one. As Bruce Bueno de Mesquita and Randolph Siverson note, "The leader—whether

48. I calculated these effects using STATA's "mfx compute" function, which holds the values of other variables at their mean in computing the marginal effect of each variable.

Reiter and Stam, *Democracy at War,* Figure 2.2, provide data on the marginal effects of increases in the democracy score but not on the relative effect of democracy compared with those of other variables. In an earlier work, Allan Stam, *Win, Lose, or Draw: Domestic Politics and the Crucible of War* (Ann Arbor: University of Michigan Press, 1996), Figures 28, 45, does. Not surprisingly, my findings about democracy's relatively small marginal effect on the likelihood of victory are similar to his.

president, prime minister, or president-for-life—who adopts policies that reduce the security of the state does so at the risk of affording their political opponents the opportunity of weakening the leader's grasp on power."[49] As this statement makes clear, the general logic of their argument applies equally to democracies and autocracies.

One could even argue that democratic leaders should be less cautious about going to war than their nondemocratic counterparts. The worst fate that a democratic leader faces is removal from office and disgrace. On the other hand, authoritarian leaders who lose wars are frequently exiled, imprisoned, or put to death. Given that fact, it seems hard to maintain that an authoritarian leader would be less wary than a democratic leader about losing a war.[50] Although the probability of democratic leaders being ousted may be higher, the costs to autocratic leaders of losing power are so great that the net result should be that both are equally wary of losing a war. Finally, if democracies are actually more selective in choosing their wars, starting only easy ones, they should engage in fewer wars than authoritarian states, because there are not likely to be many sure victories. In fact, it is widely acknowledged by scholars that democracies are at least as, if not more, war prone than other types of regimes.[51] In short, the logic undergirding the triumphalists' selection effects argument is unconvincing.

Third, the Israeli cases provide little empirical support for the selection effects explanation. Of the three wars that Israel started, just one—the 1967 Six-Day War—indisputably supports the triumphalists' claim. The 1956 Sinai War cannot be credited as a victory for Israel because Israel was forced to return captured Egyptian territory by the United States. The disastrous 1982 Lebanon war clearly demonstrates that Israel has not consistently initiated successful wars.[52]

---

49. Bueno de Mesquita and Siverson, "War and the Survival of Political Leaders," p. 853. See also William R. Thompson, "Democracy and Peace," *International Organization*, Vol. 50, No. 1 (Winter 1996), p. 149.
50. H.E. Goemans, *War and Punishment: The Causes of War Termination and the First World War* (Princeton, N.J.: Princeton University Press, 2000), pp. 39–40. For a logically rigorous argument that the incentives facing democratic and authoritarian leaders are similar, see Gordon Tullock, *Autocracy* (Dordrecht: Kluwer, 1987), p. 19.
51. See Jack Levy, "The Causes of War: A Review of Theories and Evidence," in Phillip E. Tetlock, Jo L. Husbands, Robert Jervis, Paul C. Stern, and Charles Tilly, eds., *Behavior, Society, and Nuclear War*, Vol. 1 (New York: Oxford University Press, 1989), p. 270. I thank Hein Goemans for reminding me of this point.
52. Miriam Fendius Elman, "Israel's Invasion of Lebanon, 1982: Regime Change and War Decisions," in Elman, ed., *Paths to Peace: Is Democracy the Answer?* (Cambridge, Mass.: MIT Press, 1997), p. 329.

If the triumphalists' data and approach are accepted without reservation, democracy plays one of the smallest roles in accounting for why states that start wars tend to win them. However, logical problems with the selection effects argument, and the lack of empirical support for it in what should be easy cases for them, are grounds for questioning even this modest role for democracy. In sum, democracy matters relatively little, if at all, in explaining whether states wisely select and then win their wars.

Military Effectiveness

The triumphalists offer five causal mechanisms to support their claim that democracies are better at fighting wars than nondemocracies: Democracies (1) are wealthier, (2) make better allies, (3) engage in more effective strategic evaluation, (4) enjoy greater public support, and (5) have soldiers who fight more effectively than their counterparts in authoritarian states. It is impossible to do justice to each of these arguments in the space of one article. Nevertheless, a brief assessment of these causal mechanisms suggests that none is logically compelling or has much empirical support.

DEMOCRACY AND WEALTH
Lake maintains that as a rule democracies are wealthier than authoritarian states, and because wealth is the foundation of military power, democracies are more likely to win wars.[53] This claim is based on the belief that democracies are less prone to rent seeking—that is, the governments of democratic governments are less likely to meddle in their economies, thus fostering free markets that produce greater national wealth.

---

53. Lake, "Powerful Pacifists," p. 24, and before him Frederic C. Lane, "The Economic Meaning of War and Protection" in his *Venice and History: The Collected Papers of Frederic C. Lane* (Baltimore, Md.: Johns Hopkins University Press, 1996), p. 389, n. 10, applied this argument to military power. A related argument is that liberal institutions make it easier for governments to borrow money to wage war. See Kenneth A. Schultz and Barry Weingast, "Limited Governments, Powerful States," in Randolph M. Siverson, ed., *Strategic Politicians, Institutions, and Foreign Policy* (Ann Arbor: University of Michigan Press, 1998), pp. 15–50.

For general arguments about democracies being less prone to rent seeking, see Mancur Olson, "Dictatorship, Democracy, and Development," *American Political Science Review*, Vol. 87, No. 3 (September 1993), pp. 567–576; Barry Basinger, Robert B. Ekeland Jr., and Robert Tollison, "Mercantilism as a Rent-Seeking Society," in James M. Buchanan, Robert D. Tollison, and Gordon Tullock, eds., *Toward a Theory of the Rent-Seeking Society* (College Station: Texas A&M University Press, 1980), pp. 235–268; and Mark Brawley, "Regime Types, Markets, and War: The Impact of Pervasive Rents in Foreign Policy," *Comparative Political Studies*, Vol. 26, No. 2 (July 1993), pp. 178–197.

Triumphalists maintain that democracies are better wealth creators than other types of regimes, but they provide no supporting evidence for this claim.[54] Even the large body of scholarship on the relationship between levels of democracy and levels of economic development does not provide much of a foundation for their assertion. To be sure, there is some evidence that bolsters the triumphalists' contention that democracy makes economic growth more likely,[55] but there is much more evidence for the converse proposition that wealth is a key factor in creating democracy.[56] Thus, there is no consensus in the development literature on which way the causal arrow runs.[57] Therefore there is little basis for believing the triumphalists' claim that democracies produce greater wealth than nondemocracies.

Another reason to doubt the triumphalists' assertion that democracies are superior wealth creators is that the rent-seeking logic that underpins their claim is flawed. There is no reason to think that rent seeking should be less frequent in democracies. Indeed there are compelling reasons why it should be more common.

Rent seeking is the effort by interest groups in a society to gain excess profits through nonmarket mechanisms.[58] For example, tobacco producers receive special tax breaks and subsidies as a result of political lobbying, which injects economic inefficiencies into the marketplace that slow the rest of the economy. Economists offer compelling arguments for why it is more likely that interest groups will be successful rent seekers in a democracy.[59] "Countries that have

---

54. Lake, "Powerful Pacifists," p. 28.
55. "Democracy and Growth: Why Voting Is Good for You," *Economist*, August 27, 1994, pp. 15–17; and Yi Feng, "Democracy, Political Stability, and Economic Growth," *British Journal of Political Science*, Vol. 27, No. 3 (July 1997), pp. 391–418.
For a largely theoretical argument that democracy causes growth because of the greater credibility of democratic governmental institutions, see Douglass C. North and Barry Weingast, "Constitutions and Commitment: The Evolution of Institutions Governing Public Choice in Seventeenth-Century England," *Journal of Economic History*, Vol. 49, No. 4 (December 1989), pp. 803–832.
56. John F. Helliwell, "Empirical Linkages between Democracy and Economic Growth," *British Journal of Political Science*, Vol. 24, No. 2 (April 1994), pp. 225–248; Ross E. Burkhardt and Michael S. Lewis-Beck, "Comparative Democracy: The Economic Development Thesis," *American Political Science Review*, Vol. 88, No. 4 (December 1994), pp. 903–910; Deane E. Neubaur, "Some Conditions of Democracy," *American Political Science Review*, Vol. 61, No. 4 (December 1967), pp. 1002–1009; John B. Londregen and Keith T. Poole, "Does Income Promote Democracy?" *World Politics*, Vol. 49 No. 1 (October 1996), p. 2; and Larry Diamond, "Economic Development and Democracy Reconsidered," *American Behavioral Scientist*, Vol. 35, Nos. 4/5 (March/June 1992), p. 450.
57. Mark J. Gasiorowski, "Democracy and Macroeconomic Development in Underdeveloped Countries," *Comparative Political Studies*, Vol. 33, No. 3 (April 2000), pp. 319–350.
58. James M. Buchanan, "Rent Seeking and Profit Seeking," in Buchanan, Tollison, and Tullock, *Toward a Theory of the Rent-Seeking Society*, pp. 3–15; and Robert D. Tollison, "Rent Seeking: A Survey," *KYKLOS*, Vol. 35, No. 4 (November 1982), pp. 575–602.
59. Mancur Olson, "A Theory of Incentives Facing Political Organizations: Neocorporatism and

democratic freedom of organization without upheaval or invasion the longest," Mancur Olson argues, "will suffer the most from growth-repressing organizations and combinations."[60]

Lake identifies governments, not interest groups, as the main rent seekers. But even if democratic governments are less likely to engage in rent-seeking behavior, the fact remains that interest groups in democracies are more likely to be engaged in this kind of behavior. Lake provides no evidence, however, that the lack of government interference in a democracy's economy offsets the negative effects of rent seeking by interest groups.

Moreover, although wealth is necessary for generating military might, it also is essential that a state be able to mobilize its wealth for military purposes.[61] This two-step process raises a question that Lake does not address but that might be thought essential to his position: Are democracies better able to extract resources from their societies than nondemocracies? The best available study on the subject maintains that regime type is largely irrelevant: "Politically capable governments can mobilize vast resources from the society under stress of war, but totalitarian, democratic and authoritarian regimes do not determine the level of performance."[62] In short, democracies are no better than nondemocracies at transforming economic might into military power.

Contrary to Lake's rent-seeking argument, Israel between 1948 and 1982 did not have a bigger economy, except in per capita terms, than its Arab adversaries.[63] Israeli democracy did not inhibit state rent seeking. In fact, Israel was a classic example of a state with one of the major preconditions for rent seeking:

---

the Hegemonic State," *International Political Science Review*, Vol. 7, No. 2 (April 1986), pp. 165–189; and Tollison, "Rent Seeking," p. 590.

60. Mancur Olson, *The Rise and Decline of Nations: Economic Growth, Stagflation, and Social Rigidities* (New Haven, Conn.: Yale University Press, 1982), p. 77.

61. This is a classic argument. See Alexis de Tocqueville, *Democracy in America*, Vol. 1 (New York: Vintage, 1945), p. 243. See also Reiter and Stam, "Democracy and War Initiation," p. 378; and Reiter and Stam, *Democracies at War*, pp. 117–129.

62. See Jacek Kugler and William Domke, "Comparing the Strength of Nations," *Comparative Political Studies*, Vol. 19, No. 1 (April 1986), pp. 39, 50, 66. See also Adam Przeworski and Fernando Limongi, "Political Regimes and Economic Growth," *Journal of Economic Perspectives*, Vol. 7, No. 3 (Summer 1993), pp. 51–69; Erich Weede, "The Impact of Democracy on Economic Growth: Some Evidence from Cross-National Analysis," *KYKLOS*, Vol. 36, No. 1 (February 1983), p. 35; and José Antonio Cheibub, "Political Regimes and the Extractive Capacity of Government: Taxation in Democracies and Dictatorships," *World Politics*, Vol. 50, No. 1 (April 1998), pp. 372–373.

63. According to data in the International Institute for Strategic Studies, *Military Balance* (London: IISS, various years), Israel had a gross national product of only $3.6 billion, compared with a combined Arab GNP of $16.1 billion in 1967; in 1969 and 1970, the ratio between Israel and Egypt was $4.5 billion to $6.3 billion and $5.4 billion to $6.45 billion, respectively; and in 1973 it was $8.7 billion to the combined Arab GNP of $23.53 billion. In 1982, however, Israel enjoyed an overall advantage over Syria of $21.77 billion to $16.158 billion.

The Heritage Foundation ranks Israel very high (4 on a scale of 5) in terms of the level of government intervention in the economy.[64] This is not surprising inasmuch as the economic ideology of Israel has always been socialist and collectivist. As one historian of Israel points out: "[Israel] had originally been created by East Europeans who brought with them not the ideas of Western liberal, bourgeois democracy but the collective socialism of the old Russian intelligentsia."[65] Democracy did little to constrain state intervention and did not provide Israel with more economic resources than the Arabs.

In sum, it is clear that democracies are wealthier than nondemocracies, and it is indisputable that national wealth is a key building block of military power. But contrary to what Lake and others triumphalists believe, democracy does not appear to be the source of that wealth. It seems equally plausible that states become wealthy first and then become democratic, not the other way around. Moreover, democracies enjoy no special advantage over authoritarian states in mobilizing that wealth for military purposes. Finally, even if Lake is right that state rent seeking is less of a problem in democracies, there are a number of logical reasons why rent seeking by interest groups is more of a problem in democratic political systems.

DEMOCRACY AND ALLIANCES

According Randolph Siverson and Juliann Emmons, democracies tend to form alliances with each other because they share a deep-seated commitment to two norms: cooperation and amity.[66] Some scholars argue that democratic alliances are more durable that other types of alliances.[67] This durability of democratic alliances leads Lake and others to conclude that, in war, the resulting democratic alliances are more effective than either mixed alliances or alliances com-

---

64. See Kim R. Holmes, Bryan T. Johnson, and Melanie Kirkpatrick, *1997 Index of Economic Freedom* (Washington, D.C.: Heritage Foundation and Wall Street Journal, 1997), pp. xxx, 242–244, 255–257.
65. Geoffrey Wheatcroft, *The Controversy of Zion: Jewish Nationalism, the Jewish State, and the Unresolved Jewish Dilemma* (Reading, Mass.: Addison-Wesley, 1996), p. 241.
66. Randolph M. Siverson and Juliann Emmons, "Birds of a Feather: Democratic Political Systems and Alliance Choices in the Twentieth Century," *Journal of Conflict Resolution*, Vol. 35, No. 2 (June 1991), pp. 285–300. The classic statement of the normative argument is Immanuel Kant, "Perpetual Peace," in Ted Humphrey, ed., *Perpetual Peace and Other Essays on Politics, History, and Morals* (Indianapolis: Hackett, 1983), pp. 107–145. More recent work combines normative and institutional arguments. See, for example, Kurt Taylor Gaubatz, "Democratic States and Commitment in International Relations," *International Organization*, Vol. 50, No. 1 (Winter 1996), pp. 110–111.
67. On the greater durability of democratic alliances, see William Reed, "Alliance Duration and Democracy: An Extension and Validation of 'Democratic States and Commitment in International Relations,'" *American Journal of Political Science*, Vol. 41, No. 3 (July 1997), pp. 1072–1078; and D. Scott Bennett, "Testing Alternative Models of Alliance Duration, 1816–1984," *American Journal of Political Science*, Vol. 41, No. 3 (July 1997), pp. 846–878.

prising only nondemocracies.[68] One underlying assumption that could lead to this conclusion is that democratic leaders must worry about audience costs if they renege on their alliance commitments, which should make them highly reliable allies.[69] There are reasons to suggest, however, that this is not the case.

The proposition that democracies are likely to align with each other finds little support in the historical record.[70] In fact, history offers few examples of purely democratic alliances; most have been either mixed or between nondemocracies exclusively. Siverson and Emmons's own data indicate that democratic alliances accounted for only 3.24 percent of the total in the 1920–39 period and 10.97 percent in the 1946–65 period.[71] These data can be interpreted to mean that the growth of purely democratic alliances was largely a Cold War phenomenon, where the Soviet threat, not ideological affinity, brought democracies together.[72]

The Israeli cases do not lend much support to the "birds of a feather argument" that democracies are natural and constant allies. Early in its independence, Israel experienced difficulty forming alliances with other democracies. It did, however, find significant support from the Soviet Union, Czechoslovakia, and Yugoslavia. The Soviet Union was one of the first states to formally recognize the new state of Israel. And Golda Meir concluded that "had it not been for the arms and ammunition that we were able to buy in Czechoslovakia and transport through Yugoslavia and other Balkan countries in those days at the start of the war, I do not know whether we actually could have held out until the tide changed, as it did by June of 1948."[73] More recently, Israel made common cause with such nondemocratic states as South Africa.[74] In fact, the Israeli

---

68. Lake, "Powerful Pacifists," p. 24; and Anjin Choi, "Cooperation for Victory: Democracy, International Partnerships, and State War Performance, 1816–1992," John M. Olin Institute for Strategic Studies, Harvard University, April 2002.
69. For the institutional argument that because democracies have large audience costs (e.g., leaders cannot change policies because the public is wedded to them), their commitments (or threats) are more credible, see James D. Fearon, "Domestic Political Audiences and Escalation of International Disputes," *American Political Science Review*, Vol. 88, No. 3 (September 1994), pp. 577–592; and Joe Eyerman and Robert A. Hart Jr., "An Empirical Test of the Audience Costs Proposition," *Journal of Conflict Resolution*, Vol. 40, No. 4 (December 1996), pp. 597–616.
70. Michael W. Simon and Erik Gartzke, "Political System Similarity and the Choice of Allies," *Journal of Conflict Resolution*, Vol. 40, No. 4 (December 1996), pp. 617–635; and Brian Lai and Dan Reiter, "Democracy, Political Similarity, and International Alliances, 1816–1992," *Journal of Conflict Resolution*, Vol. 44, No. 2 (April 2000), pp. 203–228.
71. Siverson and Emmons, "Birds of a Feather," p. 300.
72. This observation about the time boundedness of the democratic "birds of a feather" phenomenon is similar to the finding that the so-called democratic peace is also a recent development. On this, see Henry S. Farber and Joanne Gowa, "Polities and Peace," *International Security*, Vol. 20, No. 2 (Fall 1995), pp. 239–262.
73. Golda Meir, *My Life* (New York: G.P. Putnam's Sons, 1975), pp. 230–231.
74. Van Creveld, *The Sword and the Olive*, p. 206.

government and the South African apartheid regime were so closely aligned that they even cooperated secretly in developing each other's nuclear programs.[75] In short, democratic Israel has aligned itself with different types of regimes.

There is also little evidence to think that democratic alliances are militarily more effective than mixed or nondemocratic alliances. Large-N studies of this issue have produced contradictory findings.[76] Moreover, in the COW data set there is only one war (the debatable case of the 1956 Sinai War in which Israel, France, and Britain defeated Egypt) where the victorious alliance was composed entirely of democracies. In the overwhelming majority of other wars in which democracies won in alliance with other states, these alliances included nondemocracies.[77]

Moreover, the assumption that democracies should ally with each other is unconvincing because there are equally plausible reasons why democracies should ally with nondemocracies. Michael Simon and Eric Gartzke, for example, argue that because democracies and authoritarian states have different strengths and weaknesses (e.g., democracies have greater difficulty keeping secrets than authoritarian states), they make good allies.[78] Mancur Olson and Richard Zeckhauser suggest an alternative rationale for why different kinds of regimes attract each other. Collective action among democratic allies is likely to be difficult, they argue, because the bonds of friendship may cause democracies to contribute less than their fair share—that is, they might think that their partners will pick up any slack out of a sense of fraternal obligation. In alliances that include nondemocracies, every member is more likely to pull its own weight, because each recognizes that the others are motivated strictly by self-interest. Therefore, they will not tolerate the kind of free riding that is likely in an alliance made up solely of democracies.[79] In short, there is no good reason why democracies should prefer to ally with each other rather than with nondemocracies.

There is also reason to question the audience costs argument, which could provide the theoretical foundation for the claim that democratic alliances are

---

75. Seymor Hersh, *The Samson Option: Israel's Nuclear Arsenal and America's Foreign Policy* (New York: Random House, 1991), pp. 271–283.
76. Compare Reiter and Stam, *Democracies at War*, pp. 111–113, with Choi, "Cooperation for Victory," p. 32.
77. Reiter and Stam, "Democracy and War Initiation," p. 378.
78. Simon and Gartzke, "Political System Similarity and the Choice of Allies," pp. 617–635.
79. Mancur Olson Jr. and Richard Zeckhauser, "An Economic Theory of Alliances," in Julian R. Friedman, Christopher Bladen, and Steven Rosen, compilers, *Alliance in International Politics* (Boston: Allyn and Bacon, 1970), p. 186.

especially durable and therefore more militarily effective. Although Joe Eyerman and Robert Hart conclude that crises between democracies are resolved more easily than those between nondemocracies—and they interpret this finding as support for at least some aspects of the audience costs argument—there is still no evidence that these costs make democracies better allies.[80] The level of public support within democracies for foreign attachments varies widely; in cases where the public is not seriously engaged, there are no audience costs for failure to honor an obligation.[81] Indeed there is considerable evidence that democratic publics are not particularly attentive to international affairs, which means that more often than not audience costs play little role in the calculations of democratic leaders.[82] Even in those cases where the public strongly supports a commitment to another state, such support can evaporate quickly.[83] Finally, leaders have considerable latitude to shape public attitudes toward alliances, which means that they will sometimes be able to explain away broken promises without incurring significant audience costs. In the best available study on regime type and commitments, Kurt Gaubatz concludes that the evidence supports only the more modest conclusion that democracies are no worse than other types of regimes in making "lasting commitments."[84]

The democratic state that should have had the highest audience costs in breaking a commitment to Israel was the United States. But despite the presence of an influential pro-Israel constituency in the United States after World War II, this alignment did not become very tight until the 1970s. Indeed the

---

80. Eyerman and Hart, "An Empirical Test of the Audience Costs Proposition," pp. 597–616. But see Stephen M. Walt, "Rigor or Rigor Mortis? Rational Choice and Security Studies," *International Security*, Vol. 23, No. 4 (Spring 1999), pp. 33–35, for a discussion of the limits of this empirical support. For suggestions of other logical problems with the audience costs argument, see Kenneth A. Schultz, "Do Democratic Institutions Constrain or Inform? Contrasting Two Institutional Perspectives on Democracy and War," *International Organization*, Vol. 53, No. 2 (Spring 1999), p. 237, n. 11.
81. This point is made by Walt, "Rigor or Rigor Mortis?" pp. 33–35.
82. Ole R. Holsti, "Public Opinion and Foreign Policy: Challenges to the Almond-Lippmann Consensus, Mershon Series: Research Programs and Debates," *International Studies Quarterly*, Vol. 36, No. 4 (December 1992), p. 447; and John Mueller, *War, Presidents, and Public Opinion* (Lanham, Md.: University Press of America, 1985), p. 2.
83. Charles D. Tarlton, "The Styles of American International Thought," *World Politics*, Vol. 17, No. 4 (July 1965), pp. 584–614. This trend had become even more pronounced until September 11, 2001. The Chicago Council on Foreign Relations' most recent survey of public opinion finds that foreign policy is not even a top-ten issue for the American public. See John E. Reilly, ed., *American Public Opinion and U.S. Foreign Policy, 1999* (Chicago: CCFR, 1999), p. 7, Figure I-2, which showed that the public's concern about international problems is the lowest ever. See also John Mueller, "Eleven Propositions About American Foreign Policy and Public Opinion in an Era Free of Compelling Threats," Department of Political Science, Ohio State University, April 19, 2001.
84. Gaubatz, "Democratic States and Commitment in International Relations," p. 137.

U.S. government was ambivalent about Israeli independence in 1948,[85] opposed the democratic coalition that Israel fought beside in 1956 in Suez, and hamstrung the Israelis in 1967. Not surprisingly, once the U.S.-Israeli alliance was consolidated, the Israelis remained somewhat skeptical.[86] Other democracies such as Britain, France, and Germany were not always reliable allies either.[87] As Golda Meir recounted: "One day, weeks after the [Yom Kippur] war, I phoned [German Chancellor] Willy Brandt, who is much respected in the Socialist International, and said ' . . . I need to know what possible meaning socialism can have when not a single socialist country in all of Europe was prepared to come to the aid of the only democratic nation in the Middle East. Is it possible that democracy and fraternity do not apply in our case?'"[88] The U.S.-Israeli alliance was based not on high domestic audience costs but on the strategic interest of the United States in having allies in the Middle East to balance against the Soviet Union and later Iran and Iraq.[89] Realizing that these realpolitik considerations might someday lead to the U.S. abandonment of Israel, the Israelis and their American supporters have consistently sought to cloak the alliance in the mantle of democratic confraternity.[90] In short, democratic leaders are not necessarily constrained by alliance commitments, so there is little reason to believe that democratic alliances should be more effective than other types of alliances at winning wars.

DEMOCRACY AND SOUND STRATEGY

Some triumphalists believe that democracies are better strategic decisionmakers than nondemocracies because the voters and their representatives, not just a handful of elites, have a say in how to wage war. According to Bruce Russett, this has two positive effects: Greater public involvement in decisionmaking produces better military policies, because those who would pay the costs of going to war make the decisions about how it is conducted; and the greater the number of individuals participating in the decisionmaking process,

85. Tom Segev, *The Seventh Million: The Israelis and the Holocaust* (New York: Hill and Wang, 1993), p. 191.
86. Yoav Ben-Horin and Barry R. Posen, *Israel's Strategic Doctrine*, RAND Report 2845-NA (Santa Monica, Calif.: RAND, September 1981), pp. 9, 24. More recently, see Stephen J. Glain, "For Some Israelis, U.S. Aid Is a Burden: Some Say Strings Attached to Military Assistance Aren't Worth the Money," *Wall Street Journal*, October 26, 2000, p. A23.
87. Gilbert, *Israel*, p. 448.
88. Meir, *My Life*, p. 446.
89. Van Creveld, *The Sword and the Olive*, p. 252; and Gilbert, *Israel*, pp. 165, 225, 326, 367, 407, 445.
90. Wheatcroft, *The Controversy of Zion*, p. 308.

the lower the likelihood of strategic blunders.[91] Optimal security policies usually prevail in the marketplace of ideas, which is what Stephen Van Evera, Jack Snyder, and others argue occurs in a democratic political system.[92] On close examination, however, these claims are unpersuasive for three reasons.

First, there are no studies available that assess whether democracies or nondemocracies make better decisions about how to wage war. Indeed the triumphalists offer no systematic evidence to support this claim, but rather make their case by emphasizing the logic that underpins it. There is, however, evidence to suggest that democracies are no better at making strategy than authoritarian states.

Israeli democracy has not consistently fostered high-quality strategic evaluation and decisionmaking. Indeed Israel has made a number of major strategic blunders since 1967. The lapses in judgment that produced the 1987 Palestinian *intifada* were rooted in decisions made after Israel's 1967 victory in the Six-Day War. Even though it had been clear to many Israeli leaders early on that retaining the Occupied Territories would be more trouble than they were worth,[93] the electoral dynamics of Israeli democracy made it difficult for any leader to unilaterally withdraw from them.[94] The intelligence failures that nearly resulted in Israel's defeat in the Yom Kippur War were thoroughly documented by the 1974 Agranat Commission.[95] Both Defense Minister Moshe Dyan and Prime Minister Meir resigned after the release of the commission's report of 1974, but that has not ensured that subsequent Israeli governments have been any wiser.[96] Consider, for example, the many mistakes made by Israeli leaders that led to the 1982 Lebanon debacle.[97] Prime Minister Menachem Begin re-

---

91. Russett's reasoning follows Condorcet's jury theorem, which holds that if there is a 55 percent chance of any individual making the right decision, and 1,000 people decide using majority rule, then there is a 99.9 percent chance that such a democratic procedure will produce the right outcome. Bruce M. Russett, *Controlling the Sword: The Democratic Governance of National Security* (Cambridge, Mass.: Harvard University Press, 1990), pp. 106, 150.
92. Stephen Van Evera, "Primed for Peace: Europe after the Cold War," *International Security*, Vol. 15, No. 3 (Winter 1990/91), p. 27; Jack Snyder, *Myths of Empire: Domestic Politics and International Ambition* (Ithaca, N.Y.: Cornell University Press, 1991), pp. 18–19; and Reiter and Stam, *Democracies at War*, pp. 23–24,146, 160.
93. Gilbert, *Israel*, p. 398.
94. Ibid., p. 396; and Wheatcroft, *The Controversy of Zion*, p. 312.
95. Chaim Herzog, *The War of Atonement: October 1973* (Boston: Little, Brown, 1975), pp. 31, 278; Chaim Herzog, *The Arab-Israeli Wars: War and Peace in the Middle East from the War of Independence through Lebanon* (New York: Vintage, 1982), pp. 236–239; Amos Perlmutter, "Israel's Fourth War, October 1973: Political and Military Misperceptions," *Orbis*, Vol. 19, No. 2 (Summer 1975), pp. 434–460; and "Chief of Military Resigns in Israel, Blamed in Inquiry," *New York Times*, April 3, 1974, pp. 1, 5.
96. Gilbert, *Israel*, p. 465.
97. Schiff and Ya'ari, *Israel's Lebanon War*.

signed after the Lebanon campaign, yet the architect of that debacle, Ariel Sharon, is Israel's current prime minister.[98]

Moreover, the Israeli government has traditionally revealed very little information about its national security decisionmaking to the Israeli public.[99] Reflecting on the situation during the Yom Kippur War, former Israeli President Chaim Herzog observed that "Mrs. Meir's method of government brought about a system whereby there were not checks and balances and no alternative evaluations. Her doctrinaire, inflexible approach to problems and the government was to contribute to the failings of the government before the war. She was very much the overbearing mother who ruled the roost with an iron hand. She had very little idea of orderly administration and preferred to work closely with her cronies, creating an *ad hoc* system of government based on what was known as her 'kitchen.' But once war had broken out these very traits proved to be an asset."[100] Therefore, contradicting the marketplace of ideas argument that free and unfettered debate should produce optimal wartime policy, this undemocratic system has been effective for Israel in wartime. The fact that Israel is a democracy has not necessarily meant that it has crafted better security policies. But the lack of public input has not uniformly hindered Israeli decisionmaking either.

Second, there is no question that the public wants to avoid strategic blunders. Nobody wants to die if it can be avoided. The key issue, however, is whether there is a mechanism for translating that motivation into better wartime decisionmaking. In fact, there is not. The root of the problem is that the soldiers who fight wars hardly ever have the expertise to improve the decisionmaking process. Invariably, they have significantly less information and expertise than the civilian and military elites charged with directing the war. In the end, how well those at the top make decisions is all that matters.

Finally, a political system that gives voice to large numbers of individuals with diverse preferences may not be able to reconcile those differences and produce coherent policies. For example, Gaubatz's recent application of Kenneth Arrow's "paradox of democracy" to illustrate how national security decisions are made suggests how difficult it is to aggregate the diverse opinions

98. Gilbert, *Israel*, p. 515.
99. Van Creveld, *The Sword and the Olive*, pp. xviii, 68, 109–110; and Perlmutter, "Israel's Fourth War," p. 435.
100. Herzog, *The War of Atonement*, p. 282.

common to democracies.[101] Unfortunately, the marketplace of ideas is not necessarily an efficient producer of sound strategy.[102]

DEMOCRACY AND PUBLIC SUPPORT

According to Aaron Friedberg, democratic leaders can count on greater public support for their wars than their authoritarian counterparts because elected policymakers are accountable to the people and so will conduct wars in such a way as to ensure that public support remains high.[103] Although there is no question that democratic leaders are answerable to their constituents, it is doubtful that this link translates into greater public support for their states' wars or that it explains why they win them.

Friedberg argues that it is especially difficult for democracies to rely on coercion and centralized control to wage war while maintaining public support, because they place a high premium on the norm of consent and they usually have a limited and decentralized form of government. To maintain public support for the war effort, Friedberg maintains, democratic leaders must conduct wars while relying on the voluntary consent of the public. Doing so, in fact, is likely to increase the prospects of military success. This approach, according to Freidberg, explains why the democratic United States, rather than the authoritarian Soviet Union, prevailed in the Cold War. It is not clear, however, how much regime type affects the level of public support for a war effort.

First, there are other reasons why the United States did not become a large, intrusive, and coercive garrison state during the Cold War that could have risked losing public support in the struggle against authoritarian communism. Structural factors such as geographic isolation and possession of nuclear weapons, rather than norms and institutions, offer an equally plausible explanation for why the United States could wage the Cold War while relying on voluntary consent and with a less intrusive government than that of the Soviet Union. Therefore, the problem with Friedberg's argument is in part one of case

---

101. Kurt Taylor Gaubatz, "Intervention and Intransitivity: Public Opinion, Social Choice, and the Use of Military Force Abroad," *World Politics*, Vol. 47, No. 4 (July 1995), p. 538. Kenneth Arrow originally laid out his "paradox of democracy" argument in *Social Choice and Individual Values* (New York: Wiley, 1951).
102. For cautionary notes from an early proponent of the "marketplace of ideas," see Stephen Van Evera, "Why States Believe Foolish Ideas: Non-Self-Evaluation By States and Societies," version 3.5, January 10, 2002, p. 11, n. 21.
103. Friedberg, *In the Shadow of the Garrison State;* and Aaron Friedberg, "Why Didn't the United States Become a Garrison State?" *International Security*, Vol. 16, No. 4 (Spring 1992), pp. 109–142.

selection. His normative, institutional, and structural factors all anticipate a smaller and less coercive U.S. government relative to the Soviet Union. In the Cold War, the United States' antistatist ideas and weak governmental institutions coincided with geographical insulation and nuclear weapons. Thus it is not the best case to demonstrate that antistatist ideas and institutions were the driving force behind these strategic choices. In fact, this case could just as plausibly be interpreted as indicating that both democracy and success in war were the results of a favorable geographic location and nuclear weapons.

Second, Friedberg's assertion that the Cold War U.S. government was smaller and less intrusive than it might have otherwise been is debatable. If the comparative baseline for measuring the expansion of the U.S. Cold War state is either World War II or what some proponents of big government advocated, it was certainly smaller and less intrusive. The United States was much larger, significantly more intrusive, and somewhat more coercive, however, than it had been during the interwar period or at various times in the nineteenth century.[104] Indeed all successful states become more centralized and coercive in wartime.[105] Authoritarian Nazi Germany, which lost World War II, had remarkably little wartime centralization. On the other hand, the victors (i.e., the authoritarian Soviet Union and the democratic United States and Britain) were highly centralized.[106] This suggests that more centralized and more coercive states are more likely to win wars and also that regime type may not be the most important factor in explaining which states are able to more effectively mobilize societal resources in wartime.

Third, the triumphalists' claim about democracy and public support is not logically compelling. In particular, there is reason to believe that leaders and their publics often have different time horizons that affect their thinking about the utility of war. As Donna Nincic and Miroslav Nincic suggest, democratic publics, like consumers, tend to focus on short-term considerations when thinking about the use of force: What is the immediate payoff? In contrast, democratic leaders are inclined to think about war the way investors do: What will be the long-term payoff?[107] Given these different perspectives on the use

104. Friedberg, *In the Shadow of the Garrison State*, pp. 30–31.
105. Karen A. Rasler and William R. Thompson, *War and State-making: The Shaping of Global Powers* (Boston: Unwin and Hyman, 1989); and Bruce D. Porter, *War and the Rise of the State: The Military Foundations of Modern Politics* (New York: Free Press, 1994).
106. Overy, *Why the Allies Won*, p. 206; and Alan S. Milward, *War, Economy, and Society: 1939–1945* (Berkeley: University of California Press, 1977), pp. 99–131.
107. Donna J. Nincic and Miroslav Nincic, "Commitment to Military Intervention: The Democratic Government as Economic Investor," *Journal of Peace Research*, Vol. 32, No. 4 (July 1995), pp. 413–426.

of force, it is reasonable to expect democratic leaders and their publics to be out of step in their enthusiasm for particular wars.

Fourth, there are no comprehensive studies to support the triumphalists' claim that democracies enjoy greater public support in wartime than authoritarian states. There is actually plenty of anecdotal evidence; however, both types of regime enjoy varied levels of public support in times of conflict, and neither has an apparent advantage over the other. For example, the American public strongly endorsed U.S. participation in World War II (1941–45), but its support for the Vietnam War (1965–73) evaporated over time, leading the United States to withdraw from the conflict. Authoritarian Russia, on the other hand, saw public support for World War I disappear between 1914 and 1917, yet the Soviet Union enjoyed broad and deep public support throughout World War II.[108] The historical record thus appears to show that regime type has hardly any effect on the level of public support in wartime.

There can be little doubt that historically the state of Israel was able to count on the overwhelming support of its citizens when it went to war between 1948 and 1973. But this support was not the result of its democratic system, as the triumphalists would argue. Rather, Israelis believed that they were fighting for their very survival.[109] Golda Meir made clear why Israeli society came together in wartime despite overwhelming odds: "We couldn't afford the luxury of pessimism . . . , so we made an altogether different kind of calculation based on the fact that the 650,000 of us were more highly motivated to stay alive than anyone outside Israel could be expected to understand and that the only option available to us, if we didn't want to be pushed into the sea, was to win the war."[110] Van Creveld echoes this point: "Israeli public opinion continued to see the IDF as the one great organization standing between it and death. Even more than before, it was prepared to do its utmost to ensure the army's success by providing the necessary resources in terms of material and the very best manpower at its disposal."[111] In short, common threat, rather than shared democratic ideology, provides a more compelling explanation for why Israeli society supported Israel's war efforts so enthusiastically.

---

108. See the discussion of resurgent Russian nationalism during World War II in Overy, *Why the Allies Won*, pp. 290–293.
109. Van Creveld, *The Sword and the Olive*, pp. 125, 153, 197, 241.
110. Meir, *My Life*, p. 233. See also similar comments by David Ben Gurion and Moshe Dyan, in Dyan, *Moshe Dyan: Story of My Life* (New York: G.P. Putnam's Sons, 1975), pp. 92, 396, 441.
111. Van Creveld, *The Sword and the Olive*, p. 153.

DEMOCRACY AND FIGHTING PROFICIENCY

Reiter and Stam maintain that because democratic governments have greater legitimacy than their authoritarian counterparts, their soldiers perform better on the battlefield. They attribute this finding to the political culture of democracies, which they argue fosters greater individual initiative and better leadership among their soldiers.[112] They reject as an alternative explanation that nationalism, rather than democracy, produces superior leadership and initiative, arguing that nationalism results only in higher morale.

There is reason to think, however, that nationalism also enhances individual initiative and leadership. Many scholars believe that the French Revolution transformed warfare precisely because it democratized French society. This, they maintain, fostered a greater sense of loyalty to the regime, which in turn increased the military effectiveness of the French army in all three areas.[113] This effectiveness, however, had its roots in prerevolutionary France and survived the collapse of French democracy and the coming to power of Napoleon Bonaparte.[114] Prussia and Spain, two highly nationalistic but not democratic regimes, played important roles in defeating Napoleon by employing many of the same tactics that served revolutionary and then Napoleonic France so well.[115] Nationalism and democracy, though they sometimes reinforce each other, are not inseparable.[116] Indeed Reiter and Stam concede that nationalism, not democratic ideology, may account for combat prowess. Unfortunately, they have not systematically tested nationalism as an alternative explanation for why militaries in their data set performed well on the battlefield.[117] Thus their case rests not on explicating an unbroken chain of logical reasoning, but on showing that there is a significant statistical correlation between democracy and various combat skills.

At first glance, Reiter and Stam appear to have assembled impressive statistical support for their claim that soldiers in democratic societies display greater leadership and initiative than those from nondemocracies. On close inspection,

---

112. Reiter and Stam, "Democracy and Battlefield Military Effectiveness," pp. 259–277; and Reiter and Stam, *Democracies at War*, pp. 58–74.
113. John A. Lynn, *The Bayonets of the Republic: Motivation and Tactics in the Army of Revolutionary France, 1791–94* (Boulder, Colo.: Westview, 1996).
114. Theodore Ropp, *War in the Modern World* (New York: Collier, 1962), pp. 98–142.
115. Peter Paret, "Napoleon and the Revolution in War," in Paret, *Makers of Modern Strategy*, pp. 123–142.
116. Michael Howard, *War in European History* (Oxford: Oxford University Press, 1976), pp. 110–111, argues that democracy and nationalism are intimately related. However, Ropp, *War in the Modern World*, pp. 126, 138, reminds us that the Spanish and Prussian cases during the Napoleonic Wars demonstrate that nationalism and democracy are not necessarily linked.
117. Reiter and Stam, "Democracy and Battlefield Military Effectiveness," p. 264.

however, the Combat History Analysis Study Effort (CHASE) data set of battles, which provides the basis for these findings, is unreliable. In 1982, the Historical Evaluation and Research Organization (HERO) was commissioned to assemble this data set for the U.S. Army Concepts Analysis Agency (CAA). After receiving the initial version of the data set in 1984, CAA randomly selected 8 battles from it and submitted them for analysis to the U.S. Army Military History Institute, the U.S. Army Center for Military History, the Department of History at the U.S. Military Academy, and the U.S. Army Combat Studies Institute. A total of 159 codings were checked in the 8 cases. The results seriously called into question the data set's reliability: 106 codings (67 percent) were judged to be in error, another 29 (18 percent) were deemed questionable, and only 24 (15 percent) were ascertained to be correct by the reviewers.[118]

Despite two revisions, there is still reason to question the reliability of the 1990 version of the CHASE data set that Reiter and Stam employ. The principal problem is that the codings of certain items in the data set are imprecise. The former CAA project manager, for example, concedes that "even with our best efforts error rates of 5% to 30% are to be expected."[119] As a result of continuing conflict between CAA and HERO over the reliability of the CHASE data set, HERO was relieved of responsibility for updating that data set in 1987. Nevertheless, HERO continues to work on its own to update the 1987 version of the CHASE data set, which it calls the Land Warfare Data Base (LWDB).[120] Recently, HERO (which is now called the Dupuy Institute) compared the 1990 version of the CHASE data set with the current LWDB, focusing on 1,196 data points common to both data sets. They found that almost half (500) of the codings for those same data points were different.[121]

There were no differences between the CHASE and LWDB data sets in the "leadership" category, but the consistency between the two data sets is not evidence that the data on leadership are reliable. In its various revisions to the CHASE data set after 1987, HERO focused exclusively on relatively hard variables such as order of battles and casualties, while ignoring softer variables such as initiative and leadership. According to a HERO staff member, these

---

118. See Management and Support Directorate, *Military History: A Data Base of Selected Battles, 1600–1973,* Vol. 1, *Main Report* (Bethesda, Md.: U.S. Army Concepts Analysis Agency, September 1984), p. 21. See also Mearsheimer, "Assessing the Conventional Balance," p. 66, n. 29.
119. Robert Helmbold, "Lessons Learned Regarding Battle Data Bases," January 14, 1987, Howard Whitley Archives, Box 1, Center for Army Analysis, Ft. Belvoir, Virginia.
120. Discussions with Christopher Lawrence and Richard Anderson of the Dupuy Institute, McLean, Virginia, April 2000.
121. Letter (with attachments) to author from Christopher Lawrence, executive director, Dupuy Institute, June 8, 2000.

two variables were the "least looked at and poorest proofed section of the data base," because their codings were widely regarded as "all a judgement" anyhow.[122]

Problems with the HERO data set notwithstanding, Reiter and Stam believe that their findings are still valid on two related grounds: First, unless there is systematic bias in the codings, the fact that there is a very large number of cases should still make it possible to trust the findings. Second, because the principal architect of the original CHASE data set did not regard democracy as a key explanation for military prowess, we can be confident that the data are not biased in favor of their claims about the battlefield advantages of soldiers of democratic states.[123]

Although there may be no systematic bias in the CHASE data set, there is so much potential measurement error in the data set generally, and particularly in the leadership and initiative variables, that Reiter and Stam are left with inefficient models. Consider, for example, that if the relatively hard variables have a 5–30 percent error rate in their coding, how much more imprecise these soft variables are.[124] There is an even more serious data problem: possible bias or error in the coding of the independent variable—democracy. Ido Oren makes a convincing case that the POLITY democracy scores are highly subjective and thus unreliable.[125] The combination of problems with data for both the dependent and independent variables casts doubt on Reiter and Stam's findings that democratic armies demonstrate greater initiative and leadership skills on the battlefield.

Reiter and Stam note that there is another unbiased source of data on comparative military competence that can be used to test the triumphalists' proposition about the relationship between democracy and military performance.[126] Allen Millett, Williamson Murray, and Kenneth Watman's study of the great power militaries in World War I, the interwar period, and World War II pro-

---

122. Christopher Lawrence, telephone conversation, June 19, 2000.
123. Reiter and Stam, *Democracies at War*, pp. 71–72.
124. On the problems of bad data due to "measurement error," see William H. Williams, "How Bad Can 'Good' Data Really Be?" *American Statistician*, Vol. 32, No. 2 (May 1978), pp. 61–65. See also Gary King, Robert O. Keohane, and Sidney Verba, *Designing Social Inquiry: Scientific Inference in Qualitative Research* (Princeton, N.J.: Princeton University Press, 1994), pp. 158–163, concerning how nonsystematic error in the dependent variable reduces efficiency.
125. Ido Oren, "The Subjectivity of the 'Democratic' Peace: Changing U.S. Perceptions of Imperial Germany," *International Security*, Vol. 20, No. 2 (Fall 1995), p. 266. For a thoughtful discussion of the other limitations of the POLITY data set, see Kristian Gleditsch and Michael D. Ward, "A Reexamination of Democracy and Autocracy in Modern Politics," *Journal of Conflict Resolution*, Vol. 41, No. 3 (June 1997), pp. 361–383.
126. Reiter and Stam, "Democracy and Battlefield Military Effectiveness," p. 264.

vides indicators of their military effectiveness.[127] It offers little evidence, however, that democratic armies fight better than nondemocratic armies.[128] Given the problems with the CHASE data set and the evidence of at least one other data set, there are grounds for doubting the triumphalists' claims that democracies are more likely to win their wars because their soldiers fight better.

This conclusion is hardly surprising, given the consensus among military historians that the three most formidable armies of the twentieth century in terms of initiative and leadership were (1) Imperial Germany's army during World War I (authoritarian state),[129] (2) Nazi Germany's army during World War II (authoritarian state),[130] and (3) Israel's army between 1948 and 1973 (democratic state).[131] In the Israeli cases, necessity, rather than shared democratic ideology, accounted for the superior performance of Israeli soldiers on the battlefield between 1948 and 1973. Van Creveld attributed the combat prowess of Israeli soldiers to the fact that they had no choice but to fight well or risk death: "Nothing mattered any longer, not even fear of incurring casualties. Was not Nasser a second Hitler? Was not another Holocaust just around the corner? Thus motivated, the Israelis fought like demons."[132] Israeli troops fought so valiantly not because their democratic political system made them want to fight better but because they had to if they wanted to survive.

It is clear that ideology did play an important role in Israeli military success; that ideology, however, was not liberal democracy but rather nationalism.[133] The common Arab threat solidified the sense of Israeli national identity, which in turn increased the willingness of Israeli society to support the war effort and its soldiers to fight hard. In contrast, there is little evidence—despite much

---

127. Allan R. Millett, Williamson Murray, and Kenneth Watman, "The Effectiveness of Military Organizations," *International Security*, Vol. 11, No. 1 (Summer 1986), p. 37.
128. Lt. Gen. John H. Cushman, U.S. Army (ret.), "Challenge and Response at the Operational and Tactical Levels, 1914–45," in Williamson Murray and Allan R. Millett, eds., *Military Effectiveness*, Vol. 3, *The Second World War* (Boston: Allen and Unwin, 1988), pp. 320–340. Cross-tabulations and χ^2 for POLITY III democracy scores and Cushman's operational and tactical effectiveness grades for various countries covered in the three-volume study ($A = 4$, $B = 3$, $C = 2$, $D = 1$, $F = 0$) show no significant relationship between regime type and effectiveness.
129. Niall Ferguson, *The Pity of War: Explaining World War I* (New York: Basic Books, 1999), pp. 290–303; and Timothy T. Lupfer, *The Dynamics of Doctrine: The Changes in German Tactical Doctrine during the First World War* (Fort Leavenworth, Kans.: Combat Studies Institute, U.S. Army Command and General Staff College, 1981).
130. For a comparative discussion of the combat power of the German army, see Martin van Creveld, *Fighting Power: German and U.S. Army Performance, 1939–1945* (Westport, Conn.: Greenwood, 1982).
131. Van Creveld, *The Sword and the Olive*, p. xvii.
132. Ibid., p. 197.
133. Yehoshaphat Harkavi, "Basic Factors in the Arab Collapse during the Six Day War," *Orbis*, Vol. 11, No. 3 (Fall 1967), p. 680; and Gilbert, *Israel*, p. 174.

pan-Arab rhetoric—that the Arab-Israeli wars ever generated much nationalist sentiment in the Arab world, beyond Palestine in recent years. According to Israeli historian Benny Morris, this lack of nationalist identity put the Arabs at a distinct military disadvantage vis-à-vis the Jews: "For the average Arab villager, political independence and nation-hood were vague abstractions; his loyalties were to his family, clan, and village and, occasionally to his region. Moreover, decades of feuding had left Palestinian society deeply divided."[134] Given this lack of national consciousness, it is not surprising that the highly nationalist Israelis were generally more militarily effective than their Arab neighbors.

In sum, the triumphalists' arguments about the relationship between democracy and the economy, alliances, decisionmaking, public support, and the battlefield performance of soldiers as explanations for why democracies should do well once in war are unconvincing.

Conclusion

My skepticism about the triumphalists' argument that democracies more skillfully choose and effectively wage wars is based on two findings. First, much of the data supporting the correlation between democracy and victory are, upon closer inspection, of little value for testing the triumphalists' claim because they suffer from various shortcomings. Second, neither of the triumphalists' arguments that democracies do well because they are better at selecting wars they can win or that democracies fight better once at war are persuasive. Both rest on faulty logic and have only modest empirical support.

Therefore, if one wants to understand the sources of military effectiveness, either for one's own state or for potential allies and adversaries, whether or not that state is democratic is not the most important factor to consider. Although democracies and autocracies undoubtedly have different strengths and weaknesses that may affect some aspects of their performance in wartime, overall they seem to cancel each other out and so regime type confers no clear advantage or disadvantage. Moreover, at least until recently, military power could be produced in a variety of ways, through many different combinations of social organization, economic potential, specific doctrinal and training decisions, and

---

134. Benny Morris, *Righteous Victims: A History of the Zionist-Arab Conflict, 1881–1999* (New York: Alfred A. Knopf, 1999), p. 192. For the link between family and clan loyalty and the inability of Arabs to succeed in modern mechanized wars, see Kenneth M. Pollack, "The Influence of Arab Culture on Arab Military Effectiveness," Ph.D. dissertation, Massachusetts Institute of Technology, 1996.

strategic choices. In other words, the "recipe" for effective military performance had a lot of variability, which meant that very different regimes could produce similar levels of capability by combining other ingredients in different ways. Given this fact, it is not surprising that democracies and nondemocracies are sometimes good at fighting and sometimes bad; regime type alone does not confer a clear advantage or disadvantage in selecting or fighting wars.

One might accept that regime type was irrelevant in the past but argue that whether a state is democratic or not is now becoming more important. According to this line of reasoning, the lesson of the past eleven years is that if a state wants to have a truly cutting-edge military fully capable of taking advantage of the so-called revolution in military affairs, it cannot do this in a centralized, coercive, and information-controlled society. Specifically, if a country wants to be able to fight as successfully as the United States did in the 1991 Gulf War, it must have an open democratic society where everyone is able to freely exchange ideas and knowledge and avail themselves without restriction of computer and communication technologies.[135] The collapse of the Soviet Union at the end of the Cold War, largely because it was a centralized and coercive political system that was unable to compete militarily with the West, lends credence to this view. China, however, which remains fairly centralized and undemocratic, suggests that it may be possible for a state to reform its economy and revitalize its technology base so as to produce an effective military without political democracy.[136] Indeed, China is one of the cases that scholars need to watch to accumulate additional evidence about how much regime type may matter for military effectiveness in coming years.

My skepticism about the importance of regime type for military effectiveness stands in direct contrast to the current trends in the U.S. government, especially the intelligence community, in which there has been a renaissance of interest in the domestic-level sources of military effectiveness.[137] But if I am right, analysts should be wary about relying on monocausal theories of mili-

---

135. For the general logical underpinning of this argument, see Van Evera, "Primed for Peace," pp. 14–16; and Friedberg, *In the Shadow of the Garrison State*, p. 304.
136. For an example of how China has been able to modernize without across-the-board liberalization, see Evan A. Feigenbaum, "Who's Behind China's High-Technology 'Revolution'? How Bomb Makers Remade Beijing's Priorities, Policies, and Institutions," *International Security*, Vol. 24, No. 1 (Summer 1999), p. 119. For an argument that China does not have to match the United States across the board to pose a serious regional challenge to it, see Thomas J. Christensen, "Posing Problems without Catching Up: China's Rise and Challenges for U.S. Security Policy," *International Security*, Vol. 25, No. 4 (Spring 2001), pp. 5–40.
137. "Culture as Tool in National Security Analysis: A Roundtable," sponsored by the Strategic Assessments Group, Directorate of Intelligence, U.S. Central Intelligence Agency, McLean, Virginia, April 29, 1999.

tary effectiveness, whether they are based on regime type or some other domestic-level factor. Rather, they should look at a constellation of factors including the balance of actual and potential military power resources, the nature of the conflict, the willingness and ability of states to emulate the most successful military practices, nationalism, whether states have the common preconditions for military effectiveness and democracy, and whether their regimes are consolidated or not as indicators of how a state will do in war.[138]

The good news is that contrary to some defeatists inside and outside the U.S. government, democracy is not a liability for a state in choosing and effectively waging war. The bad news, however, is that democracy is not as large an asset as triumphalists maintain. In sum, regime type hardly matters.

---

138. See, for example, Jeffrey A. Isaacson, Christopher Layne, and John Arquilla, *Predicting Military Innovation*, documented briefing (Santa Monica, Calif.: RAND, 1999); and Ashley J. Tellis, Janice L. Bially, Christopher Layne, Melissa McPherson, and Jerry Solinger, *Measuring National Power in the Post-Industrial Age*, RAND Report 1818-A (Santa Monica, Calif.: RAND, July 1999).

Appendix. Fair Tests.

War	Misaggregation	Mixed Alliance	Gross Mismatch	Asymmetric Interests	Draw	Fair Test/Favors
Franco-Spanish			x			no
Russo-Turkish						no[a]
Mexican-American			x			no[c]
Austro-Sardinian			x			no[b]
First Schleswig-Holstein			x			no
Roman Republic			x			no[c]
La Plata			x			no
Crimean		x				no[c]
Anglo-Persian			x			no[c]
Italian Unification						no[b]
Spanish Moroccan						no[a]
Italo-Roman						no[b]
Italo-Sicilian			x			no
Franco-Mexican				x		no
Ecuador-Colombian						yes/pessimists
Second Schleswig-Holstein						yes/pessimists
Lopez		x	x			no
Spanish-Chilean						no[b]
Seven Weeks				x		no[b]
Franco-Prussian						no[a]
Russo-Turkish						no[a]
Pacific						yes/triumphalists
Sino-French			x			no[c]
Central American (1885)						yes/pessimists
Franco-Thai			x			no
Sino-Japanese						yes/triumphalists

Appendix. (continued)

War	Misaggregation	Mixed Alliance	Gross Mismatch	Asymmetric Interests	Draw	Fair Test/Favors
Greco-Turkish			x			no[c]
Spanish-American			x			no[c]
Boxer Rebellion		x				no[c]
Russo-Japanese						yes/triumphalists
Central American (1906)						yes/pessimists
Central American (1907)						no[b]
Spanish-Moroccan			x			no[c]
Italo-Turkish						no[a]
First Balkan		x				no[c]
Second Balkan		x	x			no[c]
World War I	x		x			no[c]
Hungarian			x			no[c]
Russo-Polish						yes/triumphalists[c]
Lithuanian-Polish			x			no
Greco-Turkish						no[b]
Franco-Turkish						no[b]
Sino-Soviet						yes/pessimists
Manchurian						yes/triumphalists
Chaco						yes/pessimists
Italo-Ethiopian			x			no
Sino-Japanese						yes/triumphalists
Chankufeng						yes/triumphalists
Nomohan			x			no
Russo-Finnish			x			no[c]
World War II	x	x	x			no[c]
Franco-Thai				x		no

Appendix. (continued)

War	Misaggregation	Mixed Alliance	Gross Mismatch	Asymmetric Interests	Draw	Fair Test/Favors
Israeli independence						yes/triumphalists[c]
Korea					x	no[c]
Russo-Hungarian						no[b]
Sinai			x			no[c]
Sino-Indian						yes/pessimists[c]
Second Kashmiri					?	no[c]
Vietnam				x		no[c]
Six-Day						yes/triumphalists[c]
Football						yes/triumphalists[c]
Attrition					x	no[c]
Bangladesh			x			no[c]
Yom Kippur						yes/triumphalists
Turkish-Cypriot			x			no[a]
Vietnamese-Cambodian						no[a]
Ethiopian-Somali						no[a]
Ugandan-Tanzanian						no[a]
Sino-Vietnamese (1979)						no[a]
Iran-Iraq					x	no
Falklands						yes/triumphalists[c]
Lebanon					x	no[c]
Sino-Vietnamese (1985–87)						no[a]
Gulf War			x			no[c]
Azeri-Armenian					x	no

NOTE: For further discussions of these codings, see "Assessment of the COW Universe of Interstate Wars since 1815," http://www.uky.edu/AS/PoliSci/Desch/research.htm.

a = equally democratic.
b = missing democracy scores.
c = democracy score > 6.

The Power of Democratic Cooperation

Ajin Choi

The influence of democracy on foreign policy and world politics is one of the most hotly debated issues in the field of international relations. The proposition that democracies are better equipped to win in war is no different in this respect: It is a provocative thesis that contrasts sharply with some well-received theories of international politics. In his article "Democracy and Victory: Why Regime Type Hardly Matters," Michael Desch raises several challenges to this proposition.[1] He critiques the empirical methodologies, theoretical explanations, and historical evidence that some scholars say explain democratic victory in war and concludes that, in the end, democracy has "no particular advantages or disadvantages" (p. 8).[2]

In examining five causal mechanisms offered by "democratic triumphalists" to support the military effectiveness of democracy, Desch argues that "none of them is logically compelling or has much empirical support" (p. 25). In this article I do not attempt to respond to all the points Desch raises. Instead I focus on one of his critiques—namely, whether democratic allies can be more effective than nondemocratic allies in wartime and whether they are more likely to achieve victory. My goal is twofold: first, to offer an alternative theoretical explanation for the relationship between alliance behavior, the role of domestic political institutions, and their effect on war performance to explanations that Desch attributes to democratic triumphalists; and second, to provide quantita-

Ajin Choi is a lecturer in the Department of Political Science at Yonsei University in Seoul, South Korea.

The author would like to thank Giacomo Chiozza, Peter Feaver, Christopher Gelpi, Joseph Grieco, Paul Gronke, David Lake, Dan Reiter, and Allan Stam for their suggestions. She also thanks the John M. Olin Institute for Strategic Studies at Harvard University for its support.

1. Michael C. Desch, "Democracy and Victory: Why Regime Type Hardly Matters," *International Security*, Vol. 27, No. 2 (Fall 2002), pp. 5–47. Additional cites to this article appear parenthetically in the text.
2. Studies exploring the democratic victory thesis include David A. Lake, "Powerful Pacifists: Democratic States and War," *American Political Science Review*, Vol. 86, No. 1 (March 1992), pp. 24–37; Dan Reiter and Allan C. Stam, "Democracy, War Initiation, and Victory," *American Political Science Review*, Vol. 92, No. 2 (June 1998), pp. 377–389; Dan Reiter and Allan C. Stam, *Democracies at War* (Princeton, N.J.: Princeton University Press, 2002); Christopher F. Gelpi and Michael Griesdorf, "Winner or Loser? Democracies in International Crisis, 1918–1998," *American Political Science Review*, Vol. 95, No. 3 (September 2001), pp. 633–648; and Ajin Choi, "Cooperation for Victory: Democracy, International Partnerships, and State War Performance, 1816–1992," Ph.D. dissertation, Duke University, 2001.

International Security, Vol. 28, No. 1 (Summer 2003), pp. 142–153
© 2003 by the President and Fellows of Harvard College and the Massachusetts Institute of Technology.

tive and qualitative empirical evidence in support of this alternative explanation. Although I concur with Desch that explanations for democratic victory based on "selection effects" and "military effectiveness" arguments are not completely satisfactory, I disagree with his claim that "regime type hardly matters" in determining war outcomes. Indeed my research suggests that democracies are formidable players in the international arena and that the sources of democracies' military prowess are entrenched in their political institutions.

This analysis proceeds in five steps: First, I summarize Desch's criticism of the impact of democratic alliance behavior on war performance; second, I present an alternative explanation for this behavior; third, I offer statistical evidence that supports my explanation; fourth, I illustrate the statistical findings with historical cases; and fifth, I conclude with some final observations regarding the influence of democracy on military effectiveness.

Democratic Alliance Behavior and Military Effectiveness

Democratic triumphalists, Desch argues, contend that democracies can be more reliable allies than nondemocracies because democratic leaders are more concerned than nondemocratic leaders about the possibility of punishment by the public if they fail to implement their states' international commitments. This concern is derived from James Fearon's argument that democratic leaders pay greater domestic audience costs than nondemocratic leaders when they retreat from their public declarations.[3] Triumphalists also suggest that democracies with shared norms are more likely to form alliances and are better able cooperate with one another.

Desch contends that the audience costs logic does not sufficiently explain the military effectiveness of democratic allies, noting that "there is considerable evidence that democratic publics are not particularly attentive to" foreign affairs. This suggests that audience costs do not play a significant role in the calculations of their leaders (p. 31). Therefore "democratic leaders are not necessarily constrained by" their states' international commitments (p. 32). Because of this, he concludes, there is little reason to believe that democratic allies should be more effective than other types of allies in waging war.

Desch also argues that the military effectiveness of democratic alliances based on shared norms and ideological affinity has neither theoretical nor em-

---

3. James D. Fearon, "Domestic Political Audiences and Escalation of International Disputes," *American Political Science Review,* Vol. 88, No. 3 (September 1994), pp. 577–592.

pirical support. He claims that collective action problems among democracies can be more serious in wartime because "bonds of friendship" may lead some democracies to pay less than their fair share in a war. He also points out that historically there have been few purely democratic coalitions; rather most have been mixed or exclusively nondemocratic. To support this claim further, Desch argues that Israel, a democratic state with an impressive record of wartime success, should be an easy case for the triumphalists. But according to Desch, Israel has not consistently had democratic allies, and it did not begin to develop strong ties to the United States until 1970. Moreover, Washington has maintained its political and military support for Israel not because Israel is a democracy but because of U.S. strategic interests in the Middle East (pp. 28–32).[4]

The Effectiveness of Veto Players and Transparent Political Systems

Putting aside the domestic audience costs explanation, whose validity Desch seriously doubts, why might democracies make better wartime partners and thus be more likely to win their wars? The common feature of all democratic states, regardless of whether they have presidential or parliamentary forms of government, is the existence of "veto players" and their influence in the decisionmaking process. Veto players, according to George Tsebelis, are "individual or collective actors whose agreement is necessary for a change of the status quo."[5]

Coalitions of democratic states are better able to maintain their wartime commitments because their domestic institutions include veto players whose decisionmaking processes can produce highly stable domestic preferences over time. Due to competing pref-erences and diverse interests, democratic states may find it more difficult to enter into international commitments. However, once these commitments are made and ratified through democratic institutional mechanisms, institutional constraints make them extremely difficult to reverse.[6]

---

4. In response to Desch's claims about Israel, I suggest that democratic states do not necessarily go to war either alone or with allies to save other democracies. Instead, they go to war to defend their values or national interests. The important point is that, once fully committed to fight, democratic states are better able to work with their allies when waging war. In this sense, U.S. behavior toward Israel that is motivated by American strategic interests does not contradict the triumphalists' expectation of democratic victory.
5. George Tsebelis, "Veto Players and Law Production in Parliamentary Democracies," *American Political Science Review*, Vol. 93, No. 3 (September 1999), p. 593.
6. Kurt Gaubatz, "Democratic States and Commitment in International Relations," *International Organization*, Vol. 50, No. 1 (Winter 1996), pp. 109–139.

In contrast, domestic institutions exert less influence over nondemocratic states, where a single leader or small group of leaders usually makes the major decisions. Nondemocratic states may therefore be able to respond more quickly to allies' requests than can democratic states, but for the same reason, their ability to maintain commitments may be less reliable or more unpredictable. As Mancur Olson has written, the promises of nondemocracies are "not enforceable by an independent judiciary or any other independent source of power. . . . Because of this and the obvious possibility that any dictator could, because of an insecure hold on power or the absence of an heir, take a short-term view, the promises of an autocrat are never completely credible."[7] This implies that even if a president, prime minister, or branch of government in a democratic state seeks to change the status quo in war, these efforts could be effectively blocked by other government entities. Given the role of veto players, the reliability and cohesion of democratic allies in warfare no longer relies on the audience costs logic. Rather decisions are the product of governmental checks and balances—a process embedded in democratic political institutions.

Another important element in winning a coalition war is the ability of member states to coordinate their efforts and resources. In contrast to Desch's claim that democratic coalitions are less effective because their bonds of friendship may induce an incentive to free ride, I argue that compared to nondemocracies with their closed political systems, democracies, with their open, transparent political systems, can more effectively promote wartime cooperation.

One way to achieve cooperation and reduce levels of uncertainty is to increase the quantity and quality of interstate communication. As Robert Keohane has written, states that seek cooperation with other states need "not merely information about the other government's resources and formal negotiating positions, but rather knowledge of their internal evaluations of the situation, their intentions, the intensity of their preferences, and their willingness to adhere to an agreement even in adverse future circumstances."[8] Moreover, increased communication facilitates a democratic coalition's ability to monitor its members and, if necessary, punish transgressors. From this perspective, democracies have a greater capacity than nondemocracies for communication and accessibility, and thus are more preferable partners.

---

7. Mancur Olson, "Dictatorship, Democracy, and Development," *American Political Science Review*, Vol. 87, No. 3 (September 1993), p. 571.
8. Robert O. Keohane, "The Demand of International Regimes," Stephen D. Krasner, ed., *International Regimes* (Ithaca, N.Y.: Cornell University Press, 1983), pp. 162–163.

Statistical Results of the Impacts of Democratic Allies on Victory

My research shows that the larger the number of democratic partners a state has, the more likely it is to win in war (controlling for such variables associated with war outcomes as a state's military capabilities, the capabilities of wartime allies, and which state initiates the war). According to the results of the marginal impact analysis presented in Table 1, the number of democratic partners variable increases the probability of winning a war by 62 percentage points as this variable moves from its minimum to maximum value and all other variables are set at their mean or modal values. The number of nondemocratic partners variable, on the other hand, decreases the probability of winning by 44 percentage points under the same conditions.[9]

Desch suggests that the democratic triumphalists misaggregate some wars, thus inappropriately favoring their own arguments. To avoid this, I conducted a set of sensitivity tests to determine whether my coding rules might affect the results of the statistical analysis. For these tests, I continued to use State War Performance Model 2 to determine how the coefficient on the number of democratic partners variable changes: Removal from the data set of all the U.S. and Israeli cases yields coefficients of ($b = 0.541$, $p = 0.000$) and of ($b = 0.57$, $p = 0.000$), respectively; I then eliminated Australia, Canada, and New Zealand ($b = 0.554$, $p = 0.000$); next, I removed all World War I ($b = 0.475$, $p = 0.000$) and World War II ($b = 0.591$, $p = 0.000$) cases; finally, I included only the periods after 1900 ($b = 0.851$, $p = 0.000$), 1918 ($b = 0.754$, $p = 0.000$), and 1945 ($b = 1.673$, $p = 0.003$) in the data set. As shown by the coefficients and p-values, the signs of the coefficients for the number of democratic partners variable are always positive and statistically significant across all the models, although the size of the coefficients changes slightly across the models.

To confirm these findings, I replicated Dan Reiter and Allan Stam's main model and ran their model including the number of democratic and nondemocratic partners variables.[10] According to the results of this test (presented in the right-hand column of Table 1), these two variables have a significant impact on war outcome: The number of democratic partners variable is positively related to the probability of winning ($b = 1.748$, $p = 0.005$), whereas the number of

---

9. These statistical findings were first generated in 2002 in Ajin Choi, "Democratic Synergy and Victory in War, 1816–1992," War Performance Model 2, Table 1, n. 15.
10. Reiter and Stam, "Democracy, War Initiation, and Victory," p. 385.

Table 1. Number of Coalition Partners and Its Impact on War Performance.

Variable	Choi's Model 2		Reiter and Stam's Model 5	
	Coefficient Standard Error	Marginal Impact	Coefficient Standard Error	Marginal Impact
Number of democratic partners	0.55*** (0.11)	62%	1.75** (0.62)	75%
Number of nondemocratic partners	−0.16** (0.06)	−44%	−0.34* (0.16)	−43%
Alliance contributions	1.64*** (0.48)	52%	5.65*** (1.26)	89%

NOTE: The table presents only the results of the three variables related to wartime alliances, not those of all variables in the two original models. These computations were performed using *Intercooled Stata*, version 7.0; and Michael Tomz, Jason Wittenberg, and Gary King, *CLARIFY: Software for Interpreting and Presenting Statistical Results,* Version 2.0 (Cambridge, Mass.: Harvard University Press, June 1, 2001), http://gking.harvard.edu/. All tests are two-tailed.
$*p < 0.05$
$** p < 0.01$
$*** p < 0.001$

nondemocratic partners variable continues to be negatively associated with this probability ($b = -0.338$, $p = 0.041$).

Further analysis of the statistical results reported in Table 1 yields several other interesting findings. First, according to realists, states form wartime coalitions to increase their capabilities; the importance of these capabilities to the war effort and the likelihood of victory have been a consistent theme in realist literature. In his article, Desch compares the capabilities of winning coalitions with those of their adversaries and shows that coalitions with overwhelming capabilities won their wars (pp. 10–16). Indeed, as Desch correctly points out, the difference in capabilities between wartime coalitions is an important predictor of war outcomes. As the models in Table 1 show, the probability of winning a war increases by 52 and 89 percentage points, respectively, as the alliance contribution variable moves from its minimum to its maximum value, all else being equal.

Second, this finding suggests that two coalitions with the same level of military capabilities could experience different levels of cooperation, and thus different outcomes, depending on their members' regime type. That is, in addition to capabilities, the level of cooperation among coalition members is an important determinant of victory in war; more specifically, the high level of cooperation among democratic allies significantly contributes to winning.

Third, the results in Table 1 speak to a point Desch raises regarding the effects of mixed coalitions. Desch contends that "there are cases of democracies winning wars as members of mixed alliances" (p. 10). In these cases, writes Desch, it is hard to conclude that the democratic state contributes to winning because "the nondemocracy accounted for the majority of the winning alliance's military strength" (pp. 11–12). He also states that nondemocratic members are more willing to pay their share of the war effort and that, in the case of World War II, the Soviet Union led the Allies to victory by defeating Germany on the eastern front (pp. 13–14).

If nondemocratic states were militarily stronger or if nondemocratic coalitions were less plagued with free riding, it could be that with each democratic partner added to a coalition, the probability of winning decreases, whereas with each nondemocratic partner added to a coalition, the probability of winning increases. In contrast to this expectation, however, the probability of winning increases with each democratic partner that joins a wartime coalition, whereas the probability of winning decreases with each nondemocratic partner that participates in a wartime coalition (see Table 1). In addition, if Desch's claim is right, it could be that the proportion of military victories by mixed coalitions is much lower than the proportion of victories by purely nondemocratic coalitions. According to my additional research, however, 78 percent of mixed coalitions won their wars from 1816 to 1992, whereas only 50 percent of nondemocratic coalitions won their wars.

Furthermore, given the importance of Lend Lease and other Allied aid programs during World War II, it is hard to conclude that the Soviet Union was the only major contributor to victory in 1945. As John Mueller points out, "While holding off one major enemy, it [the United States] concentrates with its allies on defeating another enemy . . . it supplied everybody. . . . If anyone was in a position to appreciate this, it was the Soviets. By various routes the United States supplied the Soviet Union with, among other things, 409,526 trucks; 12,161 combat vehicles (more than Germans had in 1939) . . . and over one-half pound of food for every Soviet soldier for every day of the war."[11] Soviet wartime aid to its Western allies, however, totaled only $2 million.[12]

---

11. John Mueller, "The Essential Irrelevance of Nuclear Weapons: Stability in the Postwar World," *International Security*, Vol. 13, No. 2 (Fall 1988), pp. 60–61.
12. Hubert Van Tuyll, *Feeding the Bear: American Aid to the Soviet Union, 1941–1945* (New York: Greenwood, 1989), p. 182.

Both Britain and the United States complained of this lack of Soviet reciprocity. On the other hand, reciprocity between the British and the Americans was significant. The United States devoted 4 percent of its domestic output to Britain through Lend Lease, while Britain provided 3 percent of its output to the United States.[13]

In sum, my findings of a strong, positive correlation between the number of democratic coalition partners and the likelihood of victory in war, in addition to a strong, negative correlation between the number of nondemocratic coalition partners and the expectation of victory, challenge Desch's conclusions about war contributions in mixed coalitions. From the statistical results presented in Table 1, it is possible to infer that coalitions with a larger number of democracies are likely to experience a synergistic effect, whereas coalitions with a larger number of nondemocracies are likely to suffer from collective action problems. The historical record confirms these findings.[14]

Historical Evidence

Given the theoretical expectation and statistical finding that democratic states are more effective partners in warfare, I conducted in-depth process tracing by examining how three variables—democracy, wartime cooperation, and war performance—interact and how they generate certain outcomes in specific historical contexts. First, I compared the level of commitment that Britain and Austria-Hungary accorded to their respective allies during World War I, and illustrated how veto players account for war performance. Second, I compared the levels of cooperation for joint air force operations among the Allies during World War II, and demonstrated how transparency and openness in democratic political systems allowed them to collaborate more closely with their coalition partners.

BRITAIN VERSUS AUSTRIA-HUNGARY IN WORLD WAR I

As World War I moved into its third year, states on both sides began to consider ways to break the stalemate through entering into separate peace negoti-

13. Richard Overy et al., "Co-operation: Trade, Aid, and Technology," in David Reynolds, Warren F. Kimball, and A.O. Chubarian, eds., *Allies at War* (New York: St Martin's, 1994), pp. 209–211.
14. Additional statistical evidence supports this finding about the number of democratic partners: First, controlling for other factors such as war duration, expected war outcome, and leadership change, democratic states are less likely than nondemocratic states to abandon their partners during war. Second, controlling for other factors such as threat, coalition size, and hegemony, democracies are better able than their nondemocratic counterparts to coordinate their war efforts and related resources. See Choi, "Cooperation for Victory."

ations. If successful, such negotiations would have seriously harmed the war's other participants, however. Believing that Germany could avoid defeat if it could convince Britain to take itself out of the war, Berlin proposed separate peace negotiations with London on September 18, 1917.

Britain's prime minister, David Lloyd George, expressed interest in the proposal despite the likely costs to its allies, arguing that "if Russia went out of the war . . . [Britain] could see no hope of the sort of victory in the war we desired. In these circumstances, it might be necessary to make a bargain with Germany."[15] Given the stalemate on the western front and the likelihood of continuing Russian withdrawal on the eastern front, he thought that Britain should consider accepting Germany's peace terms.[16]

Upon bringing the proposal to the War Cabinet, Lloyd George encountered stiff opposition. Cabinet members warned that, in accordance with the 1914 Pact of London, no negotiations should be undertaken before consulting with Britain's allies. In addition, they worried that a separate peace would leave Germany strong, allowing it to threaten Britain later on.[17] Following heated debate in the War Cabinet, the British government decided to inform its allies of Germany's proposal and to renew its commitment to remain in the war until the Germans were defeated. This decision was not made based on the will of Lloyd George or the exclusive leadership of any War Cabinet member, but rather on the bargaining that occurred between them. To ensure the survival of his government, Lloyd George had to consider the positions of the War Cabinet's members. He needed the cooperation of the Conservative Party, which at that time constituted the majority in the House of Commons.[18]

In contrast, Emperor Karl could initiate a separate peace despite some strong opposition in the Austro-Hungarian government because of the absence of effective veto players. Austria-Hungary started the war against Serbia on July 28, 1914, with German assurances of diplomatic and military support. It did not, however, anticipate that the conflict would evolve into world war or that it would last so long. When Emperor Karl ascended to the throne in November

---

15. War Cabinet minutes, May 9, 1917, CAB23/13, in David Woodward, "David Lloyd George: A Negotiated Peace with Germany and the Kuhlmann Peace Kite of September 1917," *Canadian Journal of History*, Vol. 6, No. 1 (1971), p. 80.
16. David Lloyd George, *War Memoirs of David Lloyd George*, Vol. 4 (London: Ivor Nicholson and Watson, 1934), pp. 2093–2097.
17. Woodward, "David Lloyd George," pp. 79, 88–89.
18. John Turner, *British Politics and Great War: Coalition and Conflict, 1915–1918* (New Haven, Conn.: Yale University Press, 1992), pp. 194–226; and David French, *The Strategy of the Lloyd George Coalition, 1916–1918* (Oxford: Clarendon, 1995), pp. 13–17.

1916, he began to think about separate peace negotiations, believing that the conclusion of a separate peace at the earliest possible moment—even at the expense of alliance with Germany—would preserve not only Austria-Hungary's monarchy but also its status as a great power.[19]

The majority of members in the government, however, insisted that pulling out of the alliance would result in immediate invasion by Germany and revolt in the German-populated areas of Austria-Hungary, which in turn could lead to the disintegration of the monarchy itself. Instead of agreeing to separate peace negotiations, the majority held firmly that Austria-Hungary should hold out for a comprehensive peace within the framework of continued alliance with Germany. In so doing, Austria-Hungary would strengthen its bargaining position vis-à-vis Germany in any future peace negotiations.[20] In the end, the veto players' constraints on their leader were not effective; members of the Austro-Hungarian government failed to prevent the emperor from pursuing separate peace negotiations.

ALLIED COOPERATION IN JOINT AIR FORCE OPERATIONS IN WORLD WAR II

During World War II, Europe's theaters of operation were highly dispersed. Thus one of the most effective forms of military collaboration among the United States, Britain, and the Soviet Union involved air force operations. Negotiations over the deployment of air power were a crucial element of U.S., British, and Soviet cooperation. Although the Allies attached tremendous value to the importance of coordinating their air force operations, they sometimes could not reach political consensus. As a result, they failed to implement some joint operations at critical stages of the war.

Operation VELVET, a joint Allied air force effort proposed in 1942 to destroy Germany's oil route in the eastern front, failed to materialize because of Josef Stalin's last-minute rejection. Stalin's decision was based more on political than military considerations. He worried that U.S. and British aerial domination would tarnish Soviet prestige on the eastern front. He was also concerned that if an air base were established on Russian soil, there would be great pressure for military and political interaction between Russian personnel and their Western counterparts. In Stalin's view, such a situation would have been dangerous because it could result in "political contamination" and thus

---

19. Gary W. Shanafelt, *The Secret Enemy: Austria-Hungary and the German Alliance, 1914–1918* (New York: Columbia University Press, 1985), pp. 104–115.
20. Ibid., pp. 112–116, 137–148.

threaten domestic political stability.[21] Evidence of this fear was reported by Britain's delegate to Moscow for the VELVET project, Sir Air Marshal R. Drummond, who indicated that the primary reason for the failure of VELVET was the Soviets' "almost fanatical urge to keep foreigners out of Caucasia"— not to mention Stalin's message to his Western allies that "I should be most grateful if you would expedite the dispatch of aircraft, especially fighters, but without crews."[22]

This apparent Soviet fear over contact with British and American personnel stands in stark contrast, for example, to the U.S. use of military bases in Britain. Even before the United States and Britain had decided on a plan to invade Europe, the U.S. buildup in Britain (code-named BOLERO) had begun. The buildup was based on a U.S.-British agreement at the 1942 Arcadia Conference to pursue a "Germany first strategy." As a result of BOLERO, about 3 million U.S. troops poured into Britain between 1942 and 1945—the equivalent of 7 percent of the combined prewar population of England, Wales, and Northern Ireland.[23] Using these air bases, Britain and the United States were able to engage in a successful joint offensive air-bombing operation against Germany.[24]

Conclusion

Michael Desch has raised one of the most serious challenges to the democratic victory thesis. Arguing that the thesis rests on shaky theoretical and empirical grounds, he concludes that state war performance can be fully explained without consideration of regime type as a variable. Although I disagree with Desch's main claims on this matter, his challenge makes an important contribution to the democratic triumphalists' research program: It identifies the theoretical and empirical weaknesses of their explanations and, in so doing, creates an opportunity to develop new, more persuasive explanations.

---

21. Adam B. Ulam, *Expansion and Coexistence: Soviet Foreign Policy, 1917–1973* (New York: Praeger, 1974), p. 319.
22. "Note on Moscow Conference, November 27, 1942," in Richard Lukas, *Eagles East: The Army Air Force and the Soviet Union, 1941–1945* (Tallahassee: Florida State University Press, 1970), p. 163; and Ministry of Foreign Affairs of the Soviet Union, *Stalin's Correspondence with Churchill, Attlee, Roosevelt, and Truman, 1941–1945*, Vol. 2 (London: Lawrence and Wishart, 1958), p. 45.
23. David Reynolds, *Rich Relations: The American Occupation of Britain, 1942–1945* (London: HarperCollins, 1995), pp. 89–126.
24. For the development of this operation, see Wesley Craven and James Cate, eds., *The Army Air Forces in World War II*, Vols. 1–2 (Chicago: University of Chicago Press, 1949, 1951); and Maurice Matloff and Edwin Snell, *Strategic Planning for Coalition Warfare, 1943–1944* (Washington, D.C.: Department of the Army, 1959).

In this article, I have provided new theoretical logic and empirical evidence of cooperation among democracies in war. I agree with Desch that the audience costs argument does not sufficiently explain the link between allies, democracy, and victory. Nonetheless, I contend that democracy, wartime cooperation, and war performance are connected. This connection is to be found in the very features of democratic political institutions. Due to the presence of effective veto players and to the transparency and openness of their political systems, democratic states are better able to cooperate in wartime, which in turn results in their winning more wars. This finding is supported by both statistical and historical evidence.

Scholars now know that democracy matters significantly in the international arena, and that it is a strong predictor of successful war performance. A recent study reports that despite their military competence and triumphs, democratic states are more likely to be the target of military challenges.[25] After the end of the Cold War, the international environment has not been largely favorable or peaceful. In fact, the United States has confronted a variety of serious military challenges, and it is engaged in a long and difficult war against terror. Under these conditions, the lesson of democracies' military victories over the last 200 years should not be forgotten: The strength of democracy is to be found in the power of democratic cooperation.

---

25. See Christopher F. Gelpi and Joseph M. Grieco, "Attracting Trouble: Democracy, Leadership Tenure, and Targeting of Military Challenges, 1918–1998," *Journal of Conflict Resolution*, Vol. 45, No. 6 (December 2001), pp. 794–815.

Fair Fights?

Evaluating Theories of Democracy and Victory

David A. Lake

In "Democracy and Victory: Why Regime Type Hardly Matters," Michael Desch criticizes the methods and results of several studies, mine included, that find that democracies tend to win the wars they fight.[1] After raising a number of empirical and research design issues, Desch concludes that "on balance, democracies share no particular advantages or disadvantages in selecting and waging wars. In other words, regime type hardly matters for explaining who wins and loses wars" (p. 8).

Desch does the discipline a service by challenging extant findings—skepticism is, after all, the most important trait of a social scientist. A careful review of theory and method, however, confirms the finding that democracies tend to be victorious in war. In his article, Desch separates research design from theory and thus does not provide the fair test that he claims. Scholars cannot evaluate empirical relationships outside of their theoretical context. Similarly, the concept of causality cannot be understood apart from a prior theory. Correlation may or may not exist, but causation can only be inferred. Even as the historical record highlights the distinctive nature of democracies, researchers conclude that democracy causes (at least in part) victory in war only because theory implies that it should.

The literature on the democratic peace in general, and the theory and findings on democracy and victory in particular, have contributed to a new generation of research on war as a process. With fresh attention to how war outcomes affect strategic bargaining before and during crises, scholars are moving in the direction of more synthetic and productive theories of conflict that show how attributes of states, such as democracy, interact with their choices to explain war and peace. This is one of the most promising avenues of research in contemporary security studies—and one that reinforces the need to bind empirical

David A. Lake is Professor and Chair of Political Science at the University of California, San Diego.

I am grateful to Neal Beck, Ajin Choi, Kristian Gleditsch, Dan Reiter, and Allan Stam for helpful comments and discussions.

1. Michael C. Desch, "Democracy and Victory: Why Regime Type Hardly Matters," *International Security*, Vol. 27, No. 2 (Fall 2002), pp. 5–47. All page cites in the text refer to this article.

International Security, Vol. 28, No. 1 (Summer 2003), pp. 154–167
© 2003 by the President and Fellows of Harvard College and the Massachusetts Institute of Technology.

research to theory.[2] It would be unfortunate if ill-conceived critiques of the democracy and victory literature were to abort this fruitful line of inquiry.

In this article, I review the theory that originally gave rise to the hypothesis that democracies will tend to win the wars they fight. I then discuss how Desch errs in divorcing his empirical tests from this theory, and the implications of this research strategy for his conclusions. Following this, I survey further tests of the causal mechanism propelling democracies to victory. The conclusion outlines directions for future research.

Powerful Pacifists Revisited

Desch characterizes those of us researching the relationship between regime type and victory in war as "democratic triumphalists." Contrary to the suggestion implied in this label, demonstrating the superiority of democracy was not part of my original intent in writing "Powerful Pacifists: Democratic States and War."[3] Instead, in the late 1980s I began work on a theory of state rent-seeking, with a particular emphasis on how different political regimes influence grand strategy. The key idea behind this theory is that the state is a local monopolist in producing public services and, as such, will seek to extract rents or "excess profits" from its citizens through higher than necessary taxes, bribes, or nonpecuniary transfers. Democracy, in turn, is a primary means through which society constrains the state's rent-seeking abilities. Because democratic leaders can be removed from office at less cost to citizens than autocratic leaders, they are more responsive to public opinion and less able to extract monopoly rents for themselves or their supporters. As democratic states receive smaller returns from each additional unit of territory they control, it follows that democracies possess a smaller optimal size. Conversely, because autocrats earn greater monopoly rents and receive larger returns from each unit of territory, they possess a larger optimal size and an imperialist bias in their grand strategies.

It was immediately apparent that this theory contained within it an explanation of the democratic peace, just then coming into prominence.[4] In the mutual

---

2. For a review of this budding literature, see Dan Reiter, "Exploring the Bargaining Model of War," *Perspectives on Politics*, Vol. 1, No. 1 (March 2003), pp. 27–43.
3. David A. Lake, "Powerful Pacifists: Democratic States and War," *American Political Science Review*, Vol. 86, No. 1 (March 1992), pp. 24–37.
4. Although "discovered" some years before, the relationship between democracy and war began to attract scholarly attention with Michael W. Doyle, "Liberalism and World Politics," *American Political Science Review*, Vol. 80, No. 4 (December 1986), pp. 1151–1169.

absence of this imperialist bias, two democracies would have fewer reasons to come into conflict and thus would be less likely to fight each other. As the democratic peace had already attained the status of something close to an empirical law,[5] and was therefore known to me prior to the formulation of the theory, this prediction alone could not serve as a test or distinguish my explanation from others. On further inspection, however, the theory also implied that democracies would be more likely to win the wars they did fight. According to the theory, the lower level of state rent-seeking in democracies increases the quantity of public services provided by the state, thereby providing greater benefits to citizens at lower cost and promoting loyalty. Conversely, autocracies exploit their monopoly power and, like any monopolist, restrict the supply of public services to drive up their returns and, in turn, rents. Lower levels of state rent-seeking in democracies also stimulate economic growth and higher incomes, allowing them to mobilize resources for war at lower opportunity costs. And because defeat would necessarily imply the exploitation of their citizens, democracies fight harder and, by balancing threats, come to one another's aid when challenged by potentially hegemonic autocracies.[6] This fear of exploitation produces overwhelming countercoalitions.[7] These factors combine to predict that democracies will tend to emerge victorious in the wars they do fight. As Desch correctly notes, this is a "democratic effectiveness" argument. It does not rest on democracies selecting only wars they can win, but on their fighting harder and better in those wars they engage.

The proposition that democracies tend to win the wars they fight was, at the time, a novel hypothesis. No other theory generated this prediction, although other theories have subsequently been found to contain this implication as well.[8] The historical record, in turn, confirmed this expectation. The hypothesis

---

5. Jack S. Levy, "The Causes of War: A Review of Theories and Evidence," in Philip E. Tetlock, Jo L. Husbands, Robert Jervis, Paul C. Stern, and Charles Tilly, eds., *Behavior, Society, and Nuclear War*, Vol. 1 (New York: Oxford University Press, 1989).

6. On balance of threat theory, see Stephen M. Walt, *The Origins of Alliances* (Ithaca, N.Y.: Cornell University Press, 1987).

7. Contrary to Desch's interpretation, I do not argue that democratic alliances are necessarily more effective, only that they are larger (p. 28). Similarly, nothing in my argument precludes "mixed alliances" from forming. Also note that this prediction rests not on a commitment problem, and thus audience costs (contrary to Desch, p. 30), but on a mutuality of interests against exploitative autocratic states. On democratic alliances and victory, see Ajin Choi, "The Power of Democratic Cooperation," *International Security*, Vol. 28, No. 1 (Summer 2003), pp. 142–153.

8. See Bruce Bueno de Mesquita, James D. Morrow, Randolph Siverson, and Alastair Smith, "An Institutional Explanation of the Democratic Peace," *American Political Science Review*, Vol. 93, No. 4 (December 1999), pp. 791–807; and Kenneth A. Schultz and Barry R. Weingast, "The Democratic

survived a series of demanding statistical tests, including the use of control variables for plausible rival hypotheses and the exclusion of "outliers," ambiguous cases, and so on. The democratic advantage in war also survived even more demanding and rigorous tests, with more extensive control variables, primarily conducted by Dan Reiter and Allan Stam.[9] Many scholars soon accepted this novel "fact" as one of the key traits of the democratic difference in international relations. This is how social science is supposed to work: New theories produce novel implications that are then tested against systematic evidence.

Desch on Democracy and Victory

Desch offers four substantive criticisms of the data used in the various studies of democracy and victory. As a baseline, he begins with a total of 75 wars since 1815 as coded in the Correlates of War (COW) data set, 24 of which he excludes because they involved equally democratic countries, contained missing data, ended in a draw, or were ongoing. This leaves 51 "candidate wars" for examination.

In his first criticism, Desch argues that the COW data set is misaggregated because it treats single wars (e.g., World War II) as conflicts that are more accurately described as a series of smaller wars (e.g., the Battle of France [1940], the European War [1941–45], and the Pacific War [1941–45]). Desch advocates separating these misaggregations and coding victory and loss for each. Two of the 51 candidate wars are misaggregations, according to Desch (see his appendix, pp. 45–47).

Desch then charges that the so-called democratic triumphalists count wars as democratic victories when, in fact, it was the nondemocratic members of mixed alliances that contributed most to these outcomes (e.g., the Soviet Union in the European War, pp. 13–14). Six victories, in his view, are thus incorrectly attributed to democracies.

Third, "in some cases a democracy was much more powerful than its adversary and used that advantage to overwhelm its rival. . . . Such gross mismatches," Desch argues, "should be considered only if the triumphalists can

---

Advantage: Institutional Foundations of Financial Power in International Competition," *International Organization*, Vol. 57, No. 1 (January 2003), pp. 3–42.
9. Dan Reiter and Allan C. Stam, *Democracies at War* (Princeton, N.J.: Princeton University Press, 2002).

prove that regime type caused the imbalance of power" (p. 12). Twenty-four of the 51 candidate wars, nearly half, are coded as gross mismatches.

Finally, according to Desch, "there are cases in which the belligerents' interests in the outcome of the conflict are so asymmetrical that it is impossible to ascribe the outcome to regime type and not to the balance of interests" (p. 13). These asymmetric interests are found in 4 of the candidate wars.[10]

Restricting the analysis to only wars involving states that were clearly democratic and deleting all cases of mixed alliances, gross mismatches, or asymmetric interests leaves 9 of the 51 candidate wars as what Desch calls "fair fights," of which democracies won 6 (p. 15). In statistical testing, Desch examines all participants in these 9 wars, as is standard for these tests, and disaggregates World War I into a German-Belgian war and another involving Germany, Austria, Russia, and Turkey. He also disaggregates World War II into a German-French war (1939), a German-Belgian war (1940), and a German-Dutch war (1940).[11] Thus, Desch creates a data set of 34 observations, compared with 197 in the Reiter and Stam analysis and 121 in mine. With this truncated set of observations, and such a small n, it is not surprising to find that democracy is no longer significantly related to victory in war (p. 17, Table 3). The nearly identical coefficients on the bivariate logits that Desch performs on Reiter and Stam's 197 observations (model 1) and his own set of 34 observations (model 2), suggest that the lack of significance in the latter is the result of the smaller n rather than some fundamental difference in the effect of democracy.

Desch aims to show not only that democracies are no more or less likely to win than other regimes but also that statistical tests may not be appropriate given the small number of fair fights. But Desch's research design is flawed on both theoretical and empirical grounds, negating both his substantive and methodological claims. Theoretically, deleting wars with gross mismatches and asymmetric interests is unfounded.[12] These are precisely the characteris-

---

10. The text (p. 14), but not the appendix (p. 47), describes the 1948 and 1967 Arab-Israeli wars as asymmetric conflicts.

11. The German-Austrian-Russian-Turkish war includes no democratic countries. France in 1939 is miscoded as a 0 (anocracy) when the POLITY IV data set codes it as a 10 (democracy). Dropping this war and correcting the democracy score for France does not significantly change the results Desch reports in Table 3. The Football War is missing data, and is dropped from Desch's analysis, effectively making the data set one of 8 wars not 9 as reported in the text (p. 15). Why Desch excludes the German-British war or dyad from the so-called Battle of France (discussed on p. 13 of his article), is not clear.

12. How to handle the cases of mixed alliances is less clear, but because nothing in either a democratic effectiveness or selection effects argument precludes mixed alliances from forming, there is no reason to drop them from the analysis. Moreover, including mixed alliances, and the corre-

tics that theory predicts are likely to exist in wars fought by democracies. Democratic effectiveness arguments, such as mine, predict that democracies will possess greater resources and will tend to form overwhelming counter-coalitions, producing the kind of gross mismatches that Desch wants to exclude. In addition, because democracies recognize that, if defeated, they will be exploited by autocratic victors, they are likely to enter only those wars in which they possess asymmetric interests. Selection effects theories make similar predictions: Democracies are likely to enter those wars they can win (gross mismatches) or those they are exceptionally motivated to fight (asymmetric interests). Desch's research design, thus, systematically excludes all of the wars that fit the conditions of a democratic effectiveness or selection effects theory. That no significant relationship is then found between democracy and victory is hardly unexpected—regardless of the number of observations.

Empirically, as Desch acknowledges, the appropriate test for competing explanations is to include control variables in a multivariate analysis (p. 16).[13] By excluding cases that may have an alternative cause, however, Desch de facto assumes that the rival hypotheses are deterministic but that the relationship between democracy and victory is probabilistic. This has important implications for his research design and the conclusions that follow.

Deterministic hypotheses take the traditional "if-then" form. If combatants are grossly mismatched, for instance, a deterministic hypothesis states that the more powerful country will win. A power disparity, in other words, is at least a sufficient if not also a necessary condition for victory. Probabilistic hypotheses take a "more (or less) likely" form. If states are democratic, a probabilistic hypothesis predicts that they are more likely to win. In this form, democracy is neither necessary nor sufficient for victory, but its presence increases the prob-

---

sponding autocracies in the victorious coalition, makes it less likely to find statistically significant results favoring democracies in war. Thus, if mixed alliances bias the results, it is away from rather than in support of the hypothesis.

13. Desch confuses the problem of nonindependent observations with an omitted variables problem (pp. 15–17); the latter can sometimes be addressed appropriately through a fixed effects specification, but the former cannot. Although he correctly points to a potential problem of nonindependent observations, an appropriate statistical correction is to calculate robust standard errors (a technique not available when I published my article in 1992; rerunning my models with robust standard errors does not weaken the significance of democracy). See Nathaniel Beck, Jonathan N. Katz, and Richard Tucker, "Taking Time Seriously: Time-Series-Cross-Section Analysis with a Binary Dependent Variable," *American Journal of Political Science*, Vol. 42, No. 4 (October 1998), pp. 1260–1288. Whatever dependencies that may continue to exist are exacerbated by Desch's decision to restrict the set of wars (thereby limiting variation) and disaggregate them into phases (which introduces additional dependencies).

ability of success. It is by no means clear that the theory underlying gross mismatches, asymmetric interests, and the other factors examined by Desch is in fact deterministic. Although it is not explicit in his article, I infer that he is drawing on some variant of realism. Does realism actually imply that a 2:1 power disparity must produce victory for the more powerful state regardless of any other conditions? If the usual "all else held constant" clause is necessary for the hypothesis to hold, then the hypothesis is probabilistic, not deterministic. By excluding from analysis cases of asymmetric interests, for example, Desch asserts that such asymmetries must always and everywhere lead to victory. This is dubious, even for realism. Important theoretical claims are being made here under the guise of methodology.

Even if the alternative causes identified by Desch are deterministic, testing a deterministic relationship by excluding cases from analysis is appropriate only when there are no measurement errors in either the independent or dependent variable, no random or nonsystematic errors in the environment, and no alternative causes.[14] These conditions are almost always violated, as they are in the case of victory in war. For this reason, even most deterministic hypotheses are tested as if the underlying theory implies a probabilistic relationship. Adopting a probabilistic design allows the analyst to better assess the uncertainty in his or her estimates introduced by measurement and random error or by confounding causal variables. Simply put, independent of the nature of the theory being tested and even ignoring the random error inherent in any social system, Desch's research design is invalid. Pointing to possible measurement errors in the data used by others, he cannot be confident that his own data, drawn from largely the same sources, is error free. And his strategy of excluding cases is premised on the existence of alternative explanations, which therefore requires a probabilistic, not a deterministic, design.

Importantly, unless the theory is deterministic and there are no sources of uncertainty in the estimates of the relationship between the relevant variables, a probabilistic design is preferred regardless of the number of observations. Indeed, Stanley Lieberson concludes that small-*n* research designs always "have difficulty in evaluating probabilistic theories."[15] Desch's preferred case study method is no solution. Because of the competing explanations and likely mea-

---

14. Stanley Lieberson, "Small N's and Big Conclusions: An Examination of the Reasoning in Comparative Studies Based on a Small Number of Cases," *Social Forces,* Vol. 70, No. 2 (December 1991), pp. 309–310.
15. Ibid., p. 310.

surement errors in victory, democracy, gross mismatches, and other variables of interest, a probabilistic, large-*n* research design remains the most appropriate test.

In short, although Desch raises important issues about the correct research design, his alternative of excluding cases from analysis introduces more problems than it solves.[16] By systematically discarding the very cases that the theory predicts, he selects as his observations for analysis those least likely to support the theory. Moreover, he imposes his judgment about the true causal impact of competing explanations on the data rather than testing his beliefs against the evidence. There is no doubt that further work needs to be done to probe, sharpen, and refine the effect of democracy on victory in war. But scholars cannot accept Desch's central claim that regime type does not matter.

Causal Mechanisms

Desch further asserts that alternative explanations must be "ruled out" to establish causation (p. 18). In what I regard as the more standard view, causation is an analytic concept that is prior to empirical investigation.[17] Causation is not what is left over after other explanations have been eliminated, nor does the absence of alternative explanations necessarily transform a correlation into causation. Rather, establishing "cause" requires specifying a relationship between an explanatory variable—in this debate democracy—and an outcome variable—victory in war—in which the relationship between the variables is derived from some prior theory. Because the fundamental problem of causal inference cannot be solved, all conclusions about causality are necessarily tentative.[18] Scholars conclude that one variable causes another only when theory predicts a directed relationship. They develop more confidence in their theories and causal arguments, in turn, not by dismissing alternatives but by deriv-

---

16. Desch also suggests that the relationship between democracy and victory may be spurious, depending instead on whether or not the regime is consolidated (p. 19). Desch converts the mean democracy score for victors in my study from 5.60, based on the 11-point POLITY democracy variable, to 0.59, based on POLITY's 21-point democracy-autocracy scale. The original mean in my results is just below the cutoff of 6 that I used as a threshold for democracy. Lake, "Powerful Pacifists," p. 31. Although Desch correctly notes that this average is pulled down by the autocratic victors (typically in mixed alliances with democracies), he errs in concluding that this is evidence of regimes in transition. Most of the cases in my set of wars are clustered toward the bottom and top of the democracy scale and, on further inspection, appear to be stable.
17. Gary King, Robert O. Keohane, and Sidney Verba, *Designing Social Inquiry: Scientific Inference in Qualitative Research* (Princeton, N.J.: Princeton University Press, 1994), p. 76.
18. Ibid., p. 79.

ing hypotheses and subjecting them to ever more demanding and unbiased tests. Given that many phenomena are overdetermined, or are consistent with more than one theory, scholars also gain confidence in their theories by deducing novel implications supported by empirical analysis.

As explained above, my original hypothesis on democracy and victory was composed of several steps linking state rent-seeking (more prevalent in autocracies) to lower levels of public services, lower rates of economic growth, and an imperialist bias. This is, I admit, a long and complex causal chain. Since publishing the article, I have, in association with several collaborators, subjected each of these intermediate links to further analysis. This is a form of process tracing, as often advocated by proponents of case studies, or multiplying implications and observations.[19] In either view, examining the individual links in the chain is a means of bolstering confidence in the theory. For each of the three primary links in my argument, empirical tests confirm expectations.

First, as noted above, I predict that democracies, more constrained from extracting rents from society, will provide greater levels of public services than nondemocracies. This is a direct test of the state rent-seeking hypothesis.[20] Matthew Baum and I examine this hypothesis in a set of cross-sectional models, covering seventeen different indicators of public education and health (over multiple years), and five time-series cross-sectional (TSCS) tests.[21] Using a full battery of control variables, we find a strong and robust positive relationship between democracy and the level of public services. Moreover, the effects of democracy are substantively important. Our TSCS results, to pick one example, suggest that a maximum increase in democracy decreases the rate of infant mortality by nearly 5 children per 1,000 live births.

Conversely, Desch argues that societal rent-seeking may be more common in democracies than nondemocracies, but he offers nothing more than casual speculation to support his case. Relying on Mancur Olson's *The Rise and Decline of Nations,* he largely ignores Olson's later article, "Dictatorship, Democracy, and Development," and book, *Power and Prosperity,* which produce

---

19. See ibid., pp. 85–87, 224–228.
20. Contrary to Desch's claim, the extent of the government's intervention in the economy says nothing about the magnitude of the rents extracted by the state (pp. 27–28): A large role for the government is equally consistent with high rent extraction or a high societal demand for public goods and social equality; and because rents can be accrued in nonmonetary forms, even governments with a limited role in the economy can earn large rents.
21. David A. Lake and Matthew A. Baum, "The Invisible Hand of Democracy: Political Control and the Provision of Public Services," *Comparative Political Studies,* Vol. 34, No. 6 (August 2001), pp. 587–621.

hypotheses similar to mine.[22] Claiming that "economists offer compelling arguments for why it is more likely that interest groups will be successful rent seekers in a democracy" (p. 26), Desch does not acknowledge that other economists make equally compelling arguments to the opposite effect.[23] As in my original article, I remain agnostic on the relative extent of societal rent-seeking across regimes: In democracies, many groups are likely to succeed in using state power to extract relatively small rents; in autocracies fewer groups are more likely to use state power to reap larger gains.[24] Analytically, the net effect is unclear. Most important, however, the first link in the causal chain from state rent-seeking and democracy to victory is directly tested by the public services data and clearly supported by the results.

Second, I predict that, in the relative absence of state rent-seeking, democracies will have higher rates of economic growth and levels of income. In a second article, Baum and I test this further implication of the theory and find a strong, positive but indirect effect of democracy on growth through the creation of human capital.[25] Although Desch correctly points to the conflicting evidence on this score, Baum and I argue that existing models of democracy and growth are misspecified. Most analysts begin with the neoclassical growth model—the standard in the literature, typically including only initial technology and inputs of physical and human capital and labor—and simply add democracy as one more variable. Conversely, we hypothesize that, as above, democracy constrains state rent-seeking, improves public health and education, and thereby fosters the creation of human capital and, indirectly, growth. In other words, there may or may not be a direct effect of democracy on growth, as posited by others (Baum and I find none), but we expect there will be a strong indirect effect through the creation of human capital. Using recursive regression to capture this indirect effect, and a panel of 128 countries over

---

22. Mancur Olson, *The Rise and Decline of Nations: Economic Growth, Stagflation, and Social Rigidities* (New Haven, Conn.: Yale University Press, 1982); Mancur Olson, "Dictatorship, Democracy, and Development," *American Political Science Review,* Vol. 87, No. 3 (September 1993), pp. 567–576; and Mancur Olson, *Power and Prosperity: Outgrowing Communist and Capitalist Dictatorships* (New York: Basic Books, 2000).
23. See especially Robert B. Ekelund and Robert D. Tollison, *Mercantilism as a Rent-Seeking Society: Economic Regulation in Historical Perspective* (College Station: Texas A&M University Press, 1981). For a political science contribution on this point, see Mark Brawley, "Regime Types, Markets, and War: The Impact of Pervasive Rents in Foreign Policy," *Comparative Political Studies,* Vol. 26, No. 2 (July 1993), pp. 178–197.
24. See Lake, "Powerful Pacifists," p. 34, n. 15.
25. Matthew A. Baum and David A. Lake, "The Political Economy of Growth: Democracy and Human Capital," *American Journal of Political Science,* Vol. 47, No. 2 (April 2003), pp. 333–347.

thirty years, we confirm this hypothesis as well. Specifically, we find that a maximum increase in democracy indirectly accelerates the rate of economic growth by 0.68 percentage points through increased life expectancy in countries with a gross domestic product per capita of less than $2,500 and by 0.26 percentage points through increased education in countries with a GDP per capita of more than $2,500. Thus, if the rate of economic growth were previously 2 percent per year, a maximum increase in democracy would raise that rate to 2.68 and 2.26 percent per year in relatively "poor" and "rich" countries, respectively. Again, this second important link in the causal chain between democracy and victory is supported by additional, theoretically appropriate tests.

Desch is correct, however, that countries with higher per capita income are more likely to be democratic, suggesting that the relationship between democracy and victory may be spurious.[26] Because economic growth rates themselves do not appear to be related to income levels or regime transitions in the short run, endogeneity does not appear to be a debilitating problem. But clearly the long-run relationship between income, democracy, and victory in war is worthy of further investigation.

Desch also points to the problem of mobilizing wealth for military purposes. The empirical evidence here is more mixed than Desch claims (p. 27).[27] But more important, although I predict that democracies have greater extractive capacity, it is not clear what the theory predicts about relative extraction rates.[28] If democracies enjoy greater wealth, and lower opportunity costs for extraction, should they mobilize resources at a higher rate, thereby greatly outspending their rivals and increasing their probability of victory? Or, as democracies tend to fight harder and better on the battlefield and form overwhelming countercoalitions, should they capitalize on these natural advantages in warfare and conserve resources for other valued uses? How states

---

26. See Ross E. Burkhart and Michael S. Lewis-Beck, "Comparative Democracy: The Economic Development Thesis," *American Political Science Review*, Vol. 88, No. 4 (December 1994), pp. 903–910; and John B. Londregan and Keith T. Poole, "Does High Income Promote Democracy?" *World Politics*, Vol. 49, No. 1 (October 1996), pp. 1–30.

27. In a detailed study of state extraction, Alan C. Lamborn finds that relatively more democratic Britain and France faced less resistance and were better able to mobilize resources than was relatively less democratic Germany in the lead-up to World War I. Lamborn, *The Price of Power: Risk and Foreign Policy in Britain, France, and Germany* (Boston: Unwin Hyman, 1991), especially chaps. 6–8. For a general discussion of state extraction under changing international circumstances, see Michael Mastanduno, David A. Lake, and G. John Ikenberry, "Toward a Realist Theory of State Action," *International Studies Quarterly*, Vol. 33, No. 4 (December 1989), pp. 457–474.

28. Lake, "Powerful Pacifists," p. 30.

optimize across multiple goals likely depends on the specifics of the conflict, the preferences of the citizens in each state, and the size of the coalition that is formed. This is an area where further theorizing is critical. Although I provocatively billed democracies as "powerful pacifists" in my article, the meaning of power in this context is far from clear. But because theory is still ambiguous on this point, I am hesitant to read too much into the existing findings on observed extraction rates—regardless of which way they may point.

Third, I also predict that democracies will tend to possess smaller territories. This is a relatively direct test of the hypothesis that autocracies will have an imperialist bias in their grand strategies. In another paper, Michael Hiscox and I build a model of state size, premised on the same conception of state rent-seeking, and test it empirically.[29] Controlling for federalism, which we hypothesize will be a nearly unique characteristic of democracies that allows them to capture greater economies of scale and, thus, grow to a larger optimal size, we find that democracy is negatively and significantly related to territorial size. Indeed, using 1985 as the year of observation and holding all control variables at their mean values, we predict that unitary democracies will be roughly one-fourth the size of unitary autocracies. This provides strong confirmation for this penultimate and crucial link in the causal chain between democracy and conflict propensities.

Each link in my original causal chain between democracy and victory has been isolated, subjected to rigorous empirical testing, and supported by the available evidence. This does not "prove" causation, but it does produce greater confidence in the theory as a whole and in the hypothesis on democracy and victory in particular. I see no reason to reject the hypothesis for the statistical reasons that Desch develops and that I discuss in the previous section, and I see several reasons based on these additional tests to believe that the theory is capturing a real effect. As always, further research is necessary. But theory and evidence combine to support strongly the hypothesis that democracies will tend to win the wars they fight and for the reasons I suggest.

War Outcomes and the Study of Interstate Conflict

Although I am pleased that my findings ultimately support my normative belief in the superiority of democracy as a form of government, this was not my

---

29. Michael J. Hiscox and David A. Lake, "Democracy, Federalism, and the Size of States," Harvard University and University of California, San Diego, 2002.

original purpose in taking up this research. Nor, contrary to Desch's description of the democratic triumphalists (pp. 42–43), do I or others writing on this topic claim that democracy is the most important or even a primary factor in why states win or lose wars. War outcomes are complex events, and scholars still lack good theory to explain them. My ambition was not to account for war outcomes, but to test a theory of state rent-seeking and one of its more original implications. As such, my theory does not predict that democracy will matter a lot in determining war outcomes, a little, or barely at all. It only suggests that democracy will contribute positively to victory, which it clearly does.

As the subtitle of his article indicates, Desch claims only that "regime type hardly matters," which is a statement that scholars cannot dispute without better developed theories of war outcomes than exist today. Nonetheless, there is some reason to believe that it is worthwhile to try to develop such theories. Based on my original results, the probability of victory increases from 0.34 for the most autocratic states to 0.85 for the most democratic, implying that democracies are more than 50 percentage points more likely to win the wars they fight than are autocracies. This is a greater effect than increasing military personnel or iron and steel production from their minimum to maximum values.[30] This is a big number. Even if this estimate is eventually reduced by more refined tests, as it almost certainly will be, it suggests that there are important causes of war outcomes on which scholars in the field would do well to focus attention.

The rationalist approach to war, synthesized only a decade ago, was an important step forward in conflict studies.[31] Equally important, in recent years analysts have moved beyond theories of war as a single, game-ending event modeled as a costly lottery to theories of war as a process—a sequence of events—in which when to end the conflict and on what terms are outcomes of an ongoing bargaining process.[32] Stimulated in part by the earlier work on war

---

30. See Lake, "Powerful Pacifists," p. 32, Table 2, Equation 2. Predicted probabilities calculated via Michael Tomz, Jason Wittenberg, and Gary King, *CLARIFY: Software for Interpreting and Presenting Statistical Results*, Version 1.2.1 (Cambridge, Mass.: Harvard University, June 1, 1999), http:// gking.harvard.edu.

31. James D. Fearon, "Rationalist Explanations for War," *International Organization*, Vol. 49, No. 3 (Summer 1995), pp. 379–414. See also Robert Powell, *In the Shadow of Power: States and Strategies in International Politics* (Princeton, N.J.: Princeton University Press, 1999).

32. On war as a process, see Harrison Wagner, "Bargaining and War," *American Journal of Political Science*, Vol. 44, No. 3 (July 2000), pp. 469–484; Darren Filson and Suzanne Werner, "A Bargaining Model of War and Peace: Anticipating the Onset, Duration, and Outcome of War," *American Journal of Political Science*, Vol. 46, No. 4 (October 2002), pp. 819–837; and Branislav L. Slantchev, "The Power to Hurt: Costly Conflict with Completely Informed States," *American Political Science Review*, Vol. 97, No. 1 (February 2003), pp. 123–133.

outcomes, this is a potentially revolutionary advance in understanding conflict. It would be a tragedy if flawed critiques of the small but growing literature on war outcomes were to divert the field from this promising line of research.

Ultimately, and as implied by these new theories of war, the democratic effectiveness and selection effects arguments will need to be synthesized. Democracies may well be more careful in choosing the wars they enter, but it is their greater capabilities, ability to fight harder, and oversized coalitions that allow them to be more selective. And knowing their propensity for victory, opponents will choose more carefully as well. All of this has implications not only for which wars get fought but also for those international bargains negotiated in the shadow of war.

If I am in any way a triumphalist, it is about the ability of social scientists to eventually develop empirically verified theories that help all scholars to understand and, hopefully, mitigate violent conflict. Although I fully recognize the difficulty of the road ahead, it is only through careful attention to theory and research design that scholars can hope to realize this ambition.

Understanding Victory

Dan Reiter and
Allan C. Stam

Why Political Institutions Matter

In our book, *Democracies at War*, we asked the question: Why do democracies tend to win the wars they fight? We confirmed this pattern, first noted by David Lake in his "Powerful Pacifists" article, using statistical tests and numerous historical cases.[1] Notably, this phenomenon confounds the traditional realpolitik fear that democratic liberalism is a luxury that states may be unable to afford. Our basic answer to the question is that democracies tend to win because they put themselves in a position to do so. The constraints that flow from democratic political structures lead the executives of liberal democracies to hesitate before starting wars, particularly wars where victory on the battlefield appears to be less than clear-cut.

Democracies' willingness to start wars only against relatively weaker states says nothing about the actual military efficiency or capacity of democratic states. Rather, it says that when they do start a fight, they are more likely to pick on relatively weaker target states. We also find, however, that in addition to this "selection effects" explanation of democratic success, democratic armies enjoy a small advantage on the battlefield.

Michael Desch, a prominent realist scholar, reviews these claims in his article "Democracy and Victory: Why Regime Type Hardly Matters."[2] His assertion that regime type is irrelevant to the probability of military victory is consistent with the broader realist agenda, which argues that domestic politics matters little in the formation of foreign policy or the interactions between states. Desch makes a valuable contribution in advancing the debate over this question. There are many points about which Desch and we agree. Democracies do

Dan Reiter is Professor of Political Science at Emory University. Allan C. Stam is Associate Professor of Government at Dartmouth College.

We received helpful comments and suggestions from David Lake, Bruce Russett, Andrew Stigler, and William Wohlforth. Thanks also to Michael Desch for kindly sharing his data with us. Any remaining errors are ours.

1. Dan Reiter and Allan C. Stam, *Democracies at War* (Princeton, N.J.: Princeton University Press, 2002); and David A. Lake, "Powerful Pacifists: Democratic States and War," *American Political Science Review*, Vol. 86, No. 1 (March 1992), pp. 24–37.
2. Michael C. Desch, "Democracy and Victory: Why Regime Type Hardly Matters," *International Security*, Vol. 27, No. 2 (Fall 2002), pp. 5–47. Further references to this article appear parenthetically in the text.

International Security, Vol. 28, No. 1 (Summer 2003), pp. 168–179
© 2003 by the President and Fellows of Harvard College and the Massachusetts Institute of Technology.

not win wars because of economic superiority, greater public support during wartime, or having more or better allies. "Realist" factors such as military strategy, leadership, and industrial capabilities directly account in part for the war outcomes we observe. Regarding the irrelevance of regime type, however, we fail to find Desch's claims convincing because his analysis contains serious errors in both logic and method. In the next section, we review our central argument and explain how best to test it empirically. In the following section, we take up and rebut Desch's critiques of our primary data set. We then address some of Desch's other critiques.

Our Theory and Our Tests

How have democracies managed to compile their impressive wartime record? Generally, there are three possible answers. First, they might simply pick on weaker states, thereby arriving on the battlefield with more of the material determinants of victory. We call this argument the "selection effects" explanation. Because democratic political institutions make elected leaders more vulnerable to losing office, their foreign policies are different from those of leaders of other kinds of states. Specifically, they tend to represent more closely the values and preferences of the people than do the foreign policies of more authoritarian states. States where the elites must generate the contemporaneous consent of the people tend to avoid starting wars in which they have only a moderate or low chance of victory, so that when they go to war, they win.

Second, democracies might be materially more powerful. Their armed forces might fight with greater efficiency or better strategies, they might be able to extract relatively more resources from their societies, or they might choose allies that are relatively more reliable or powerful. The third possibility is that both of these explanations are partly correct.

The theories are disarmingly simple. Testing them is also quite straightforward. To test our theories, we examined all interstate wars that occurred between 1815 and 1985. We also looked closely at a number of specific cases to be sure that we were not spuriously attributing causal power to the correlations we uncovered. To test the selection effects theory, we looked in particular at the record of all war initiators, as the theory implies that democratic war initiators, being more risk averse, should be more likely than autocratic war initiators to win the wars they start. We conducted two kinds of statistical tests, cross tabulations and regression analysis, where we controlled for many of the material factors that determine the outcomes of wartime battles.

Desch carefully and accurately replicates these analyses. Unfortunately, he then makes a crucial error in logic and research design. After demonstrating that our basic findings are robust, he returns to the data set to judge whether, in his view, the cases are "fair tests." In doing so, he excludes more than 85 percent of the cases in which democracies were the initiators because he says these were not "fair fights." He then reestimates the statistical models and finds that democracy no longer correlates with victory. This is not surprising, but it is completely beside the point. By dropping all of the "unfair fights" from the data set, he has excluded many of the cases that our theory predicts will be the wars that democracies should win. As we explain below, our selection effects theory predicts that we should see precisely the type of cases that Desch omits—instances where the democratic war initiator is much more powerful than the target state.

To test the second theory—that democracies win because they fight wars more effectively—we again took two statistical approaches, cross tabulations and regression analysis. In our cross tabulations, we focused on target states (i.e., those states that find themselves at war when other states initiate attacks against them). If democracies are more powerful than other kinds of states, as well as more prudent, then we would expect them to do better than other kinds of states even when targeted. This was exactly what we found. Democratic targets are more likely to win than are other kinds of targets. To ensure that this finding was not a result of some other determinant of military victory randomly correlated with democracy levels, we also estimated a sophisticated regression model that allowed us to control for numerous material and ideational factors such as military capabilities, alliance contributions, and military strategy. Even with a large set of control variables, we still found that democratic targets outfought autocratic targets.

These results created something of a puzzle, because even when we controlled for many of the factors that military historians argue account for war outcomes, we still found a small military advantage that we could attribute to a state's level of democratization. In other words, even when we control for the number of troops, measures of technology, strategy, alliance contributions, and war matériel, democracies are still more likely than other kinds of states to win wars. Even when other states target them, democracies seem to be more powerful than their opponents; but there is no single factor in our war data set to which we could attribute their winning record. Therefore, we looked to other sources of data that would let us investigate democracies' performance at the level of the individual soldier. There, we found that soldiers of democracies are more likely to take the initiative and enjoy somewhat better leadership; also, they are less likely to surrender on the battlefield.

We remain confident in our findings. Importantly, an array of studies using a variety of data sets has produced findings consistent with our theoretical perspective that the constraints imposed by democratic political institutions push elected leaders to fight short, successful, low-cost wars and to avoid potential foreign policy disasters. Specifically, these studies have found that public opinion is generally reasonable, rational, and stable, thus providing a reliable foundation for foreign policy debate. Following military defeat, democratic leaders lose power faster than do autocratic leaders. Fear of eroding public support leads democracies to tend to fight shorter wars with fewer casualties. Moreover, this fear causes democracies to choose military strategies that promise rapid victory with low costs. Closely related to our findings about wartime performance, the aversion to foreign policy failure means that when democracies initiate foreign policy crises short of war, they usually emerge with diplomatic success. Our work on democratic wartime performance is also consistent with the institutional explanations of democracies' tendency not to fight each other.[3] Last, our research findings have withstood sophisticated methodological challenges.[4]

3. Benjamin I. Page and Robert Y. Shapiro, *The Rational Public: Fifty Years of Trends in Americans' Policy* (Chicago: University of Chicago Press, 1992); Bruce Bueno de Mesquita and Randolph M. Siverson, "War and the Survival of Political Leaders: A Comparative Study of Regime Types and Political Accountability," *American Political Science Review,* Vol. 89, No. 4 (December 1995), pp. 841–855; Reiter and Stam, *Democracies at War;* Scott Sigmund Gartner and Gary Segura, "War, Casualties, and Public Opinion," *Journal of Conflict Resolution,* Vol. 42, No. 3 (June 1998), pp. 278–300; Dan Reiter and Curtis Meek, "Determinants of Military Strategy, 1903–1994: A Quantitative Empirical Test," *International Studies Quarterly,* Vol. 43, No. 2 (June 1999), pp. 363–387; Christopher F. Gelpi and Michael Griesdorf, "Winners or Losers? Democracies in International Crisis, 1918–94," *American Political Science Review,* Vol. 95, No. 3 (September 2001), pp. 633–661; D. Scott Bennett and Allan C. Stam, "Research Design and Estimator Choices in the Analysis of Interstate Dyads: When Decisions Matter," *Journal of Conflict Resolution,* Vol. 44, No. 5 (October 2000), pp. 653–685; Brett Ashley Leeds and David R. Davis, "Beneath the Surface: Regime Type and International Interactions, 1953–1978," *Journal of Peace Research,* Vol. 36, No. 1 (January 1999), pp. 5–21; Dan Reiter and Erik R. Tillman, "Public, Legislative, and Executive Constraints on the Democratic Initiation of Conflict," *Journal of Politics,* Vol. 64, No. 3 (August 2002), pp. 810–826; and Bruce M. Russett and John R. Oneal, *Triangulating Peace: Democracy, Interdependence, and International Organizations* (New York: W.W. Norton, 2001).
4. David H. Clark and William Reed, "A Unified Model of War Onset and Outcomes," *Journal of Politics,* Vol. 65, No. 1 (February 2003), pp. 69–91; and William Reed and David H. Clark, "War Initiators and War Winners: The Consequences of Linking Theories of Democratic War Success," *Journal of Conflict Resolution,* Vol. 44, No. 3 (June 2000), pp. 378–395.

Desch's Specific Data Critiques

Desch begins his critique by successfully replicating our results. He then sets out to disprove our claims in two steps. First, he claims that more than two-thirds of our observations do not qualify as wars suitable for testing our propositions. Then, having eviscerated the data set, he tries to show that our findings about democracy and victory fail under scrutiny. In this section, we examine Desch's claims and rebut each of them, restoring confidence in the power of our original results.

REASSEMBLING OUR DATA

Desch outlines six potential problems with our data leading him to drop the majority of observations from his subsequent analysis. We disagree both with his general reasoning, as we pointed out above, and with these supposed problems.

Desch's main assertion is that we should drop wars between mismatched opponents. The existence of these cases, however, proves our selection effects theory that democracies seek out gross mismatches and avoid conflicts where their chances of victory are lower, starting wars only when they are confident they will win. Hence, they must be included in any fair analysis. By taking into consideration an array of control variables for material factors, we can safely include power mismatches in the analysis and still reliably estimate the independent effect of democracy.

Second, Desch asserts that our disaggregation of the data—that is, separating single large wars such as World War II into multiple smaller ones to improve coding accuracy—can bias the test in favor of our theory (pp. 13–14, n. 25). In principle, this might be true. As Desch notes, however, more of the cases that we create through disaggregation oppose our theory than favor it.

Third, Desch notes that democracies have sometimes won wars with powerful autocratic allies and, apparently assuming that wars are won only by single causes, concludes that these democracies should not receive any credit for these victories. We maintain that war outcomes are determined by multiple outcomes, which demands the inclusion of multiple independent variables. We controlled for alliance contributions in our regression models, finding that belligerents are more likely to win when they are democratic as well as when they have powerful allies.

Fourth, Desch claims that some of our codings are questionable, thereby weakening our case. In particular, he takes issue with our interpretation of Israel's performance in the 1969–70 War of Attrition and the 1982 Lebanon War. We coded the war outcomes in our data based on whether a side achieved its immediate military aims, a standard that matches well with our theory's predictions. We recognize that doing so involves making difficult judgment calls, but we stand by our coding decisions. In the War of Attrition, Israel won because it successfully repelled Egyptian attacks, and in the process suffered far fewer casualties than its opponent (260 Israeli dead and about 5,000 Egyptian dead).[5] In the Lebanon War, Israel accomplished its immediate goals of occupying southern Lebanon and evicting the Palestinian Liberation Organization.[6] Following Desch's standard of judging war outcomes based on long-term political gains or losses, we should then also consider many autocratic victories to be defeats, including the German victory over Belgium in World War I, all of the German victories in World War II, and the Japanese victories in East Asia in the 1930s.

Fifth, Desch argues that asymmetric interests explain war outcomes better than does regime type. Analyzing similar data, however, Allan Stam found in his book *Win, Lose, or Draw* that a control variable for asymmetric interests (or issue area) was not statistically significant. He also found that democracies were significantly more likely to win when controlling for the relative salience of the issue at stake for each side in a war.[7] For democracies, leaders are aware that when lesser interests are at stake, public support is likely to erode quickly. They therefore either avoid these wars or design strategies for short, low-cost wars, such as those in the 1999 Kosovo campaign, the 1991 Gulf War, and the 1982 Lebanon War. When democratic leaders occasionally underestimate how quickly they can attain victory, as in Vietnam, the decline in public support pushes them to accept a draw rather than holding out for elusive victory.[8]

Sixth, Desch argues that in some instances states that we might think of as "democratic" are not true democracies. Unfortunately, he does not indicate which cases he means. In our analyses, we treat a state's level of democracy

---

5. Michael Clodfelter, *Warfare and Armed Conflicts: A Statistical Reference to Casualty and Other Figures, 1500–2000,* 2d ed. (Jefferson, N.C.: McFarland and Company, 2002), p. 638.
6. Eric Silver, *Begin: A Biography* (London: Weidenfeld and Nicolson, 1984), especially p. 229. On Israel meeting its military goals, see Ze'ev Schiff and Ehud Ya'ari, *Israel's Lebanon War* (New York: Simon and Schuster, 1984), pp. 230, 306.
7. Allan C. Stam III, *Win, Lose, or Draw: Domestic Politics and the Crucible of War* (Ann Arbor: University of Michigan Press, 1996), pp. 114–115.
8. D. Scott Bennett and Allan C. Stam III, "The Declining Advantages of Democracy: A Combined Model of War Outcomes and Duration," *Journal of Conflict Resolution,* Vol. 42, No. 3 (June 1998), pp. 344–366; and Reiter and Stam, *Democracies at War,* chap. 7.

as a continuous variable, and we draw our democracy measures from the POLITY III data set, a standard source that correlates highly with other measures of regime type.[9]

Why Including Control Variables Is Better Than Dropping Cases

Desch's research design approach is to look at "crucial cases" to devise "fair tests" (p. 13, n. 23), proposing that "a fair test of a theory involves identifying crucial cases that clearly rule out alternative explanations" (p. 13). To get such a "fair test," Desch eliminates 54 of the 75 wars that we studied, arguing that the factors that determined the outcomes of those excluded cases are not relevant to the central democracy hypothesis, meaning that we should ignore these cases. In taking this approach, Desch creates some confusion with his use of the term "crucial cases" in his discussion of our statistical analysis compared with how the term is usually used in the context of qualitative research. Harry Eckstein has defined a case as crucial if it *"must closely fit* a theory if one is to have confidence in the theory's validity."[10] Eckstein has high standards for what would qualify as a true crucial case, noting that such cases "will not commonly occur and that one will need unusual luck to find them out economically if they exist."[11] Importantly, no single crucial case can decisively prove or disprove our theory, as we recognize that regime type affects war outcomes only probabilistically, and in every war there are always several other factors (some measurable and some not) that also determine outcomes. The combination of probabilistic hypotheses and multiple alternative factors makes statistical analysis preferable.

More generally, in qualitative case study research, the benefits of being able heuristically to develop new hypotheses and to examine closely the causal mechanisms in question offset the cost of limited generalizability imposed by a reduced sample size. Desch's decision to truncate the data set before conducting statistical analysis incurs the costs of a reduced sample without reaping these qualitative research benefits. This truncation is both unnecessary and dangerous. It is unnecessary because the standard practice of including control variables in statistical analysis enables one to discern accurately the individual

9. Keith Jaggers and Ted Robert Gurr, "Tracking Democracy's Third Wave with the Polity III Data," *Journal of Peace Research*, Vol. 32, No. 4 (November 1995), pp. 469–482.
10. Harry Eckstein, *Regarding Politics: Essays on Political Theory, Stability, and Change* (Berkeley: University of California Press, 1992), p. 157 (emphasis in original).
11. Ibid.

and independent effects of separate variables without having to delete large numbers of cases. It is dangerous because the omission of so much information is likely to bias the results, especially when the omission means removing most cases successfully predicted by one of the theories being tested, most notably cases in which democracies initiated mismatches that they won. Overall, if our choice is to address potentially confounding factors through control variables and other techniques or simply to eliminate most of our available data, the former approach is clearly superior; the latter is disingenuous and likely produces biased results.

Perhaps most surprising is that even if we use Desch's truncated data set, his null result that democracies are no more likely to win quickly collapses. In Table 1, we replicate the results for his models 1 and 2 (p. 17). Oddly, Desch included no control variables, only the variable "Democracy," despite his realist claim that material factors such as power explain war outcomes completely. Estimating a misspecified model by excluding relevant control variables leads to biased results. We agree with Desch that material factors matter. We therefore reran his analysis with a measure of military-industrial capabilities as a control. These results are in model 3.

When we include the military-industrial capabilities variable, democracy becomes statistically significant, and the coefficient increases more than two and a half times. Hence, even if one accepts all of Desch's critiques about the composition of the data set, his finding of no relationship between democracy and victory evaporates with the most basic improvement of his statistical model. We next consider some of Desch's narrower critiques of our work.

Desch's Other Critiques

Desch presents a series of other critiques of our claims. Space limitations force us to address only his primary points.

CORRELATION VERSUS CAUSATION?

Desch notes that correlation does not mean causation. In our book, we work through a number of historical illustrations to demonstrate the plausibility of our theory, looking at the 1971 Bangladesh War, the 1991 Gulf War, the 1941 Japanese attack on Pearl Harbor, and others, demonstrating that the trappings of democracy directly affected the war outcomes in question.

In challenging our causal claims, Desch mentions the different military performances of Britain and France, noting in particular that France won 58 per-

Table 1. Reanalysis of Desch's "Fair Fights" Data Set.

Variables	Model 1 (full democra-cies at war data set)	Model 2 (Desch "fair fights > 6" data set)	Model 3 (Desch "fair fights > 6" data set)
constant	0.1410283	−0.3440138	−1.508952**
	(0.097201)	(0.227655)	(0.4491125)
democracy	0.0359429**	0.0364302	0.0942184*
	(0.0137452)	(0.0313352)	(0.0397805)
military-industrial capabilities	—	—	3.028452*
			(0.9551511)
pseudo R^2	0.0248	0.0332	0.2998
log likelihood	−133.04446	−21.342535	−15.456298
N	197	34	34

NOTE: We thank Michael Desch for sharing his data. Standard errors are robust; significance tests are two-tailed.
*$p < 0.05$
**$p < 0.001$

cent of its wars before 1815 and 56 percent of its wars after; Britain, on the other hand, won 81 percent of its wars before and 89 percent after. The problem with this summary is that, for France in the post-1815 period, it includes wars fought while France was both democratic and autocratic. If one looks only at wars fought since 1815 in which France was democratic, the French record is 4 wins (Roman Republic, Boxer Rebellion, World War I, and Gulf War), 1 loss (World War II), and 2 draws (Sinai and the Sino-French War). This win/loss record of about 80 percent matches the overall record of democratic performance for our data set.

SUBSTANTIVE EFFECTS: SIGNIFICANT OR MARGINAL?
Desch next argues that even if democracy has a statistically significant effect on war outcomes, its substantive effect is so small that it should be considered irrelevant. Here he confuses the selection effects argument—that democracies win because they initiate war only when material conditions such as strategy, capabilities, and terrain are very favorable—and the war-fighting argument— that democracies prosecute wars more effectively. We agree with Desch that material and ideational factors such as industrial capabilities, technology, and

military strategy are the critical factors determining outcomes. However, this misses the point of our selection effects theory, where we argue that democracies are better or more willing to judge the effects of these factors. Using reliable estimates of their relative power, democracies start only those wars they are likely to win—that is, those wars in which they will enjoy an "unfair fight." In *Democracies at War*, we demonstrate the very real advantages of democracies (and democratic initiators in particular). Table 2 reports these advantages.[12]

Desch provides some statistical analysis that the marginal effect of democracy is minor. However, his simplistic approach of looking at unit changes in individual variables to assess substantive significance is misguided, both because unit changes have different meanings for different variables (as the ranges in values of the variables differ widely), and because the presence of interaction variables makes it inappropriate to isolate the effects of individual variables. The estimates from our data predict that a democratic initiator (POLITY score of 10) has a 93 percent chance of winning, whereas a mostly authoritarian regime (POLITY score of -7) has only a 62 percent chance of winning, if one appropriately codes all democracy, initiation, and interaction variables and holds other values at their means.

DEMOCRACIES ARE BETTER DECISIONMAKERS
Desch disputes the proposition that democracies are better strategic decisionmakers. Our point is a bit more sophisticated than this, as we argue that democracies are both better and more risk-averse decisionmakers—each of these factors contributing to democratic success in war. His claim that there is no evidence that democracies make better choices is incorrect. Our study shows that among states that start wars, democracies are much more likely to win; relatedly, Jack Snyder found that democracies are less likely to engage in imperial overexpansion, and Christopher Gelpi and Michael Griesdorf found that, when democracies initiate international crises, they are more likely to win.[13]

Relatedly, Desch's claim that Israel is a poor decisionmaker in the context of its foreign policy is like claiming that the glass is one-quarter empty instead three-quarters full. Like all states, democratic or otherwise, Israel has made its share of poor decisions. In the big picture, however, it is hard to consider

12. Reiter and Stam, *Democracies at War*, p. 29, Table 2.1.
13. Reiter and Stam, *Democracies at War*; Jack Snyder, *Myths of Empire: Domestic Politics and International Ambition* (Ithaca, N.Y.: Cornell University Press, 1991); and Gelpi and Griesdorf, "Winners or Losers?"

Table 2. Winning Percentages for War Initiators and Targets by Regime Type.

	Dictatorships	Oligarchs	Democracies	Total
	War Initiators			
Wins	21	21	14	56
Losses	14	15	1	30
Winning percentage	60%	58%	93%	65%
	Targets			
Wins	16	18	12	46
Losses	31	27	7	65
Winning percentage	34%	40%	63%	41%

Israel's survival and prosperity while it is surrounded by extremely hostile neighbors anything other than a success. Israel has won every war it has fought since independence. It has also gone beyond military victory, concluding a peace agreement with one neighbor (Egypt) and taking long strides toward reconciliation with another (Jordan).

In comparison, consider the strategic record of Israel's neighbors. They have lost every war fought with Israel. Iraq initiated two disastrous invasions of neighbors, Iran and Kuwait. In 2003, it lost a third war because of its embrace of strategically unnecessary weapons of mass destruction. Desch describes Israel's national soul-searching following its victory in the 1973 Yom Kippur War as indicative of failure, but he again misses the point. This willingness to confront past errors to improve future policies explains why democracies make better decisions.

DEMOCRACIES ON THE BATTLEFIELD
Like Desch, we are skeptical that democracies win wars because of greater wealth, more or better allies, or greater societal commitment during wartime. We do argue that democratic armies enjoy higher levels of battlefield military effectiveness because their soldiers fight with higher initiative, their small-unit leadership is better, and democratic soldiers are less likely to surrender to the enemy. There is a considerable amount of historical scholarship supporting these points, especially in the Middle East.[14] Even in Germany's greatest World

14. Kenneth M. Pollack, *Arabs at War: Military Effectiveness, 1948–1991* (Lincoln: University of Nebraska Press, 2002); Michael B. Oren, *Six Days of War: June 1967 and the Making of the Modern Middle*

War II victory, its rapid conquest of France, studies have shown that unit for unit French forces fought as well as German forces. Studies of the European campaign also indicate U.S. tactical superiority.[15] Desch presents an extensive critique of the HERO data that we use, ultimately concluding that there is likely some measurement error, which would result in the attenuation of correlations found there. HERO is the only available quantitative data set on battles, however, and it suggests democratic superiority. We agree with Desch that the HERO data set is imperfect and that much work remains.

Conclusion

Winston Churchill put it best when he remarked, "Democracy is the worst form of government, except for all the others." The historical record bears out Churchill's point. Despite occasional errors, democracies enjoy much better wartime and foreign policy performance than do authoritarian states. Scholars can still be confident that democracies are more likely to win wars, especially the wars they initiate. They should not view democracy as a costly luxury that inevitably endangers the state, but rather as a set of institutional characteristics that improve a state's foreign policy decisionmaking and battlefield performance.

East (Oxford: Oxford University Press, 2002); and Risa Brooks, "Institutions at the Domestic/International Nexus: The Political-Military Origins of Strategic Integration, Military Effectiveness, and War," Ph.D. dissertation, University of California, San Diego, 2000.
15. Don W. Alexander, "Repercussions of the Breda Variant," *French Historical Studies*, Vol. 8, No. 3 (Spring 1974), especially pp. 462–465; John Sloan Brown, "Colonel Trevor N. Dupuy and the Mythos of Wehrmacht Superiority: A Reconsideration," *Military Affairs*, Vol. 50, No. 1 (January 1986), pp. 16–20; Stephen E. Ambrose, *D-Day, June 6, 1944: The Climactic Battle of World War II* (New York: Simon and Schuster, 1994); and Stephen E. Ambrose, *Citizen Soldiers: The U.S. Army from the Normandy Beaches to the Bulge to the Surrender of Germany—June 7, 1944–May 7, 1945* (New York: Simon and Schuster, 1997).

Democracy and Victory

Fair Fights or Food Fights?

Michael C. Desch

$\mathbf{A}$jin Choi, David Lake, and Dan Reiter and Allan Stam have each provided useful rejoinders to the critique of democratic triumphalism in my recent article "Democracy and Victory: Why Regime Type Hardly Matters."[1] In response, I begin by summarizing our arguments and pointing out several issues where we have little or no disagreement. I then examine our two major areas of contention: how best to test whether democracy matters much in explaining military outcomes, and whether the democratic triumphalists' proposed mechanisms convincingly explain why democracies frequently appear to win their wars.

The Arguments

Democratic triumphalists argue that democracies are more likely to achieve victory in warfare because of the nature of their domestic regimes. According to the triumphalists, democracies (1) start only wars they can win easily, and (2) enjoy important wartime advantages such as greater wealth, stronger alliances, better strategic thinking, higher public support, and more effective soldiers.

After examining the data and methods that underpin these findings, I concluded that "whether a state is democratic is not the *most important* factor to consider" in determining a state's likelihood of victory in war—hence the subtitle of my article "Why Regime Type *Hardly* Matters" (p. 42, emphasis added). I do not argue that regime type plays no role—only that it appears to be modest compared with other factors.

Michael C. Desch is Professor and Director of the Patterson School of Diplomacy and International Commerce at the University of Kentucky.

The author greatly appreciates the advice and comments of Eugene Gholz, Douglass Gibler, John Mearsheimer, and Scott Wolford as well as research support from Glenn Rudolph.

1. Michael C. Desch, "Democracy and Victory: Why Regime Type Hardly Matters," *International Security*, Vol. 27, No. 2 (Fall 2002), pp. 5–47. Subsequent references to this article and the rejoinders appear parenthetically in the text.

International Security, Vol. 28, No. 1 (Summer 2003), pp. 180–194

POINTS OF AGREEMENT

The triumphalists and I recognize the striking empirical regularity that democracies have been on the winning side of most wars since 1815. If one looks at all the cases of wars involving democratic states (with a score of 6 or better on the 21-point [−10 to 10] POLITY Democracy Index), democracies win more than 80 percent of the time. If one broadens the range of cases to look at whether the more democratic state prevailed irrespective of whether it had a democracy score of 6 or higher, the more democratic state prevails more than 70 percent of the time. And even my restrictive subset of "fair fight" cases still credits democracies with winning more than 60 percent of the time. Given this, the triumphalists and I also agree that on balance democracy is not an obstacle to the successful conduct of war.

We also accept that victory is likely to be the result of a variety of factors, including material power and other, nonmaterial, variables (Desch, p. 23; Lake, p. 166; and Reiter and Stam, p. 172). Where we part company is over how much relative influence these factors have in explaining whether a state wisely chooses and then effectively wages war.

There are two other points where the triumphalists suggest that we have major disagreements when in fact we do not. First, Lake and Reiter and Stam argue that I advance a deterministic theory of military victory (i.e., "the more powerful side always wins its wars"), whereas they offer a more theoretically and empirically defensible probabilistic theory of victory (i.e., "more democratic states are more likely to win their wars")[2] (Lake, pp. 158–161, and Reiter and Stam, p. 174). This is not an accurate portrayal of my position. When listing the alternative theories that I think better explain victory, I qualified almost every one with words such as "often," "sometimes," "likely," and "could" (Desch, p. 7). Indeed there are few, if any, deterministic theories in the field of international relations.[3]

---

2. Contrary to Lake's claims, multiple case studies can be used to test probabilistic theories because they provide a great deal of information. Not only does this make them an efficient means of testing theories, but it also makes it possible to determine whether or not particular cases are outliers.
3. One exception is the "democratic peace," whose proponents treat it as "the closest thing we have to an empirical law in the study of international relations." See Jack S. Levy, "Domestic Politics and War," in Robert I. Rotberg and Theodore K. Rabb, eds., *The Origin and Prevention of Major Wars* (New York: Cambridge University Press, 1989), p. 88; and James Lee Ray, "Wars between Democracies: Rare or Non-existent?" *International Interactions*, Vol. 18, No. 3 (Spring 1988), pp. 251–276.

Second, Lake argues that causation can be inferred only in the context of a prior theoretical framework and criticizes me for not having one. But elsewhere Lake and Reiter and Stam identify me as a realist based not only on my previous work but also, undoubtedly, on my conclusion that materialist variables such as power play a more important role in explaining victory than does democracy. In other words, they recognize that my explanation of military victory is not divorced from an overarching theoretical framework. Thus, we are in basic agreement about the importance of theory in guiding empirical research.

Points of Contention

Despite these areas of agreement, this exchange has highlighted important differences between the triumphalists and me (as well as among the triumphalists themselves). These include (1) how scholars should gauge democracy's role in explaining victory; and (2) what makes democracies more likely to select winnable wars and fight more effectively. To support their claim that democracy is the key to victory, the triumphalists need to provide both a convincing test of their hypothesis as well as a compelling set of causal mechanisms explaining why democracies should be more likely to win their wars. They have done neither.

DISAGREEMENT #1: HOW TO TEST THE TRIUMPHALISTS' ARGUMENT
In this section I lay out the triumphalists' claims that democracy is the key to victory and show that, with one exception (Choi, pp. 144–145), they fail to demonstrate that regime type is the most important factor in this regard. I then discuss their criticisms of my approaches for measuring democracy's role in explaining military victory.

In their book *Democracies at War*, the most recent and comprehensive statement of the democratic triumphalists' case, Reiter and Stam make strong claims about the role of democracy in explaining why democratic states have been on the winning side of so many wars. They conclude that "democracy has . . . been the *surest* means to power in the arena of battle" and that "democratic political institutions hold the *key* to prudent and successful foreign policy."[4] Choi similarly argues that "democracies are formidable players in the

---

4. Dan Reiter and Allan C. Stam III, *Democracies at War* (Princeton, N.J.: Princeton University Press, 2002), pp. 197, 205 (emphasis added).

international arena and that the sources of democracies' military prowess are entrenched in their political institutions" (Choi, p. 143).

Instead of measuring the size of the relative effect of democracy and other variables that might explain victory, Lake and Reiter and Stam are content merely to show that democracy is statistically significant in their models (Lake, pp. 165–167, and Reiter and Stam, pp. 175–176). But there is more to interpreting the results of models besides the statistical significance of the variables. Their practical significance, which is a function not only of the variables' statistical significance but also of the size of the coefficient and their signs, must also be considered. "Too much focus on statistical significance," one standard text on econometrics reminds us, "can lead to [the] false conclusion that a variable is 'important' . . . even though its estimated effect is modest."[5] In short, democracy may be statistically significant but still not be the key to victory.

To measure the practical significance of democracy, I first sought to isolate its effects by focusing on cases in which democracies fought without the advantages of power imbalances, nondemocratic allies, or an asymmetry of interests in their favor. I characterized such cases as "fair fights." My rationale for focusing on these "fair fights" was that because alternative theories of military victory rely on these variables, it is important to find cases where democracies won without these advantages to see if democracy has much independent effect.

To illustrate the effects of focusing on this small subset of cases that I argue are fair tests of the democracy and victory proposition, I reported the results of two simple probit models estimating the relationship between level of democracy and likelihood of victory using the full Reiter and Stam data set ($N = 197$) and my "fair fight" data set ($N = 34$). The models show that by itself the democracy variable becomes insignificant using my data set. Reiter and Stam take me to task for not including a control variable for power, which if included would make the democracy variable significant. They ignore, however, that in Table 4 of "Democracy and Victory" I report that I ran their model 4 of the selection effects argument—including their control variables for military capability—using only the "fair fight" data set and found no significant relationship between their democracy*initiation variable and victory (see Table 1). This suggests that the statistical significance of democracy is highly sensitive to model specification.

---

5. Jeffrey M. Wooldridge, *Introductory Econometrics: A Modern Approach* (Mason, Ohio: South-Western College Publishing, 2000), p. 131.

Table 1. Selection Effects with Fair-Fight Data (win/lose).

Variable	Model 1
constant	−11.97823
	(3.840502)
democracy*initiation	0.1070775
	(0.0710027)
initiation	10.33682**
	(3.606029)
democracy*target	1.167899**
	(0.3995539)
capabilities	2.870727
	(1.803471)
allies capabilities	−1.039488
	(1.670907)
quality ratio	0.49914*
	(0.2532648)
pseudo R^2	0.5227
log likelihood	−10.535901
N	34

NOTE: Standard errors are robust; significance tests are two-tailed.
*$p < 0.05$
**$p < 0.01$
***$p < 0.001$

Second, I calculated the marginal effects (the impact of a one-unit increase in each independent variable on the likelihood of change in the dependent variable) of democracy and other variables in the triumphalists' models. My reasoning here was that because we all agree that wars are complex and perhaps overdetermined processes, it is necessary to find a means of determining which of these variables play the most crucial role in explaining victory.

Lake and Reiter and Stam object that my dropping of cases that are not "fair fights" is methodologically unsound because it deprives scholars of lots of useful information. There is, however, no methodological problem with focusing on "fair fight" cases, because doing so has an effect similar to adding control variables in a multivariate equation. The purpose of control variables is to account for variation in the dependent variable that may be wrongly attributed to the independent variable. Adding them thus avoids this "omitted variable bias."[6] Although adding control variables is the standard approach in the observational social sciences, my dropping cases where the winner had an overwhelming power advantage is akin to running a controlled experiment in the natural sciences. Both approaches are scientifically valid menas of controlling for the effects of potentially perturbing variables.[7]

The real issue is whether limiting consideration to just "fair fights" provides an equitable test of the democratic triumphalists' theories. Lake argues that my looking at only "fair fights" is an unfair test of his theory because he believes that one of the wartime advantages of democracies is their ability to generate more wealth, which in turn gives them greater military resources with which to wage war. Because wealth is one of the sinews of military power, it is not surprising that wealthier states tend to win their wars. Excluding such cases, in Lake's view, eliminates those cases on which his theory depends.

There are two problems with Lake's argument: First, his theory is impossible to test against the most likely alternative theory: that states win when they have a preponderance of power. Second, as Lake acknowledges (p. 164), it is possible that the relationship between democracy and victory is spurious, inasmuch as wealth may explain both democracy and military success. Lake's subsequent effort to establish the causal chain from democracy to wealth by showing that democracies are more likely to provide public services does not shed much light on the question of whether democracies are likely to produce greater wealth.[8]

Reiter and Stam also challenge Lake's claim that democratic states are better able to generate wealth.[9] Instead they maintain that democracies start only wars they can easily win. But like Lake, they maintain that focusing on "fair fights" is not an adequate test of their theory because they claim that the main advantage of democracies is finding unfair fights.

The problem here is the inconsistency between Reiter and Stam's response to me and their critique of Lake and others who argue that democracies win wars

---

6. Gary King, Robert O. Keohane, and Sidney Verba, *Designing Social Inquiry: Scientific Inference in Qualitative Research* (Princeton, N.J.: Princeton University Press, 1994), pp. 168–184.
7. For discussion of the similarities of the two approaches, see Alan Agresti and Barbara Finlay, *Statistical Methods for Social Sciences,* 3d ed. (Upper Saddle, N.J.: Prentice Hall, 1999), pp. 359–362.
8. David A. Lake and Matthew A. Baum, "The Invisible Hand of Democracy: Political Control and the Provision of Public Services," *Comparative Political Studies,* Vol. 34, No. 6 (August 2001), pp. 587–621.
9. Reiter and Stam, *Democracies at War,* pp. 114–143.

because they are wealthier. After testing Lake's wealth argument, Reiter and Stam conclude that they "can reject two hypotheses: that democracies in general win their wars because they have higher capabilities, and that democratic initiators win wars because they have significantly higher capabilities than do other kinds of initiators." Elsewhere they note that their selection effects argument "does not imply that [democracies] win because they are more powerful, rather that they are better at avoiding wars they would have gone on to lose had they fought them."[10] It is difficult to reconcile these arguments that power imbalances do not explain victory with their claim in their response to me that democratic initiators tend to win because they are better at selecting "unfair fights" in which they have a decided advantage in military power (Reiter and Stam, p. 170).

Even if the core of Reiter and Stam's argument is that democracies' main advantage is their ability to select weak adversaries, regime type may not play the key role here. In "Democracy and Victory," I calculated the marginal effects of the variables in the triumphalists' selection effects model—using all their cases—and concluded that of all of the variables democracy had one of the smallest impacts on the likelihood of victory.

Reiter and Stam respond that this approach is inappropriate for assessing their selection effects argument for two reasons. First, because the democracy variable is included as part of an interaction term with an initiation variable, it cannot be gauged in this fashion. Second, because the democracy and power variables have different scales, calculating their marginal effects is like comparing apples and oranges. Their democracy*initiation interaction variable, however, is just 1 or 0 (whether the state initiated the war or not) multiplied by its democracy score. There is little reason to think that marginal effects cannot be calculated for it in the same way they would be for the straight democracy variable. Although it is true that the democracy and power variables have different scales, if one calculates elasticities (the effects of a percentage increase in the independent variables which makes them more comparable) rather than marginal effects, democracy still has a relatively small impact on the likelihood of victory.[11]

In sum, the triumphalists' research design does not convincingly demonstrate that democracy is the main reason states win their wars.

---

10. Ibid., pp. 20, 138.
11. The elasticity for democracy*initiation is −0.0486938; for the state's capabilities, it is 0.9587111; and for its allies' capabilities, it is 0.6250279.

DISAGREEMENT #2: RESERVATIONS ABOUT THE CAUSAL MECHANISMS

In my article I criticized the triumphalists' arguments that democracies are more likely to win their wars because of selection effects and greater wartime effectiveness. The triumphalists offer various defenses of these causal mechanisms, but many problems remain.

SELECTION EFFECTS. Although in Reiter and Stam's data set democracies win 94 percent of the wars they start, it is questionable whether this success is due to the fact that democratic political leaders are more careful because they fear electoral retribution if they lose.[12] The triumphalists have only a handful of cases of democracies starting wars; the coding of many of them is questionable; and in many of the others, the triumphalists' causal mechanisms do not operate.

There are only 16 cases of democracies starting wars since 1815, and half of these involve the same three countries: Britain, Israel, and the United States. Given that only three states account for such a large proportion of the winners, it is useful to ask how generalizable the triumphalists' findings are (see Table 2).

Moreover, the coding of 6 of these cases is questionable because democracies did not initiate the wars. Reiter and Stam and other triumphalists credit Britain, France, and the United States with initiating the Boxer Rebellion in 1900, ignoring the fact that diplomats and citizens of those countries were already under attack by the Boxers when the Western powers sent their relief expedition to China.[13] The 1919 Czech-Hungarian War is widely considered to have begun with Hungarian Communist Bela Kun's attack on Slovakia rather than with democratic Czechoslovakia attacking Hungary as the triumphalists maintain.[14] Finally, they count Britain and the United States as having initiated the 1941 through 1945 phase of World War II in Europe, despite the fact that Germany began that war in 1939 with its attack on Poland. In sum, there are only 10 clear cases of democracies starting wars since 1815.

These 10 cases must be closely examined to see whether the triumphalists' causal mechanisms really explain why these democracies won their wars. De-

---

12. I arrive at slightly different numbers than Reiter and Stam, because in Appendix 2.2 they list the United States as initiating the European phase of World War II, but in my copy of their data set they do not. Ibid., pp. 52–57.

13. Michael Clodfelter, *Warfare and Armed Conflicts: A Statistical Reference to Casualty and Other Figures, 1618–1991* (Jefferson, N.C.: McFarland, 1992), p. 643.

14. George Childs Kohn, *Dictionary of Wars*, rev. ed. (New York: Facts on File, 1999), p. 264; and R. Ernest Dupuy and Trevor N. Dupuy, *The Harper Encyclopedia of Military History: From 3500 B.C. to the Present* (New York: HarperCollins, 1993), p. 1091.

Table 2. Assessment of Triumphalists' Selection Effects Cases.

Status	Cases
Questionable coding	France/Boxer Rebellion (1900) United Kingdom/Boxer Rebellion United States/Boxer Rebellion Czech-Hungarian War (1919) United Kingdom/World War II (1941–45) United States/World War II (1941–45)
Process tracing supports selection effects	Spanish-American (1898) First Balkan (1912–13) Six Day War (1967) Bangladesh (1971) Turkey/Cyprus Invasion (1974)
Process tracing does not support selection effects	Mexican War (1846–48) Greco-Turkish War (1897) Russo-Polish War (1919–20) Sinai War (1956) Lebanon War (1982)

mocracies, according to the triumphalists, should be better at finding easy fights because they cannot start wars if there is substantial public opposition; they should enjoy free and high-quality debate about war; democratic leaders will suffer or prosper depending on how the war goes; and so the primary concern of democratic leaders in starting wars should be their influence on the fate of the leader.[15] Rather than directly testing these propositions in detailed case studies, the triumphalists are largely content with establishing a correlation between the democracy, initiation, and victory variables and inferring that the selection effect explains it.

A closer examination of the 10 remaining cases shows that the triumphalists' causal mechanisms do not explain many instances of democratic victory. To be sure, their propositions appear to be at work in 5 cases: the 1898 Spanish American War, the First Balkan War of 1912–13, the 1967 Six Day War, the 1971 Bangladesh War, and Turkey's 1974 invasion of Cyprus. However, in the other 5—the Mexican War of 1846–48, the 1897 Greco-Turkish War, the Russo-Polish War of 1919–20, the 1956 Sinai War, and Israel's Lebanon War of 1982—democracy does not seem to be the explanation for why these countries did or did not launch successful wars. Instead of democracies winning a very impressive 94

15. Reiter and Stam, *Democracies at War*, pp. 10, 19–20, 21, 23, 144, 146, 162.

percent of the wars they start, democratic selection effects actually explain only 50 percent of these victories.

The 1846–48 U.S.-Mexican War does not support many of the triumphalists' propositions. For example, President James Polk started the war without substantial public and congressional support: Many Americans opposed the annexation of Texas, fearing it would upset the delicate balance between free and slave-holding states.[16] Nor was there much open debate about what came to be called "Mr. Polk's War," which the president initiated by secretly sending U.S. forces into a disputed area of the border, where they were sure to be attacked by the Mexicans.[17] It is difficult to ascertain from this case whether victory helped the domestic fortunes of the Polk regime: He did not run for re-election after the war. His party—the Whigs—lost the election to Zachary Taylor, whose campaign was clearly aided by his role in the victorious war. But overall, Polk's motive appears to have been not domestic political gain but rather territorial consolidation of U.S. control of North America.[18]

The 1897 Greco-Turkish War does not support the triumphalists' argument because after Greece's loss to the Turks, Greek Crown Prince Constantine—the military commander—was not punished for losing the war. Indeed he would again command Greek military forces in the First Balkan War and eventually become king of Greece.[19]

The Russo-Polish War of 1919–20 also provides little support for the triumphalists' causal mechanisms. Despite Poland's democratic institutions and widespread opposition to war with Russia, the Polish military leader Jozef Piłsudski was not deterred from launching it. "Piłsudski," as Richard Watt concludes "kept his own counsel and made his own plans, paying no great attention to the desires of the Sejm, the majority of whose members, both on the Left and on the Right, wanted peace with Soviet Russia."[20] The triumphalists' selection effects argument assumes that both the public and civilian government

---

16. Thomas A. Bailey, *A Diplomatic History of the American People,* 10th ed. (Englewood Cliffs, N.J.: Prentice-Hall, 1970), p. 257; and Robert H. Ferrell, *American Diplomacy: A History* (New York: W.W. Norton, 1975), p. 190.
17. Bailey, *A Diplomatic History of the American People,* p. 256.
18. John J. Mearsheimer, *The Tragedy of Great Power Politics* (New York: W.W. Norton, 2001), pp. 242–244.
19. C.M. Woodhouse, *Modern Greece: A Short History* (London: Faber and Faber, 1986), p. 193.
20. Richard M. Watt, *Bitter Glory: Poland and Its Fate, 1918–1939* (New York: Hippocrene, 1982), p. 105. Also cf. Serge Michiel Shewchuk, "The Russo-Polish War of 1920," Ph.D. dissertation, University of Maryland, 1966, p. 146; Adam Zamoyski, *The Battle for the Marchlands* (New York: Columbia University Press, 1981), p. 6; and Viscount D'Abernon, *The Eighteenth Decisive Battle of the World* (London: Hodder and Stoughton, 1931), p. 39.

officials have some input into, and knowledge about, the conduct of wars. Piłsudski, however, had complete authority for running the war and kept the rest of the government in the dark.[21] One foreign ministry official wondered in 1919: "'Where are we going? This question intrigues everybody. . . . To the Dnieper? To the Dvina? And then?'"[22] Nor is it clear that the changing course of the war had much effect on Piłsudski's political fortunes. During the bleakest days before the miraculous turnabout in August 1920, there was no discernable loss of support for Piłsudski or any effort to reduce his substantial authority. Finally, there is little evidence that Piłsudski thought much about the war in terms of his political fortunes in the Polish democratic political system. The only consistent elements in Piłsudski's life were Polish nationalism and the struggle for Poland's independence.[23]

Nor does Israel's successful 1956 Sinai campaign support the selection effects argument. Neither the Israeli public nor the cabinet strongly supported the war, as evidenced by the great lengths to which Israeli Prime Minister David Ben-Gurion went to keep preparations for the operation a secret.[24] Only ten Israeli civilians besides Ben-Gurion knew about the war before October 25, 1956, and the Israeli cabinet was informed only the night before military operations began.[25] Indeed Ben-Gurion alone made all the decisions about the war.[26] As one of his colleagues famously quipped: "'Ben-Gurion the defense minister consulted with Ben-Gurion the foreign minister and received the green light from Ben-Gurion the prime minister.'"[27] The results of the war were mixed (Israel reduced the *fedayeen* threat from Egypt but was forced to unilaterally withdraw from the Sinai; it also became increasingly isolated internationally).

21. Watt, *Bitter Glory*, p. 125; Michael Palij, *The Ukrainian-Polish Defensive Alliance, 1919–1921: An Aspect of the Ukrainian Revolution* (Edmonton: Canadian Institute of Ukrainian Studies Press, 1995), p. 118; Norman Davies, *White Eagle, Red Star: The Polish-Soviet War, 1919–20* (New York: St. Martin's, 1972), p. 31; and Robert Szymczak, "Bolshevik Wave Breaks at Warsaw," *Military History*, Vol. 11, No. 6 (February 1995), p. 56.
22. Quoted in Watt, *Bitter Glory*, p. 99.
23. Zygmunt J. Gasiorowski, "Joseph Pilsudski in the Light of British Reports," *Slavonic and East European Review*, Vol. 50, No. 121 (1972), p. 566.
24. Hugh Thomas, *Suez* (New York: Harper and Row, 1966), p. 16; and Avner Yaniv and Robert J. Lieber, "Personal Whim or Strategic Imperative? The Israeli Invasion of Lebanon," *International Security*, Vol. 8, No. 2 (Fall 1983), p. 140.
25. Michael Brecher, *Decisions in Israel's Foreign Policy* (New Haven, Conn.: Yale University Press, 1975), pp. 232–234; and Zeev Schiff, *A History of the Israeli Army: 1847 to the Present* (New York: Macmillan, 1985), p. 93.
26. Howard M. Sachar, *A History of Israel: From the Rise of Zionism to Our Time*, 2d ed. (New York: Knopf, 1996), p. 478; and Avi Shlaim, *The Iron Wall: Israel and the Arab World* (New York: W.W. Norton, 2001), p. 149
27. Moshe Sharett, quoted in Shlaim, *The Iron Wall*, p. 150.

Moreover, their effects on Ben-Gurion's political career remain unclear.[28] But there is little evidence that domestic political concerns were of much importance to Ben-Gurion, who seemed to make strategic decisions overwhelmingly in terms of what he thought was best for Israel's survival and prosperity.[29]

Israel's 1982 war against Syria in Lebanon also provides no support for the triumphalists' selection effects argument. Prime Minister Menachem Begin and Defense Minister Ariel Sharon began the war despite overwhelming opposition from the cabinet and the Israeli public, who opposed transforming a retaliatory raid against Palestine Liberation Organization forces in south Lebanon into a war with Syria.[30] Nor did Israeli democracy foster thoughtful debate via the marketplace of ideas. Begin refused to listen to the objections of military and intelligence professionals, and Sharon kept the cabinet in the dark about the scope of the Israeli operation and thwarted efforts by Israel's press to inform the public about their plans.[31] All the major decisions about the war were made by a small group of individuals: Begin, Sharon, Foreign Minister Yitzak Shamir, and Chief of Staff Gen. Rafael Eitan.[32] Despite Israel's tactical victory over Syria, the general impression among Israelis was that the war was a failure.[33] Syria was not ejected from Lebanon, and by the mid-1980s it had again become the dominant power in that country. Although Israel's failure in Lebanon did exact some political costs, neither Begin nor Sharon suffered serious political punishment: Begin resigned in 1983, largely for personal reasons;[34] and Sharon lost his defense portfolio (largely because of the massacres

---

28. Ibid., pp. 183–185; Sachar, *A History of Israel*, pp. 513–514; and Thomas Baylisss, *How Israel Was Won: A Concise History of the Arab-Israeli Conflict* (Lanham, Md.: Lexington, 1999), p. 125.

29. Brecher, *Decisions in Israel's Foreign Policy*, p. 65.

30. Zeev Schiff and Ehud Ya'ari, *Israel's Lebanon War* (New York: Simon and Schuster, 1984), pp. 34, 55, 57, 127, 163; Schiff, *A History of the Israeli Army*, pp. 239–240; Martin Gilbert, *Israel: A History* (New York: William Morrow, 1998), p. 504; Lieber and Yaniv, "Personal Whim or Strategic Imperative?" p. 137; and Trevor N. Dupuy and Paul Martell, *Flawed Victory: The Arab-Israeli Conflict and the 1982 War in Lebanon* (Fairfax, Va.: HERO Books, 1986), pp. 60, 148.

31. Schiff and Ya'ariv, *Israel's Lebanon War*, pp. 33, 39, 41, 58, 97, 100, 101, 103, 113, 266–268, 303, 304; Shlaim, *The Iron Wall*, p. 397; Schiff, *A History of the Israeli Army*, p. 246; Richard Gabriel, *Operation Peace for Galilee: The Israeli-PLO War in Lebanon* (New York: Hill and Wang, 1984), p. 68; and Dupuy and Martell, *Flawed Victory*, pp. 96, 142.

32. Schiff and Ya'ariv, *Israel's Lebanon War*, pp. 43, 164, 181, 212–213; and Gabriel, *Operation Peace for Galilee*, p. 158.

33. Schiff and Ya'ariv, *Israel's Lebanon War*, p. 293; Benny Morris, *Righteous Victims: A History of the Zionist-Arab Conflict, 1881–1999* (New York: Knopf, 1999), p. 590; Anthony H. Cordesman, *The Arab-Israeli Military Balance and the Art of Operations* (Washington, D.C.: American Enterprise Institute, 1987), pp. 55–56; and Dupuy and Martell, *Flawed Victory*, p. 154.

34. Schiff and Ya'ariv, *Israel's Lebanon War*, p. 284; Sachar, *A History of Israel*, p. 920; and Shlaim, *The Iron Wall*, p. 419.

at the Sabra and Shatila refugee camps), but he remained in the cabinet and eventually became Israel's prime minister.[35] Shamir also later became prime minister. There is little evidence that their domestic political fates affected either Begin's or Sharon's calculations about the advisability of the war with Syria. For Begin, intervention in Lebanon was largely about saving the Christian community from slaughter at the hands of the Muslims.[36] For Sharon, the war was about advancing his plan to fundamentally reshape the Middle East to increase Israel's security.[37]

Finding cases where democracies did not start losing wars is as important for proving the triumphalists' selection effects argument as identifying cases they did start and win. Reiter and Stam briefly mention only two candidates, however: the 1898 Fashoda crisis and the 1911 Moroccan crisis. In neither instance do they show that the specific mechanisms of democracy led France to avoid war. Indeed, in the latter case they quote French Prime Minister Joseph Caillaux endorsing Napoleon's advice not to go to war unless the chances of victory were higher than 70 percent. Because Napoleonic France was an autocracy, it is not true that only democracies are selective about their wars.[38] The triumphalists need to do much more work to identify cases of democracies not going to war because they thought they would lose, and demonstrate that this assessment was the result of the specific mechanisms of the triumphalists' selection effects argument.

In sum, the causal mechanisms that the triumphalists believe encourage democratic leaders to be more cautious about the wars they wage do not explain victory in a large number of cases. This weakens scholars' confidence in their selection effects theory of democratic victory.

WARTIME EFFECTIVENESS. In "Democracy and Victory," I offered two reasons why the triumphalists' wartime effectiveness arguments are unpersuasive. First, the logic behind their causal mechanisms is flawed; and second, their evidence is weak. In their responses, the triumphalists are far less vigorous in defending their causal mechanisms. Reiter and Stam, for example, concede that the evidence undergirding their argument that soldiers in democracies fight better is problematic (p. 179); and Choi also admits that the audience costs ar-

---

35. Shlaim, *The Iron Wall*, p. 417; and Schiff, *A History of the Israeli Army*, pp. 238, 240.
36. Schiff and Ya'ariv, *Israel's Lebanon War*, pp. 25, 34, 39, 220; Sachar, *A History of Israel*, pp. 900, 913, 916, 920; and Dupuy and Martell, *Flawed Victory*, p. 15.
37. Schiff and Ya'airv, *Israel's Lebanon War*, p. 230.
38. Reiter and Stam, *Democracies at War*, pp. 12–13. John J. Mearsheimer, *Conventional Deterrence* (Ithaca, N.Y.: Cornell University Press, 1983), examines cases of both democracies (Britain, France, India, and Israel) and nondemocracies (Egypt, Germany, North Korea, the Soviet Union, and Vietnam) that started wars only when they thought they had a high chance of victory.

gument does not provide a solid foundation for the claim that democracies make better allies (p. 143). In fact, these various battlefield effectiveness arguments are controversial among the triumphalists themselves. In their book, Reiter and Stam reject Lake's theory that the lack of state rent-seeking makes democracies better allies, and they conclude that realist theories of alliance behavior seem to best explain the alignment decisions of both democracies and nondemocracies.[39]

In her response, Choi seeks to rescue the triumphalist alliance argument by offering a new theory that holds that democracies, because they are more transparent and have checks and balances, enjoy an advantage in war because they form cohesive and reliable alignments. I have four reservations about her argument. First, the number of democratic partners in an alliance—the measure for the independent variable in her quantitative analysis—does not really capture her causal mechanisms: that the presence of checks and balances and greater transparency in democracies makes them more reliable allies.

Second, there is reason to question whether greater transparency makes democratic alliances more reliable. Often, rather than clarifying what is going on in a democracy, transparency overwhelms outside observers with excessive and contradictory information.[40] Also, the number of checks and balances can vary from democracy to democracy, and it is possible to have them in nondemocracies too. In other words, it is not clear that transparency and checks and balances make democracies better allies.

Third, historically there have been few all-democratic alliances, which presents something of a puzzle for Choi's argument. If a larger number of democracies make an alliance more reliable, cohesive, and thereby more effective, then the most effective alliances should be those composed purely of democracies. In the COW data set, though, there is only one purely democratic alliance—Britain, France, and Israel in the Suez War—and that alliance was not a model of cohesiveness and reliability.[41] Moreover, if nondemocracies are less attractive allies than democracies, why have most alliances involved a combination of both democracies and nondemocracies?

Finally, Choi does not solve the problem of determining which state made the greater contribution to victory in a mixed alliance. Everything she says about the relations between Britain and the United States being more harmoni-

39. Reiter and Stam, *Democracies at War*, p. 105.
40. Bernard I. Finel and Kristin M. Lord, "The Surprising Logic of Transparency," *International Studies Quarterly*, Vol. 43, No. 2 (June 1999), pp. 315–339.
41. Shlaim, *The Iron Wall*, pp. 162–185.

ous than relations between the democracies and the Soviet Union during World War II may be true; nevertheless, the Soviet Union played the principal role in the defeat of Germany. "Few would now contest," notes historian Richard Overy, "that the Soviet war effort was the most important factor . . . in the defeat of Germany."[42]

Conclusion

Democracies seem to have won most of their wars over the last 200 years and so the triumphalists have identified an intriguing empirical regularity. Yet there remain significant problems both with how the triumphalists test their claim that democracy is the key to victory and with the causal mechanisms that they suggest explain why democracies have so often been on the winning side of wars. In other words, the democratic triumphalists have not yet provided a compelling explanation for the correlation between democracy and victory. Given this, I hope that our exchange spurs more research by both proponents and critics of the proposition that democracy is the key to military victory.

---

42. Richard Overy, *Russia's War: A History of the Soviet War Effort, 1941–1945* (Harmondsworth, U.K.: Penguin, 1997), p. xi.

How Smart and Tough Are Democracies?

Alexander B. Downes

Reassessing Theories of Democratic Victory in War

$\mathbf{T}$he argument that democracies are more likely than nondemocracies to win the wars they fight—particularly the wars they start—has risen to the status of near-conventional wisdom in the last decade. First articulated by David Lake in his 1992 article "Powerful Pacifists," this thesis has become firmly associated with the work of Dan Reiter and Allan Stam. In their seminal 2002 book, *Democracies at War*, which builds on several previously published articles, Reiter and Stam found that democracies win nearly all of the wars they start, and about two-thirds of the wars in which they are targeted by other states, leading to an overall success rate of 76 percent. This record of democratic success is significantly better than the performance of dictatorships and mixed regimes.[1]

Reiter and Stam offer two explanations for their findings. First, they argue that democracies win most of the wars they initiate because these states are systematically better at choosing wars they can win. Accountability to voters gives democratic leaders powerful incentives not to lose wars because defeat is likely to be punished by removal from office. The robust marketplace of ideas in democracies also gives decisionmakers access to high-quality information regarding their adversaries, thus allowing leaders to make better decisions for war or peace. Second, Reiter and Stam argue that democracies are superior war fighters, not because democracies outproduce their foes or overwhelm them with powerful coalitions, but because democratic culture produces soldiers who are more skilled and dedicated than soldiers from non-

Alexander B. Downes is Assistant Professor of Political Science at Duke University.

For constructive feedback on previous versions of this article, the author would like to thank Jonathan Caverley, Kathryn McNabb Cochran, Michael Desch, Matthew Fuhrmann, Christopher Gelpi, Sarah Kreps, Matthew Kroenig, Jason Lyall, Sebastian Rosato, Joshua Rovner, Elizabeth Saunders, John Schuessler, Todd Sechser, Jessica Weeks, seminar participants at Harvard University's John M. Olin Institute for Strategic Studies and the Massachusetts Institute of Technology's Security Studies Program, as well as the anonymous reviewers.

1. David A. Lake, "Powerful Pacifists: Democratic States and War," *American Political Science Review*, Vol. 86, No. 1 (March 1992), pp. 24–37; and Dan Reiter and Allan C. Stam, *Democracies at War* (Princeton, N.J.: Princeton University Press, 2002), pp. 28–33. See also Bruce Bueno de Mesquita and Randolph M. Siverson, "War and the Survival of Political Leaders: A Comparative Study of Regime Types and Political Accountability," *American Political Science Review*, Vol. 89, No. 4 (December 1995), p. 852; and Bruce Bueno de Mesquita, Alastair Smith, Randolph M. Siverson, and James D. Morrow, *The Logic of Political Survival* (Cambridge, Mass.: MIT Press, 2003), chap. 6.

International Security, Vol. 33, No. 4 (Spring 2009), pp. 9–51
© 2009 by the President and Fellows of Harvard College and the Massachusetts Institute of Technology.

democratic societies.[2] Because democratic culture emphasizes individualism, and because democratic citizens are fighting for a popular government, soldiers from democracies are more likely to take the initiative, exhibit better leadership, and fight with higher morale on the battlefield than soldiers from repressive societies.

Reiter and Stam test these selection effects and war-fighting arguments using quantitative methods. In an analysis of interstate war outcomes from 1816 to 1990, for example, they find that war initiators and targets are significantly more likely to prevail as they become more democratic. Using a data set of battles in wars from 1800 to 1982, Reiter and Stam also find that soldiers from democracies exhibit greater initiative and leadership in battle, but not necessarily higher morale.

Although many scholars have criticized Reiter and Stam's arguments and findings,[3] few have challenged the data analysis of interstate war outcomes that constitutes the chief piece of evidence for their argument that democracies win wars more frequently than autocracies.[4] In this article, I reexamine the quantitative evidence on war outcomes that Reiter and Stam offer in support of their arguments. I begin by laying out in greater detail the selection effects and war-fighting theories of democratic effectiveness, and briefly summarize some of the main criticisms that have been leveled against these arguments.

In the next section, I reanalyze Reiter and Stam's data on war outcomes and show that their statistical results are not robust. I suggest two alterations to the analysis. First, Reiter and Stam code all states that do not initiate wars as targets even when these states joined the war later on the initiator's side. In actu-

---

2. For the democratic power argument, see Lake, "Powerful Pacifists"; and Bueno de Mesquita et al., *The Logic of Political Survival,* pp. 232–236, 257–258. Bueno de Mesquita and colleagues also argue that democracies start wars only if they are "near certain of victory" (p. 240).
3. Michael C. Desch, "Democracy and Victory: Why Regime Type Hardly Matters," *International Security,* Vol. 27, No. 2 (Fall 2002), pp. 5–47; Michael C. Desch, "Democracy and Victory: Fair Fights or Food Fights?" *International Security,* Vol. 28, No. 1 (Summer 2003), pp. 180–194; Michael C. Desch, *Power and Military Effectiveness: The Fallacy of Democratic Triumphalism* (Baltimore, Md.: Johns Hopkins University Press, 2008); Ajin Choi, "The Power of Democratic Cooperation," *International Security,* Vol. 28, No. 1 (Summer 2003), pp. 142–153; Ajin Choi, "Democratic Synergy and Victory in War, 1816–1992," *International Studies Quarterly,* Vol. 48, No. 3 (September 2004), pp. 663–682; Risa A. Brooks, "Making Military Might: Why Do States Fail and Succeed? A Review Essay," *International Security,* Vol. 28, No. 2 (Fall 2003), pp. 149–191; Stephen Biddle and Stephen Long, "Democracy and Military Effectiveness: A Deeper Look," *Journal of Conflict Resolution,* Vol. 48, No. 4 (August 2004), pp. 525–546; and Jasen J. Castillo, "The Will to Fight: National Cohesion and Military Staying Power," unpublished manuscript, Texas A & M University, July 2008.
4. The exceptions are Desch, "Democracy and Victory: Why Regime Type Hardly Matters"; Desch, "Democracy and Victory: Fair Fights or Food Fights?"; and Desch, *Power and Military Effectiveness.* I discuss Desch's critique further below.

ality there are three categories of states: initiators, targets, and joiners. I code a variable for war joiners and add it to the analysis. Second, Reiter and Stam exclude draws from their analysis of war initiation and victory with little justification given that costly stalemates can threaten the tenure of democratic leaders. I therefore add wars that ended in draws to the data set. I use ordered probit and multinomial logit models on a dependent variable consisting of wins, draws, and losses to show that democracies of all types—initiators, targets, and joiners—are not significantly more likely to win wars.

I then analyze the relationship between democracy and victory from another angle to provide a more complete understanding of how democratic politics might affect decisions to go to war in ways less positive than those outlined in the selection effects argument. I do so through an in-depth case study of the Johnson administration's decision to begin bombing North Vietnam in 1965 and then to send large numbers of U.S. ground troops into action in South Vietnam later that year. Vietnam is an anomaly for the selection effects argument in two ways. First, the United States initiated the interstate phase of the war, but at best emerged from Vietnam in 1973 with a costly draw; others code the war as a loss for the United States.[5] Second, President Lyndon Johnson and his key subordinates chose to fight in Vietnam even though they understood that the prospects for a quick and decisive victory were slim. The selection effects argument, however, implies that because democratic leaders are cautious in selecting only those wars they are highly likely to win, democratic elites who choose to initiate or enter wars should be confident of victory.

The case study investigates this anomalous case for selection effects and the puzzle of U.S. escalation to develop new theories of how democracy affects leaders' choices to go to war.[6] First, I briefly make the case for coding the United States as either the initiator or a joiner of the Vietnam War. Although Reiter and Stam code the Vietnam War as being initiated by North Vietnam, in fact the United States was the first state to use interstate force when it began the bombing of North Vietnam in February 1965.[7] More important, Vietnam was a war of choice for the United States, meaning that the war did "not result

---

5. Reiter and Stam code the war as a draw; the Correlates of War data set codes it as a defeat for the United States. See Reiter and Stam, *Democracies at War*, p. 56; and Correlates of War Interstate War Data, ver. 3.0, http://www.correlatesofwar.org.
6. On the use of case studies for this purpose, see Alexander L. George and Andrew Bennett, *Case Studies and Theory Development in the Social Sciences* (Cambridge, Mass.: MIT Press, 2005), pp. 20–21, 111–115.
7. See, for example, the Correlates of War (COW) data set.

from an overt, imminent, or existential threat to a state's survival."[8] The choice to intervene in Vietnam, in other words, was the same type of decision as the choice to initiate war, and thus should be governed by the same factors highlighted in the selection effects argument.

Second, I document the prevailing pessimism about the likelihood of victory in Vietnam among the key decisionmakers in Johnson's administration in 1964 and 1965. These men, including the president himself, were deeply pessimistic about the military and political situation in South Vietnam. Moreover, they were not optimistic that bombing the North or introducing U.S. ground troops in the South would coerce Hanoi to stop supporting its Vietcong allies or allow U.S. and South Vietnamese forces to defeat the insurgency. Civilian and military officials alike warned that the war would require several years and hundreds of thousands of U.S. troops with no guarantee of victory. Despite this widespread pessimism in Washington, the president chose to take the United States into Vietnam.

Third, I argue that democratic politics was an important factor in explaining why Johnson decided to fight in Vietnam even though victory appeared unlikely. The case suggests that under certain circumstances democratic processes can compel leaders to embark on wars even when the prospects of winning are uncertain. In Vietnam, for example, President Johnson appeared to believe that he had more to lose from not fighting than from entering a costly and protracted war in Southeast Asia. Johnson judged that his treasured domestic reform agenda, the Great Society, would be killed if he did not stand firm and prevent the "loss" of South Vietnam. Leaders of democracies may thus face pressure to fight abroad to protect their legislative agendas or programs at home. The recent case of Iraq suggests a second mechanism whereby democracy can lead to risky war choices: it may prompt leaders to downplay or minimize the potential costs of conflict to obtain public consent for wars they want to fight for other reasons. Leaders of democracies have incentives not to plan for the postwar era if the costs of regime change, occupation, and nation building are potentially high because divulging those costs to the people beforehand might dampen public ardor for war. Failing to plan for the day after the initial victory, however, increases the likelihood that things will go wrong later and that democracies will blunder into costly quagmires.

---

8. Elizabeth N. Saunders, "Wars of Choice: Leadership, Threat Perception, and Military Intervention," unpublished manuscript, George Washington University, October 2008, p. 1 n. 1.

The article concludes that more research is necessary not only to parse the varied effects of democracy on military effectiveness, but also to investigate alternative factors that may provide superior explanations for military outcomes.

Selection Effects, War Fighting, and Democratic Victory

This section sketches the selection effects and war-fighting arguments that together make up the theory of democratic victory in war.

SELECTION EFFECTS

The selection effects argument posits that democracies are better than nondemocracies at selecting winnable wars. Two facets of democratic institutions provide the reason: electoral accountability and the marketplace of ideas. The first mechanism, electoral accountability, focuses on the political consequences of policy failure. In democracies, it is easier to remove leaders than in nondemocracies: people simply have to go to the polls and cast their votes for a competing candidate in sufficient numbers, and the incumbent is forced out of office. In autocracies, by contrast, it typically takes a violent coup, revolt, or rebellion to oust unpopular leaders because elections are either rigged or nonexistent.

Moreover, the argument maintains that losing a war is a major policy disappointment that is likely to turn the public against the leader responsible. Defeat is costly not only in money expended but also in human terms, namely the nation's sons and daughters whose lives are lost in a failed cause. National pride may also suffer depending on the depth of the humiliation caused by the adversary. Public anger is likely to be more intense if the leader who lost the war is also the one who started it.

This combination of ease of removal in democracies and the likelihood that policy failure—in the form of losing the war—will turn the public against the leader and increase the likelihood that he or she will suffer defeat in the next election induces a healthy dose of caution in democratic elites. As Reiter and Stam put it, "Because democratic executives know they risk ouster if they lead their state to defeat, they will be especially unwilling to launch risky military ventures. In contrast, autocratic leaders know that defeat in war is unlikely to threaten their hold on power. As a result, they will be more willing to initiate risky wars that democracies avoid. . . . Simply put, compared to other kinds of states, democracies require a higher confidence of victory before they are willing to launch a war. . . . The prediction that follows is that democracies are especially likely to win wars that they initiate."[9]

The second causal mechanism in the selection effects argument posits that democratic leaders are able to make better decisions because they have access to high-quality information. Democratic policymakers' "estimates of the probability of winning," in other words, "are more accurate representations of their actual probabilities of victory."[10] The main reason for this information advantage is the freewheeling, competitive marketplace of ideas in democracies. Freedom of speech and freedom of the press, for example, permit the expression of a multiplicity of voices and viewpoints on foreign policy, which in turn leads to better policymaking. "The proposition that the vigorous discussion of alternatives and open dissemination of information in democratic systems produce better decisions," write Reiter and Stam, "is an idea at the core of political liberalism." An energetic press also limits the ability of the government to misrepresent the facts or purvey "unfounded, mendacious, or self-serving foreign policy arguments," as does the presence of opposition parties hoping to displace the current regime and gain power for themselves.[11] The virtues of this public discussion are augmented by the unvarnished and outstanding military advice that policymakers receive from a professional and meritocratic officer corps.[12] The marketplace of ideas thus improves the overall quality of information available, encourages healthy debate among a variety of alternatives, and places limits on political actors' ability to mislead the public.

WAR FIGHTING

The second part of Reiter and Stam's theory contends that democracies are superior war fighters once engaged in conflict. Some scholars argue that democracies prevail because they are wealthier than autocracies and mobilize a greater share of that wealth in wartime.[13] Others maintain that democracies are more likely to come to each other's defense—thereby forming overwhelm-

---

9. Reiter and Stam, *Democracies at War*, p. 20.
10. Ibid., p. 23.
11. Ibid.
12. Ibid., pp. 23–24. Militaries in authoritarian states, by contrast, sometimes represent the greatest threat to the survival of the regime and are thus kept weak and staffed by incompetent (but loyal) officers. This practice undermines both the quality of military advice and the army's ability to fight. See, for example, Stephen Biddle and Robert Zirkle, "Technology, Civil-Military Relations, and Warfare in the Developing World," *Journal of Strategic Studies*, Vol. 19, No. 2 (June 1996), pp. 174, 179–180.
13. Lake, "Powerful Pacifists"; and Bueno de Mesquita et al., *The Logic of Political Survival*.

ing countercoalitions—and are also better able to cooperate in prosecuting the war.[14] Reiter and Stam reject these claims and advance an alternative explanation for democratic battlefield success. They put forward three related propositions regarding the influence of democracy on tactical military effectiveness. First, the popular nature of democratic governments engenders trust, loyalty, and consent among soldiers, which translates into an army that fights with higher morale than an autocratic army. Second, the emphasis on individual rights and freedoms in liberal democracies produces soldiers willing to take the initiative in battle. Third, amicable civil-military relations in democracies mean that the cream rises to the top in democratic militaries rather than incompetent regime loyalists as in many autocracies. Because of this meritocratic system, soldiers from democracies display better leadership skills in battle.[15]

EVIDENCE

Reiter and Stam test their arguments using quantitative methods. Examining a data set of interstate wars between 1816 and 1990, they find that democracies won 93 percent of the wars they started, compared with 60 percent for dictatorships and 58 percent for mixed regimes. Democratic targets also won a majority of their wars (63 percent), although at a lower rate than democratic initiators, but still higher than other regime types (34 percent for dictatorships and 40 percent for oligarchies). In a multivariate probit model controlling for traditional variables such as relative material capabilities, material contributions from allies, terrain, and choice of military strategy, both democratic initiators and democratic targets correlate positively and significantly with victory. Reiter and Stam also perform a separate analysis on a sample of battles taken from the Historical Evaluation and Research Organization (HERO) data set. They find that although democracy is not associated with higher morale, democratic armies did display greater levels of initiative and leadership in battle.[16]

Reevaluating the Statistical Evidence for Democracy and Victory

A measure of how tremendously influential Reiter and Stam's work has been in the field is the substantial amount of criticism it has elicited. Several critics

---

14. Lake, "Powerful Pacifists"; and Choi, "Democratic Synergy and Victory in War."
15. Reiter and Stam, *Democracies at War*, pp. 60–71. Reiter and Stam also argue that because democracies do not abuse enemy prisoners of war, enemy soldiers are more willing to surrender to democratic armies rather than fight to the death. They do not test this argument, however.
16. Ibid., pp. 29, 45, 78–81.

of the selection effects argument, for example, cite evidence from the George W. Bush administration's decision to attack Iraq in 2003 to argue that the democratic marketplace of ideas does not operate as theorized and is incapable of providing much of a constraint on powerful executives.[17] Others target the electoral accountability mechanism, arguing that authoritarian leaders may actually be more cautious about going to war than democrats because autocrats who lose wars are sometimes exiled or killed, whereas democrats may be removed from office but are rarely punished.[18] Indeed, one statistical study finds no evidence that defeat in crisis or war raises democratic leaders' risk of losing office, or that prevailing in such conflicts lowers the risk of removal.[19] Case studies have also turned up tepid support for electoral accountability, finding instead that democratic leaders often keep their own counsel and initiate wars that lack broad public support.[20]

Scholars have also questioned the arguments and evidence for the warfighting explanation of democratic victory. Some contend that alternative explanations—such as gross imbalances of power or alliances with nondemocratic states in which the autocracy made up the bulk of the alliance's military power—account for many democratic victories. Once these "unfair fights" are removed, democracies perform only slightly above average.[21] Others point out that the HERO data set used by Reiter and Stam in their analysis of battlefield

---

17. Chaim Kaufmann, "Threat Inflation and the Failure of the Marketplace of Ideas: The Selling of the Iraq War," *International Security,* Vol. 29, No. 1 (Summer 2004), pp. 5–48; Jon Western, "The War over Iraq: Selling War to the American Public," *Security Studies,* Vol. 14, No. 1 (January 2005), pp. 106–139; and Jane Kellett Cramer, "Militarized Patriotism: Why the U.S. Marketplace of Ideas Failed before the Iraq War," *Security Studies,* Vol. 16, No. 3 (July 2007), pp. 489–524.
18. Desch, "Democracy and Victory: Why Regime Type Hardly Matters," pp. 9–16. See also H.E. Goemans, *War and Punishment: The Causes of War Termination and the First World War* (Princeton, N.J.: Princeton University Press, 2000). For an argument that leaders in certain types of autocracies are just as accountable domestically as leaders in democracies, see Jessica L. Weeks, "Autocratic Audience Costs: Regime Type and Signaling Resolve," *International Organization,* Vol. 62, No. 1 (Winter 2008), pp. 35–64.
19. Giacomo Chiozza and H.E. Goemans, "International Conflict and the Tenure of Leaders: Is War Still *Ex Post* Inefficient?" *American Journal of Political Science,* Vol. 48, No. 3 (July 2004), p. 613.
20. Desch, "Democracy and Victory: Fair Fights or Food Fights?" pp. 189–192; and Desch, *Power and Military Effectiveness,* pp. 50–51, 72–76, 97–104.
21. Desch, "Democracy and Victory: Why Regime Type Hardly Matters," pp. 9–16. Desch deemed 54 of the 75 wars in Reiter and Stam's data set to be unfair fights. Of the 21 remaining conflicts, he credited the more democratic side with 12 wins and 9 losses for a winning record of 57 percent. This move sparked a serious methodological dispute regarding the legitimacy of dropping cases versus using control variables. See David A. Lake, "Fair Fights? Evaluating Theories of Democracy and Victory"; Dan Reiter and Allan C. Stam, "Understanding Victory: Why Political Institutions Matter"; and Desch, "Democracy and Victory: Fair Fights or Food Fights?" all in *International Security,* Vol. 28, No. 1 (Summer 2003), pp. 154–167, 168–179, and 180–194, respectively.

effectiveness is plagued by measurement error and dubious intercoder reliability, and that the data set's leadership, initiative, and morale variables rely on subjective, unverifiable judgments rather than hard data.[22] A third critique suggests that the relationship between democracy and victory is spurious: wealthy countries are more likely to become democratic, and wealth also explains war outcomes.[23]

Although many of these critiques raise valid issues and warrant further research, few of them directly confront Reiter and Stam's large-*n* evidence that democratic war initiators and targets are significantly more likely to prevail. In this section, therefore, I revisit Reiter and Stam's core statistical analysis. Although others have criticized their coding of particular war initiators or outcomes and questioned their approach of using regression analysis with control variables—and I share some of these concerns—coding issues are not the main basis of my critique. Nor do I reject Reiter and Stam's multivariate regression approach. Rather, I argue that the statistical significance of their results is critically affected by two debatable decisions: the choice to code all noninitiators of wars as targets, and the decision to omit wars that ended in draws. I reanalyze their data after dividing states into initiators, targets, and joiners, and adding draws to the data set.[24] After making these changes, I find that democratic initiators, targets, and joiners are not significantly more likely to win wars. The fragility of the statistical evidence to reasonable alternative coding and modeling choices raises doubts about the robustness of the finding that democracies are better at choosing and fighting wars.

---

22. Desch, "Democracy and Victory: Why Regime Type Hardly Matters," pp. 39–40. Brooks also notes that the initiative variable in HERO—which Reiter and Stam use as an indicator of individual battlefield initiative—actually codes which side attacked first. Brooks, "Making Military Might," pp. 181–182.

23. Desch, *Power and Military Effectiveness*, pp. 171–173. Similarly, Biddle and Long find that after controlling for factors such as amicable civil-military relations and human capital, democracy actually reduces battlefield effectiveness. Biddle and Long, "Democracy and Military Effectiveness." Several recent studies also find that democracy makes no difference or actually reduces the likelihood of victory in counterinsurgency wars. See Jason Lyall and Isaiah Wilson III, "Rage against the Machines: Explaining Outcomes in Counterinsurgency Wars," *International Organization*, Vol. 63, No. 1 (January 2009), pp. 67–106; Jason Lyall, "Do Democracies Make Inferior Counterinsurgents? Reassessing Democracy's Impact on War Outcomes and Duration," *International Organization* (forthcoming); and Jonathan D. Caverley, "Democracies Will Continue to Fight Small Wars . . . Poorly," paper presented at the annual meeting of the International Studies Association, New York, February 2009, p. 31.

24. Reiter and Stam report that adding draws did not much change the results. They also recoded as initiators states that joined wars later, reporting that this weakened the results. They do not appear to have tested these two changes simultaneously, however. Reiter and Stam, *Democracies at War*, p. 213 n. 62, and p. 40.

REITER AND STAM'S APPROACH

Reiter and Stam gathered data on war outcomes for interstate wars occurring between 1816 and 1990. Arguing that their hypotheses "relate only to the likelihood of victory," Reiter and Stam dropped all conflicts that ended in draws, leaving 197 cases in the data set.[25] The authors then coded each belligerent's level of democracy as adjudicated by the widely used Polity index, a 21-point scale normalized around zero (ranging from −10 to +10), and whether that belligerent initiated the war or was a target. The coding of initiation relies primarily on the Correlates of War (COW) project, which defines a war initiator as the first state to use force in an interstate dispute that resulted in at least 1,000 battle deaths.[26] Reiter and Stam note two instances in which they changed COW's initiation coding: the Crimean War (1853–56, initiator changed from Turkey to Russia) and the First Balkan War (1912–13, changed to include as initiators all three members of the Balkan League, which declared war simultaneously).[27] All states that are not coded as war initiators are coded as targets. This includes states that were actually attacked by the initiator, as well as states that joined the war later on either side, including as allies of the initiator. Britain and France, for example, are coded as targets in the Crimean War even though they joined the conflict a year after it started. Turkey is also coded as a target in the eastern theater of World War I despite having joined the side of the states Reiter and Stam coded as the war's initiators, Germany and Austria. Reiter and Stam also coded several other variables that should affect the likelihood of victory, including states' relative power and the power of their allies, the quality of their troops, the two sides' military strategies, and the roughness of the terrain over which the war was fought.

A pair of interaction terms—between states' Polity scores and war initiation, on the one hand, and Polity score and war target, on the other—tests Reiter and Stam's selection effects and war-fighting arguments. Each of these terms is

---

25. Ibid., p. 39. Reiter and Stam "coded war outcomes . . . based on whether a state achieved its immediate military aims." By contrast, wars that "end in what is essentially the prewar status quo" are considered draws. Reiter and Stam, "Understanding Victory," p. 173; and Reiter and Stam, *Democracies at War*, p. 211 n. 41.
26. This is a rather narrow conception of war initiation, and it is unclear if it always captures the logic of the selection effects argument, which is about how wisely states choose war as a policy option. Using force first in a dispute is not always a good indicator of whether a government chose to go to war, or had war thrust upon it by the choice of another state.
27. Reiter and Stam also changed COW's coding of the initiator of the Russo-Polish War (1919–20), Changkufeng (1938), and Vietnam (1965–73); did not code an initiator in the Vietnamese-Cambodian War (1975–79); and divided the Yom Kippur War (1973) into two parts, crediting Israel with two victories. Reiter and Stam changed a few war outcome codings as well, but because restoring them to their original values does not affect the results, I defer discussion of these cases to an online appendix, http://www.duke.edu/~downes.

positive and significant in a probit regression controlling for other determinants of victory, signifying that both initiators and targets become more likely to win as they become increasingly democratic. I was able to replicate Reiter and Stam's main model from *Democracies at War;* the results appear in model 1 of table 1.[28] The coefficients and standard errors for all variables are nearly identical to those reported in the book.[29]

Previous appraisals of Reiter and Stam's work have criticized some of their codings of war initiation and outcomes. Michael Desch, for example, contends that Reiter and Stam misidentify the initiator in three wars—the Boxer Rebellion (1900), the Hungarian War (1919), and World War II in Europe (1940–45), initiated by autocracies rather than democracies—and mistakenly code the outcome in another two—Israeli draws rather than victories in the War of Attrition (1969–70) and the First Lebanon War (1982).[30] Making these suggested coding changes and rerunning the analysis renders the Polity × target interaction statistically insignificant, as shown in model 2, but does not alter Reiter and Stam's finding that war initiators are more likely to win as they become more democratic. The utility of this critique is limited, however, because it does not address the more fundamental issues in Reiter and Stam's analysis, that is, their amalgamation of war joiners and war targets and their decision to exclude an entire class of war outcomes. I take up these issues below.[31]

---

28. Reiter and Stam, *Democracies at War,* p. 45, table 2.2, model 4. Replication data are available at http://dvn.iq.harvard.edu/dvn/dv/stam.
29. The significance levels, however, differ slightly. The *p*-value for democratic targets, for example, is only 0.02, not <0.01 as stated in the text. Similarly, the *p*-value for war initiator is 0.008, rather than <0.001.
30. Desch, "Democracy and Victory: Fair Fights or Food Fights?" p. 187; and Desch, "Democracy and Victory: Why Regime Type Hardly Matters," p. 14. Desch also credits Israel with only one victory in the Yom Kippur War. Desch, "Democracy and Victory: Why Regime Type Hardly Matters," p. 16. Because Reiter and Stam exclude draws, Desch's three suggested outcome changes leave 192 cases in the analysis.
31. That said, I find some of Reiter and Stam's changes to the COW initiation and war outcome codings to be unconvincing. Clearly Britain and the United States did not initiate World War II in Europe in 1940, for example. I code Germany and Italy as initiators. Even if the two democracies had started the war, this would not support the selection effects argument because they could not have been optimistic in 1940 about the war's eventual outcome. Other changes made by Reiter and Stam that I return to their original values include the United States as the initiator of the Vietnam War; Vietnam as the initiator of the war with Cambodia in 1975; a single war between Israel and the Arab states in 1973; and a draw (rather than an Israeli win) in the War of Attrition. That said, I leave Reiter and Stam's coding of the Vietnam War as a draw unchanged. Coding the war as a loss for the United States and South Vietnam (as the COW data set does) further weakens the result reported below for democratic initiators.

Table 1. Probit and Ordered Probit Models of War Outcomes

	Model 1 Reiter and Stam Replication (probit, dependent variable is win/lose) †	Model 2 Reiter and Stam's Model with Desch's Suggested Changes (probit, dependent variable is win/lose) ‡	Model 3 Tripartite Coding of Belligerents, Tripartite Coding of War Outcomes (ordered probit, dependent variable is win/draw/lose)
Polity (−10 to +10) × Initiation	0.068** (0.030)	0.082*** (0.031)	—
Polity (−10 to +10) × Target	0.064** (0.028)	0.042 (0.030)	—
Polity (1 to 21)	—	—	0.030 (0.045)
Initiation	0.91*** (0.34)	1.00** (0.49)	0.37 (0.56)
Target	—	—	−0.16 (0.60)
Polity (1 to 21) × Initiation	—	—	−0.009 (0.047)
Polity (1 to 21) × Target	—	—	−0.008 (0.048)
Relative capabilities	3.73*** (0.52)	3.99*** (0.79)	2.35*** (0.49)
Alliance contribution	4.72*** (0.68)	4.99*** (1.09)	3.00*** (0.74)
Quality ratio	0.05 (0.03)	0.06 (0.04)	0.04* (0.02)
Terrain	−10.93*** (2.94)	−13.11*** (3.92)	−1.89* (1.13)
Strategy × Terrain	3.56*** (0.97)	4.17*** (1.30)	0.50 (0.36)
Strategy 1	7.24** (2.89)	8.80** (3.73)	−0.47 (1.30)
Strategy 2	3.48* (1.99)	3.81 (2.70)	−2.55*** (0.88)
Strategy 3	3.36** (1.43)	4.06** (1.83)	−0.21 (0.65)
Strategy 4	3.07** (1.25)	3.43** (1.46)	1.11 (0.73)
Constant	−5.52*** (1.70)	−6.20*** (2.18)	—

Table 1. *(continued)*

N	197	192	233
Log Pseudo-LL	−64.89	−62.78	−168.46
Wald Chi^2	96.62***	76.26***	95.89***

Robust standard errors in parentheses (model 1; clustered on each war, models 2–3).
*$p < 0.10$; **$p < 0.05$; ***$p < 0.01$.
† Dan Reiter and Allan C. Stam, *Democracies at War* (Princeton, N.J.: Princeton University Press, 2002), table 2.2, model 4.
‡ Initiator changed to China in the Boxer Rebellion, Hungary in the Hungarian War, and Germany and Italy in World War II in Europe; outcome of the War of Attrition and First Lebanon War changed from Israeli victories to draws (and thus dropped from the analysis, which includes only wins and losses); and Israel credited with only one win in the Yom Kippur War. Michael C. Desch, "Democracy and Victory: Fair Fights or Food Fights?" *International Security*, Vol. 28, No. 1 (Summer 2003), p. 187; and Michael C. Desch, "Democracy and Victory: Why Regime Type Hardly Matters," *International Security*, Vol. 27, No. 2 (Fall 2002), p. 14.

THE MISSING WAR JOINERS

As described above, Reiter and Stam use two interaction terms to test their democracy arguments: the first multiplies states' Polity scores by war initiator, and the second does the same for Polity and war target. Puzzlingly, Reiter and Stam equate "targets" with "noninitiators" even when a state later joined a war on the initiator's side. The way they code the variable conflates actual targets with states that join wars after they have started.

To take into account this third category, I created a variable called "war joiners" coded 1 if a state entered a war after it began and did not have an alliance or defense pact with the side it joined.[32] Twenty-five states fall into this category. "War initiators," by contrast, are coded as those states that either attacked first or had a preexisting alliance with the attacker and entered the war within one week.[33] Many states, for example, such as Austria in the Second Schleswig-Holstein War (1864), are listed in the COW data set as entering a war on the same date as the "initiator" (in this case, Prussia) but are not coded

32. An alternative way to code joiners would be to define them as states that enter a war after a certain amount of time has passed. I employ a purely temporal definition of belligerency status as a robustness check below.
33. In practice, states tend to enter wars within a few days or wait at least several months, meaning that the one-week cutoff is not sensitive to minor changes. I chose one week because states that wait longer than a few days to enter a conflict cannot reasonably be described as initiators or targets.

as initiators by COW or Reiter and Stam. Austria and Prussia, however, concluded an alliance on January 16, 1864, to enforce the Treaty of London, which granted Schleswig and Holstein autonomy within Denmark, and went to war together two weeks later.[34] This is also the case for many states that enter wars within a few days of the conflict's onset. Italy in the Austro-Prussian War (1866) and Britain and France in the Sinai/Suez War (1956), for instance, concluded prewar alliances with the technical initiators of these conflicts (Prussia and Israel, respectively). In the Suez case, the short delay in the Anglo-French attack was actually part of the agreed plan.[35] In all cases, therefore, formal allies of the state that attacked first which entered the war within seven days are coded as initiators.[36] "War targets," finally, consist of states that are the object of an attack by the war initiator(s) at the outset of the conflict or countries that have an alliance with the victim state and join the war within a week. A choice to attack a state that is part of an alliance ultimately accepts the risk that the allies will fight on behalf of their partner, and thus such allies should also be considered targets.

To clarify these categories, consider World War I in the western theater. Germany (allied with Austria) attacked France to initiate the conflict. Germany and Austria are coded as initiators and France as a target. Several other states, though, including Bulgaria, Greece, Turkey, and the United States were not allied with either the initiators or the targets when the war started, and did not join the war for several months or years. I code these states as joiners. Italy and Romania are also coded as joiners: although members of the Triple Alliance, these states did not go to war in 1914, instead choosing to join the Entente in 1915 and 1916, respectively. The trickiest case is Britain, which—although it was a member of the Triple Entente and entered the war within a few days— never clearly committed to fight on France's behalf in case of a German attack. I nevertheless code Britain as a target by virtue of its alliance membership.[37]

---

34. A.J.P. Taylor, *The Struggle for Mastery in Europe, 1848–1918* (Oxford: Oxford University Press, 1954), pp. 142–145. Taylor's account also indicates that the two powers attacked at the same time: "The two Great Powers then declared that they would act alone; on 1 February their forces crossed the frontier into Sleswick" (p. 146).
35. Jack S. Levy and Joseph R. Gochal, "Democracy and Preventive War: Israel and the 1956 Sinai Campaign," *Security Studies*, Vol. 11, No. 2 (Winter 2001/02), p. 38.
36. The COW initiator variable, which Reiter and Stam follow, is supposedly coded as the first state(s) to use force. Several cases in the COW data set and in *Democracies at War*, however, appear to be coded on the basis of joint declarations of war rather than first uses of force, including the allies in the Boxer Rebellion, the First Balkan War, the eastern and western theaters of World War I, the Hungarian War, and the Palestine War. My coding thus simply applies this rule consistently rather than haphazardly.
37. The details of the case largely support coding Britain as a joiner rather than as a target. Britain's Triple Entente membership did not include a commitment to go to war if France (or Russia)

THE MISSING DRAWS

A second debatable choice in Reiter and Stam's analysis concerns their treatment of draws. In their investigation of democracy, war initiation, and victory, Reiter and Stam contend that their hypotheses bear only on the likelihood of winning, and thus they omit draws. Later in their book, they perform a separate analysis that includes draws where they find that democracies settle for stalemates more quickly than autocracies in wartime.[38]

The omission of draws from the analysis of war initiation and victory, however, is questionable. The selection effects argument stipulates that democratic leaders, fearing possible removal from office for initiating costly or losing wars, choose to start only those wars they think they can win. Given this logic, the justification for excluding draws is tenuous because all outcomes that are not victories are potentially damaging to the incumbent's chances of retaining office. Excluding draws would perhaps be warranted if democratic leaders were never punished for settling for these outcomes, but history demonstrates otherwise. Harry Truman, for example, was politically crippled by the stalemate in Korea and gave up on seeking another term as president because he had little chance of winning re-election. Fifteen years later Lyndon Johnson—faced with plummeting support for the war in Vietnam following the Tet Offensive—similarly ended his bid for another term in office. In between these two U.S. cases, British Prime Minister Anthony Eden resigned shortly after Soviet and American pressure forced him to cut short his joint attack with France and Israel on Egypt after Gamal Nasser nationalized the Suez Canal. Finally, although George W. Bush narrowly managed to defeat a strong challenge from Senator John Kerry in 2004, the president's sky-high approval ratings following the September 11 terrorist attacks plunged when the easy ouster of Saddam Hussein in 2003 gave way to a protracted Sunni insurgency in Iraq. In short, democratic leaders are frequently punished for draws in the same way they are punished for losses. According to the selection effects logic, therefore, democracies should avoid wars they foresee may become protracted stalemates, just as they should shun wars they fear they may lose.[39]

---

was attacked by Germany. The British Cabinet had rejected a more formal alliance in June 1914, similarly rejected a commitment to defend France in July, and engaged in a fierce debate over whether to enter the war when it broke out. Changing the coding of Britain to a joiner does not substantially change any of the results shown below.

38. Reiter and Stam, *Democracies at War*, pp. 164–192. This analysis, it should be noted, includes many of the same variables that Reiter and Stam earlier argued applied only to the likelihood of victory, such as regime type, war initiation, strategy, terrain, and the balance of material capabilities.

39. For a similar approach to dealing with draws, see Jessica L. Weeks, "Leaders, Accountability, and Foreign Policy in Non-Democracies," Ph.D. dissertation, Stanford University, 2008, pp. 17–26;

CHANGES TO THE STATISTICAL ANALYSIS

Introducing a third category of belligerency and war outcomes complicates the statistical analysis, necessitating several changes to Reiter and Stam's model. First, additional variables and interaction terms must be added to the equation. Because all states for Reiter and Stam were either initiators or targets, they were able to omit the Polity term from their regressions (including it results in multicollinearity). To capture the effects of change in democracy on likelihood of victory for initiators and targets, they simply included (in addition to a dummy variable for war initiator) two interaction terms: Polity × initiator and Polity × target. With three categories of states, however, five terms are needed to capture all the relevant effects: β_1Polity, $\beta_2$initiator, $\beta_3$target, β_4Polity × initiator, and β_5Polity × target.[40] β_1 represents the effect of a one-unit increase in Polity score when the state is not an initiator or a target (i.e., it is a joiner). The other terms in the equation have no substantive meaning in isolation, but must be interpreted together. β_1, β_2, and β_4 in combination determine the effect of a one-unit increase in Polity for war initiators; β_1, β_3, and β_5 do the same for war targets.

Because the effects and significance of interaction terms are not immediately obvious from examining coefficients, the substantive impact of these variables must be examined over the range of values they may plausibly take.[41] In this case, given that states that score at both extremes of the Polity index (full autocracies and full democracies) are present in the data set, it is necessary to examine the effect of level of democracy and type of belligerent, as the former varies from its lowest to highest value. To be able to interpret the interaction terms, however, the Polity scores need to be rescaled so that they are not symmetrical around zero. In this case, I have added 11 to each state's Polity score, such that the variable now ranges between 1 and 21.[42]

---

and Lyall and Wilson, "Rage against the Machines." One might argue that leaders in these cases simply got it wrong: information available to them ex ante pointed toward an easy victory, but unforeseeable events that occurred later resulted in a reversal of fortune. This is not true, however, for Korea and Vietnam. Indeed, even Reiter and Stam conclude with regard to Vietnam that "the outlook for the conflict was not promising" in the early stages and that "the American leadership did not in 1965 foresee an imminent victory." See Reiter and Stam, *Democracies at War*, pp. 173, 174.

40. Methodologists have shown that it is imperative to include all constituent variables when using interaction terms. Thomas Brambor, William Roberts Clark, and Matt Golder, "Understanding Interaction Models: Improving Empirical Analyses," *Political Analysis*, Vol. 14, No. 1 (Winter 2006), pp. 66–69; and Bear F. Braumoeller, "Hypothesis Testing and Multiplicative Interaction Terms," *International Organization*, Vol. 58, No. 4 (Fall 2004), p. 809.

41. Reiter and Stam, *Democracies at War*, pp. 52–57.

42. Many of the variables used in Reiter and Stam's analysis were originally coded for this earlier book. Allan C. Stam III, *Win, Lose, or Draw: Domestic Politics and the Crucible of War* (Ann Arbor:

Second, to test whether the omission of draws has any influence on Reiter and Stam's finding that democracies are more likely to win in war, I added to the data set those wars that Reiter and Stam coded as ending in draws. There were sixteen draws involving thirty-seven states.[43] Owing to a missing Polity score for India in the First Kashmir War (1947–48), these additions result in a total of 233 cases in the new data set. I then obtained data for the relevant independent variables for these cases from Stam's *Win, Lose, or Draw* data set.[44]

Third, because the dependent variable is now trichotomous, I use an ordered probit model to estimate the effects of the independent variables on war outcomes (win = 2, draw = 1, loss = 0). This model is specifically designed for dependent variables where outcomes are clearly ordered and ranked (e.g., disagree, agree, strongly agree), but the distances between the categories are unclear.[45] An ordered model is justified in this case because the underlying variable—military effectiveness—is continuous, and these three outcomes— win, draw, or lose—represent ranked and ordered levels of military proficiency.

STATISTICAL RESULTS

Model 3 in table 1 shows the results of an ordered probit regression of war outcomes. Included are the five variables needed to test the joint effects of democracy and war initiator/target/joiner, as well as the same slate of control variables used by Reiter and Stam in *Democracies at War*: relative power, allies' contribution, troop quality, difficulty of terrain, four combinations of attacker/ defender military strategy, and an interaction between strategy and terrain.[46] The coefficients for the key variables are quite different from Reiter and Stam's results in model 1, but this is not very informative given the presence of multiple interaction terms in the model. The real difference in the results becomes apparent only when the substantive effects for regime type on different war outcomes are shown graphically. Figures 1, 2, and 3 plot the marginal effects of

University of Michigan Press, 1996). The data are available at http://dvn.iq.harvard.edu/dvn/dv/ stam.
43. J. Scott Long, *Regression Models for Categorical and Limited Dependent Variables* (Thousand Oaks, Calif.: Sage, 1997), p. 114.
44. Interaction terms cannot be interpreted as marginal effects. Brambor, Clark, and Golder, "Understanding Interaction Models," pp. 71–73.
45. I calculate robust standard errors clustered on each war on the presumption that observations within a war are correlated with each other, but observations across different wars are not. Reiter and Stam also used robust standard errors but do not seem to have clustered them on each war.
46. For detailed descriptions of these variables, see Reiter and Stam, *Democracies at War*, pp. 41–44.

Figure 1. Marginal Effect of Change in Democracy on Probability of Victory for War Initiators (ordered probit, with 95 percent confidence interval; dependent variable = win/draw/lose)

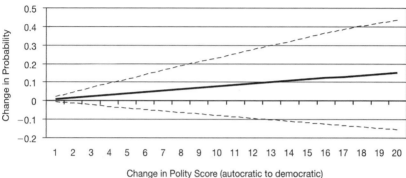

changes in democracy on the likelihood of victory for war initiators, targets, and joiners.[47] Figure 1, for example, shows that moving from the most autocratic to the most democratic regime type increases the likelihood of an initiator winning by about 15 percent. The problem is that the 95 percent confidence interval for this effect includes zero across its entire range, indicating that it is not statistically significant. Figures 2 and 3 similarly demonstrate that as war targets and joiners become increasingly democratic, they become more likely to win (16 and 22 percent, respectively), but the wide confidence intervals show that these estimates are also statistically insignificant. Table 2 summarizes the marginal effects of changing Polity from least to most democratic on the likelihood of victory, draws, and defeat for war initiators, targets, and joiners, with the accompanying 95 and 90 percent confidence intervals. Although in each case the effect is in the expected direction—increasing democracy makes victory more likely and lowers the probability of defeat (the effect on draws is negative but minimal)—in no instance is it statistically significant. In other words, in statistical terms the substantive effects of democracy are indistinguishable from zero, meaning one cannot reject the null hypothesis that de-

47. Marginal effects were calculated using CLARIFY with continuous variables set to their means and binary variables set to their modes. Michael Tomz, Jason Wittenberg, and Gary King, *CLARIFY: Software for Interpreting and Presenting Statistical Results*, ver. 2.1, http://gking.harvard.edu/stats.shtml.

Figure 2. Marginal Effect of Change in Democracy on Probability of Victory for War Targets (ordered probit, with 95 percent confidence interval; dependent variable = win/draw/lose)

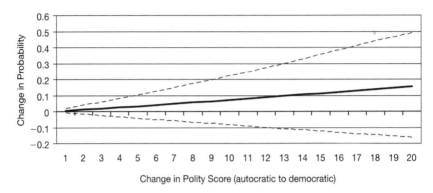

Figure 3. Marginal Effect of Change in Democracy on Probability of Victory for War Joiners (ordered probit, with 95 percent confidence interval; dependent variable = win/draw/lose)

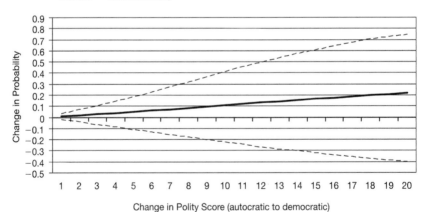

Table 2. Summary of Marginal Effects of Changing Democracy from Minimum to Maximum on All War Outcomes for War Initiators, Targets, and Joiners (from table 1, model 3)

Type of Belligerent	War Outcome	Change in Probability	Standard Error	95 Percent Confidence Interval	90 Percent Confidence Interval
Initiator	Win	0.151	0.154	−0.157, 0.436	−0.112, 0.394
Initiator	Draw	−0.036	0.041	−0.125, 0.035	−0.113, 0.024
Initiator	Lose	−0.114	0.120	−0.346, 0.119	−0.314, 0.085
Target	Win	0.156	0.170	−0.163, 0.487	−0.125, 0.437
Target	Draw	0.003	0.028	−0.056, 0.064	−0.041, 0.051
Target	Lose	−0.159	0.173	−0.490, 0.179	−0.436, 0.136
Joiner	Win	0.218	0.309	−0.401, 0.703	−0.289, 0.695
Joiner	Draw	−0.032	0.054	−0.164, 0.055	−0.135, 0.041
Joiner	Lose	−0.185	0.284	−0.688, 0.397	−0.622, 0.300

mocracy has no effect on the likelihood of victory and defeat for any type of belligerent.[48]

One objection to these results is that I have defined joiners too broadly.[49] In particular, one could argue that belligerent status should be defined solely on the basis of when states enter wars rather than on the basis of prewar alliances. Only states that enter a war well after it has started—more than one week, for example—constitute joiners, whereas those that quickly enter on the attacker's or victim's side should be considered initiators or targets, respectively. To check for the possibility that this different conception of war joining affects the results, I recoded initiators, targets, and joiners along the lines just suggested and reanalyzed the model.[50] The results for initiators become somewhat stronger: shifting from lowest to highest levels of democracy increases the probability of victory 22 percent. This increase, however, still remains insignificant at the most generous level of confidence (90 percent confidence interval = −0.062, 0.478). Results for targets and joiners are weaker than with the previous definition of belligerency.[51] Thus, I conclude that a purely temporal definition of war joiners does not affect the results.

48. It is worth noting that the marginal effects shown in figures 1–3 and table 2 represent the maximum possible effects that democracy can exert on war outcomes. More realistic changes in regime type—such as from an autocracy to a mixed regime, or a mixed regime to a democracy—result in smaller changes in the likelihood of victory and defeat.
49. Details on all robustness and specification checks, as well as the data used in the analysis, may be found on the author's website, http://www.duke.edu/~downes.
50. There are now 18 joiners instead of 25.
51. The increase in the probability of victory that results from moving from least to most democratic is 9 percent for targets and 17 percent for joiners. Neither is significant.

Table 3. Marginal Effects of Changing Democracy from Minimum to Maximum on Probability of Victory for War Initiators, Targets, and Joiners (from model 3 reestimated with multinomial logit)

Type of Belligerent	Change in Probability of Victory	Standard Error	95 Percent Confidence Interval	90 Percent Confidence Interval
Initiator	0.037	0.230	−0.415, 0.467	−0.343, 0.400
Target	0.122	0.211	−0.291, 0.519	−0.232, 0.467
Joiner	0.258	0.265	−0.243, 0.736	−0.173, 0.689

A second possible objection to this analysis is that ordered probit is the wrong statistical model to analyze war outcomes. Ordered logit/probit assumes that each independent variable exerts an effect of the same magnitude and in the same direction on each category of the dependent variable. If a variable increases the likelihood of a draw relative to a loss, in other words, it also increases the probability of a win relative to a draw by a similar amount. These assumptions may be incorrect, however. Multinomial logit allows the effects of the independent variables to differ for each category of the dependent variable. Coefficients are calculated for two war outcomes compared with a third that is omitted (losses and draws compared to wins, for example). To see if the type of statistical estimator influences the results, I reran the analysis using multinomial logit and calculated marginal effects as discussed above. Table 3 shows that the results are consistent with those produced by ordered probit: democratic initiators, targets, and joiners are not statistically more likely to prevail in war.

I performed two other robustness checks. First, if democracies are cleverer about choosing which wars to enter, the selection effect should apply to war joiners as well as initiators. To test this argument, I created a variable coded 1 if a state was either an initiator or a joiner. I interacted this variable with the Polity score and reanalyzed model 3.[52] The results represent only a slight improvement on those for initiators alone: moving from least to most democratic improves the likelihood of victory 19 percent for states that started or joined ongoing wars, but the effect is insignificant (90 percent confidence level = −0.063, 0.426).

Second, given my argument above that draws have been politically damaging to democratic leaders—and hence that draws might be the "functional

---

52. In this case only three variables are needed to estimate the joint effect of democracy and belligerent status: Polity score, initiator/joiner, and Polity × initiator/joiner. The Polity score term picks up the effect of democratic targets.

equivalent" of losses—one could simply recode draws as losses (technically "not wins") and retain a dichotomous dependent variable rather than switch to a tripartite coding of war outcomes. Reestimating the model using probit after making this adjustment confirms the earlier results: democratic initiators, targets, and joiners are not significantly more likely to prevail in war.[53]

In summary, the statistical significance of Reiter and Stam's core results depends on two debatable choices: first the decision to amalgamate war targets and joiners, and second the decision to exclude draws. When targets and joiners are disaggregated, and draws are included in the analysis, no type of belligerent—initiators, targets, or joiners—is significantly more likely to win (or less likely to lose) as it becomes more democratic. This finding is robust to a variety of specification checks.

One might concede that my reanalysis shows that Reiter and Stam's findings do not hold for the time period they analyzed but point out that the last war in their original data set—the Second Sino-Vietnamese War (1985)—began more than twenty years ago. By my count, nine interstate wars (possibly ten, depending on how one counts the Russo-Georgian conflict of 2008) have been fought since that time—Chad-Libya (1987), Iraq-Kuwait (1990), Persian Gulf (1991), Armenia-Azerbaijan (1992–94), Ethiopia-Eritrea (1998–2000), India-Pakistan (1999), the Kosovo War between NATO and Yugoslavia (1999), United States-Afghanistan (2001), and United States-Iraq (2003)—several of them with positive outcomes for democratic belligerents.[54] It is possible that the addition of these post–Cold War contests may improve the statistical significance of the relationship between democracy and victory.[55] What is unclear is whether such a finding would really bolster confidence in democratic military effectiveness. The process by which the United States went to war with Iraq, for example, was deeply flawed, and this defective process contributed to the insurgency that has plagued the country ever since.[56] Similarly, few predicted that the Taliban would be defeated so quickly, and the political goals of the

---

53. Changes in probability of victory that result from shifting from least to most democratic (with 90 percent confidence intervals) are as follows: initiators, 0.008 (−0.327, 0.344); targets, 0.124 (−0.209, 0.461); joiners 0.247 (−0.244, 0.703).

54. Each of these wars—with the possible exceptions of Iraq's conquest of Kuwait and the Kosovo War—exceeded the 1,000 battle-death threshold.

55. I cannot test this conjecture because I do not have data for all of Reiter and Stam's variables for recent wars.

56. See, for example, Kaufmann, "Threat Inflation and the Marketplace of Ideas"; James Fallows, "Blind into Baghdad," *Atlantic Monthly*, Vol. 293, No. 1 (January/February 2004), pp. 52–74; and Thomas E. Ricks, *Fiasco: The American Military Adventure in Iraq* (New York: Penguin, 2006).

war—the destruction of al-Qaida and stable democracy for Afghanistan—have not been realized, as the writ of Hamid Karzai's government remains limited to Kabul, and al-Qaida (and the Taliban) has acquired a new sanctuary in the mountainous tribal areas of Pakistan. President Bill Clinton began a war with Yugoslavia essentially without a strategy, believing that a few days of bombing would cause Slobodan Milošević to give up, and President George H.W. Bush decided to confront Iraq in 1991 when many believed doing so would be a costly bloodbath.[57] Democratic India repelled Pakistan's cross-border intrusion into Kashmir in 1999, but Pakistan—which initiated the war—at that time was also a democracy. Given the shaky statistical foundation of the democracy and victory findings, process tracing of the decisions to start wars and the sources of victory in wars becomes all the more important for testing theories. Others have begun this task,[58] and I provide my own contribution with the U.S. decision to fight in Vietnam below.

Choosing Stalemate in Vietnam

The previous section showed that the apparent statistical correlation between democracy and victory for both war initiators and targets is tenuous, and may well be illusory. This section turns to the causal logic of the theory, specifically of the selection effects argument. Reiter and Stam posit that electoral accountability and the marketplace of ideas cause leaders of democracies to choose only winnable wars. In Vietnam, however, we observe a democratic leader entering a war even though he knew the odds of victory were slight. Although there are several plausible explanations for this, one of the most prominent is that the need to protect his domestic legislative agenda caused Lyndon Johnson to perceive the costs of fighting a losing war in Vietnam as lower than the costs of staying out. In other words, in certain circumstances democracy may propel leaders to engage in wars with a low probability of success, thus causing democracies to suffer the occasional draw or loss. I do not claim that these circumstances are common; in the conclusion I suggest some potential cases for further research. My claim is merely that domestic politics played an important role in at least one important case—Vietnam—and that this

57. Ivo H. Daalder and Michael E. O'Hanlon, *Winning Ugly: NATO's War to Save Kosovo* (Washington, D.C.: Brookings Institution Press, 2000), pp. 91–96; and Jacob Weisberg, "Gulfballs: How the Experts Blew It, Big-Time," *New Republic*, March 25, 1991, p. 19.
58. See, for example, Desch, *Power and Military Effectiveness*.

represents an alternative mechanism through which democracy can lead to war.

Some readers might object that a single case study cannot provide definitive evidence against a probabilistic theory such as Reiter and Stam's. But it is not my purpose to "falsify" their argument. Rather, I endeavor to make two points. First, I show that democracies sometimes go to war when they are unsure of the prospects of victory, and that this can help explain cases of democratic stalemate and defeat. Second, I propose an alternative mechanism inherent in democratic politics that can cause democracies to choose war under less-than-ideal circumstances.

Below, I first show that Vietnam was a war of choice for the United States. A strong argument exists on technical grounds to code the United States as the initiator of the war, but the more important point is that the Johnson administration had a choice about whether to enter or escalate the conflict and selected itself into the war. I then demonstrate that at both critical decision points—to bomb North Vietnam in February 1965 and to send U.S. combat troops in June—key policymakers advocated war despite being deeply pessimistic about the prospects for victory, understanding that the United States faced a long and costly war of attrition. This contradicts the expectation from the selection effects argument that leaders of democracies choose war only when they have high confidence in victory.[59] Lastly, I argue that democratic politics actually pushed Johnson to fight even in the face of long odds.

THE VIETNAM WAR: A WAR OF CHOICE FOR THE UNITED STATES

The question of who started the interstate phase of the Vietnam War is difficult to answer. A civil war began in 1960 in South Vietnam between the government, supported by the United States, and the National Liberation Front, better known as the Vietcong (VC), supported by North Vietnam. In 1965 a separate interstate conflict broke out pitting North Vietnam against the United States and South Vietnam. According to the COW definition of war initiator as

---

59. Reiter and Stam, for example, cite the case of French leaders in 1911 declining to go to war against Germany when the general staff rated their chances of victory at less than 70 percent. Reiter and Stam, *Democracies at War*, p. 13. Of course, this "70 percent rule" is undermined by both its reputed source—Napoleon Bonaparte, hardly a democrat—and Reiter and Stam's reference in the same paragraph to the belief of some U.S. government officials that U.S. chances of victory in Vietnam rose only to 70 percent by 1968 with the commitment of several hundred thousand troops. One would think that a 70 percent chance of winning would be the bare minimum for a leader to risk his political life; more likely he would require an 80 or a 90 percent likelihood of victory.

the first state actor to use force against another state, the United States should be coded as the initiator of the conflict. The reason is that the United States was the first state to use interstate force when President Johnson ordered Flaming Dart I, a set of air raids on army barracks in southern North Vietnam, on February 7, 1965. Four days later, Johnson ordered another air strike (Flaming Dart II). Soon thereafter, the president formally approved Rolling Thunder, an ongoing bombing campaign described by National Security Adviser McGeorge Bundy as "a program of sustained reprisal against the North."[60]

Skeptics of this coding would point out that both Flaming Dart raids were launched in response to VC attacks that killed U.S. servicemen, first at Pleiku and then at Qui Nhon.[61] Because the VC was a nonstate actor, however, to code these strikes as the war's beginning presumes that the VC was fully controlled by—or doing the bidding of—Hanoi. Subsequent research has shown that the attack on Pleiku was in fact not ordered by the North.[62] Nor does the presence of North Vietnamese troops in South Vietnam before 1965 necessarily point to the North as the initiator. Although Hanoi began to send individual soldiers south in late 1963 or early 1964, and the 325th Division of the People's Army of Vietnam (PAVN) left for South Vietnam in mid-November 1964, PAVN regiments did not openly enter the fray until the communist spring offensive began in May 1965.[63] Until then Hanoi's troops fought as part of VC units.

Further evidence indicates that on December 1, 1964, Johnson had in principle already approved a policy of bombing North Vietnam, desiring only increased political stability in Saigon (and a suitable provocation) before implementing it.[64] Phase I of the plan began with bombings of infiltration routes in Laos (Operation Barrel Roll) on December 14. The VC strike on Pleiku was used as a pretext to put into effect a policy that had already been decided upon. As Bundy remarked to a reporter, "'Pleikus are streetcars'—meaning

60. Bundy to Johnson, "The Situation in Vietnam," Memorandum, February 7, 1965, in *Foreign Relations of the United States* (hereafter *FRUS*), 1964–1968, Vol. 2, doc. 84. All *FRUS* documents cited in this article are available online at http://www.state.gov/r/pa/ho/frus.
61. Eight Americans died in the attack on Pleiku; twenty-three were killed at Qui Nhon.
62. According to historian Mark Moyar, Vietnamese sources indicate that the "attack had been conceived and ordered by the local commander of the Viet Cong forces in Pleiku province, who . . . was simply trying to hurt his adversaries." Moyar, *Triumph Forsaken: The Vietnam War, 1954–1965* (Cambridge: Cambridge University Press, 2006), p. 486 n. 14.
63. Ibid., p. 372. South Vietnamese and U.S. forces also began covert and naval incursions into the North under the aegis of OPLAN-34A at about the same time that Hanoi covertly began sending troops south.
64. Fredrik Logevall, *Choosing War: The Lost Chance for Peace and the Escalation of War in Vietnam* (Berkeley: University of California Press, 1999), p. 270.

that the United States had been waiting for such an incident and planned to make use of it." "Pleiku," remarks historian Brian VanDeMark, "seemed a propitious moment to begin the bombing of North Vietnam which Bundy and others had already embraced as a remedy to South Vietnam's political failure." As David Kaiser puts it, "Having lost the war in the South Vietnamese countryside, they [Johnson and his advisers] had chosen to begin a new war against the government, the society, and the regular army of North Vietnam." William Gibbons concurs: "Thus, on February 8, 1965 the decision was made by the President to begin waging war on the Democratic Republic of Vietnam."[65]

Although the weight of the evidence probably favors labeling the United States as the initiator of the war in the technical sense, the more important point is that President Johnson and his team of advisers chose to fight a war against North Vietnam that was not forced upon them.[66] It was not as if PAVN armored divisions were pouring across the Rio Grande from Mexico; there was no attack on the U.S. homeland and no existential threat to the survival of the United States. Even those who argued that U.S. entry into the war was necessary to prevent Asian dominoes from falling understood that withdrawing from Vietnam would have had only a marginal impact on U.S. security in the short term. The president may have paid a political price for pulling up stakes, but that is not the same as saying the United States had no choice but to fight. If not the initiator of the war, then, the United States chose to join an ongoing war that had only a peripheral impact on its security. The selection effects argument should therefore apply to the U.S. decision to fight in Vietnam.

THE VIETNAM WORKING GROUP AND THE DECISION TO BOMB

The selection effects hypothesis implies that because democratic leaders start only those wars they think they can win, they should be confident about their military prospects when deciding to go to war.[67] As the evidence below dem-

---

65. Bundy, quoted in Robert L. Gallucci, *Neither Peace nor Honor: The Politics of American Military Policy in Viet-Nam* (Baltimore, Md.: Johns Hopkins University Press, 1975), p. 46; Brian VanDeMark, *Into the Quagmire: Lyndon Johnson and the Escalation of the Vietnam War* (New York: Oxford University Press, 1991), p. 63; David Kaiser, *American Tragedy: Kennedy, Johnson, and the Origins of the Vietnam War* (Cambridge, Mass.: Belknap, 2000), p. 411; and William Conrad Gibbons, *The U.S. Government and the Vietnam War: Executive and Legislative Roles and Relationships*, Pt. 3: January–July 1965 (Princeton, N.J.: Princeton University Press, 1989), p. 68.
66. This is a central theme of Logevall, *Choosing War.*
67. One might expect leaders to cite a favorable military balance, superior military technology or strategy, or the outstanding fighting qualities of their troops. Interestingly, Reiter and Stam do not identify a particular factor that leaders examine when deciding whether a war is winnable. For ex-

onstrates, this was emphatically not the case: U.S. policymakers understood just how dire the political and military situation was in South Vietnam, and they knew that using force against North Vietnam would probably not improve matters on the ground. I focus on the two key decisions for escalation: the decision to bomb North Vietnam, and the choice to send a large number of U.S. ground troops to South Vietnam.[68]

By the time the Johnson administration began seriously debating escalation in Vietnam in November 1964, its most prominent members agreed that the situation there was ominous. Since the overthrow of South Vietnamese President Ngo Dinh Diem a year earlier, much of the news out of Saigon had been grim.[69] Summing up the thoughts of many, the deputy U.S. ambassador in Saigon, David Nes, wrote in February 1964, "I do not see in the present military regime or any conceivable successor much hope in providing the real political and social leadership or the just and effective country-wide administration so essential to the success of our counter-insurgency program."[70] The president shared his subordinates' pessimism, telling Secretary of Defense Robert McNamara in late April, "We're not getting it done. . . . We're losing."[71] In September a draft paper by Assistant Secretary of Defense for International Security Affairs John McNaughton argued that pacification in the South could not succeed without a strong central government. Unfortunately, he continued, "It has become apparent that there is no likelihood that a government

---

ample, they find no evidence that democracies won "because they were bigger or had more allies, better troops, or better strategy choices." Reiter and Stam, *Democracies at War*, p. 37.

68. I do not address the role of the marketplace of ideas in the Vietnam decisions, except to note that Johnson adopted a "policy of minimum candor" in his dealings with the public, the media, and congressional leaders. He refused to announce major decisions publicly, concealed the level of escalation, and endeavored to squelch congressional debate on the wisdom of fighting in Vietnam. The case is thus not supportive of marketplace of ideas arguments. See Stanley Karnow, *Vietnam: A History* (New York: Viking, 1983), p. 430; George C. Herring, *America's Longest War: The United States and Vietnam, 1950–1975*, 2d ed. (Philadelphia, Pa.: Temple University Press, 1986), p. 133; and John Schuessler, "Doing Good by Stealth? Democracy, Deception, and the Use of Force," Ph.D. dissertation, University of Chicago, 2007, chap. 3.

69. For selected examples of this reporting, see Robert McNamara, "Vietnam Situation," Memorandum for the President, December 21, 1963, in Senator Mike Gravel, ed., *The Pentagon Papers: The Defense Department History of United States Decisionmaking on Vietnam*, Vol. 3 (Boston: Beacon, 1971), p. 494; Special National Intelligence Estimate (SNIE), No. 50–64, "Short-Term Prospects in Southeast Asia," summarized in Carroll to McNamara, Memorandum, February 12, 1964, in *FRUS, 1964–1968*, Vol. 1, doc. 42; McNamara to Johnson, "South Vietnam," Memorandum, March 16, 1964, in *FRUS, 1964–1968*, Vol. 1, doc. 84; and William Bundy, "Next Courses of Action in Southeast Asia," Memorandum, August 13, 1964, in *FRUS, 1964–1968*, Vol. 1, doc. 313.

70. Memo to Ambassador Lodge, February 17, 1964, quoted in Logevall, *Choosing War*, p. 113.

71. Quoted in Logevall, *Choosing War*, p. 145.

sufficiently strong to administer a successful pacification program will develop. It follows that our current U.S. policy, which is based on such a program, will not succeed."[72] These gloomy appraisals of the political situation in South Vietnam were reinforced by negative reports on military conditions, as well as pessimism regarding the utility of using force against North Vietnam to improve the situation in South Vietnam.[73]

Despite these downbeat assessments, by September 1964 the president and his advisers generally agreed that a greatly expanded U.S. presence would be required if South Vietnam was to be saved.[74] Two factors, however, led Johnson to hold off on escalation. First, he was worried that the rickety government in South Vietnam remained too fragile to withstand a major escalation in the war. As he put it in a White House meeting in early September, "With a weak and wobbly situation it would be unwise to attack until we could stabilize our base."[75] Second, the president, having established his hawkish credentials in August by launching retaliatory air strikes and obtaining the Gulf of Tonkin resolution, tacked back toward the doves to present himself as the candidate of peace in the upcoming presidential election. Johnson wished to avoid any major escalation in Vietnam before November, having previously rebuffed a recommendation in March by the Joint Chiefs of Staff to attack the North by telling them "he did not want to start a war before November."[76] At the same time, Johnson did not want to be tagged as the man who lost Vietnam; he preferred to avoid a major decision either way until after he was safely reelected.

---

72. John McNaughton, "Plan of Action for South Vietnam," 2d draft, September 3, 1964, in Gravel, *The Pentagon Papers*, Vol. 3, p. 556. McNaughton reiterated this judgment two months later: "Progress inside SVN is important, but it is unlikely despite our best ideas and efforts." John McNaughton, "Action for South Vietnam," 3d draft, November 7, 1964, in Gravel, *The Pentagon Papers*, Vol. 3, p. 602.
73. See, for example, the memos by George Ball and William Bundy (October 5 and 19, respectively), discussed in Kaiser, *American Tragedy*, pp. 349–353. Two Pentagon war games on Vietnam conducted in April and September (Sigma I and II) suggested that bombing North Vietnam would do nothing to staunch the insurgency in the South and might actually increase the North's will to resist. "Sigma I-64: Final Report," D-6, and "Sigma II-64: Final Report," D-2, D-7, D-14, both in *Declassified Document Reference Service.*
74. Logevall, *Choosing War*, p. 235.
75. Memorandum of a Meeting, White House, September 9, 1964, in *FRUS*, 1964–1968, Vol. 1, doc. 343.
76. Quoted in Kaiser, *American Tragedy*, p. 304. The chiefs' proposal may be found in Joint Chiefs of Staff to McNamara, "Vietnam," Memorandum, March 2, 1964, in *FRUS*, 1964–1968, Vol. 1, doc. 66. In fact, during the presidential campaign against Republic Senator Barry Goldwater, an advocate of escalation in Vietnam, President Johnson repeatedly told voters that he would not "send American boys 9 or 10,000 miles away from home to do what Asian boys ought to be doing for themselves." Quoted in VanDeMark, *Into the Quagmire*, p. 19.

According to McGeorge Bundy, 1964 was the "year off. It was so under Johnson, and it would have been under [a surviving Kennedy] as well. Neither man wanted to go to the election as the one who either made war or lost Vietnam. If you could put it off you did." There was little doubt, however, which way Johnson was leaning.[77]

Shortly before Election Day, when his victory appeared assured, Johnson tasked a National Security Council working group to explore the options available to the United States in Vietnam. The group, chaired by Assistant Secretary of State William Bundy, quickly coalesced on three alternatives: Option A, which would continue the status quo; Option B, a "systematic program of military pressures against the north," escalating at a "fairly rapid pace"; and Option C, which supplemented current policies with a strategy of graduated military pressure against the North.[78] The group would eventually settle on Option C, but not because its members believed it stood a good chance of succeeding.[79] This attitude is evident in McNaughton's November 19 draft "Courses of Action in Southeast Asia." McNaughton argued, for example, that "the United States could not be certain of securing its maximum objectives in South Vietnam even if it undertook all-out war with North Vietnam or China."[80] What is more, even the best-case outcome of pursuing Option C did not promise much of an accomplishment. A redraft of this paper by William Bundy classified possible outcomes as follows: "At best: To avoid heavy risk and punishment, the DRV [North Vietnam] might feign compliance and settle for an opportunity to subvert the South another day. That is, a respite might be gained. At worst: South Vietnam might come apart while we were pursuing the course of action. In between: We might be faced with no improvement in the internal South Vietnam situation and with the difficult decision whether to

---

77. Quoted in Logevall, *Choosing War,* p. 108. Johnson's desire to delay a decision until after the 1964 election might be construed as support for Kurt Taylor Gaubatz's argument that democratic executives are more likely to initiate wars early in their terms of office when the political risk is low rather than later on. This only applies, however, if the leader believes the war will be short, and thus he or she can reap the political benefits of victory come re-election time. As Gaubatz notes, his argument "would not explain behavior if a conflict was expected to be long and drawn out." This is exactly what Johnson believed about Vietnam. See Gaubatz, *Elections and War: The Electoral Incentive in the Democratic Politics of War and Peace* (Stanford, Calif.: Stanford University Press, 1999), p. 54.
78. "Courses of Action in Southeast Asia," November 21, 1964, in Gravel, *The Pentagon Papers,* Vol. 3, pp. 656–666.
79. Logevall, *Choosing War,* p. 259. See also National Security Council Working Group on Vietnam, "Intelligence Assessment: The Situation in Vietnam," November 24, 1964, in Gravel, *The Pentagon Papers,* Vol. 3, pp. 651–652.
80. Kaiser, *American Tragedy,* p. 363.

escalate on up to major conflict with China."[81] The best-case scenario, in other words, was a return to the status quo in which the VC insurgency continued with reduced support from the North but Hanoi simply waited for another opportunity to subvert the South.

In late November, the U.S. ambassador to Saigon, Maxwell Taylor, arrived in Washington to participate in meetings where the working group's conclusions would be debated with the president in attendance. Taylor came armed with a dismal report on the situation in South Vietnam. The report concluded, for example, that "the counterinsurgency program country-wide is bogged down and will require heroic treatment to assure survival," and "it is impossible to foresee a stable and effective government under any name in anything like the near future."[82] Yet rather than call for a reappraisal of the U.S. commitment to the South, Taylor was adamant that the United States should take strong action (in the form of bombing) against Hanoi. Johnson and his wise men agreed: escalation "should be undertaken regardless of the political picture in Saigon, either to reward the GVN [Government of Vietnam] or to keep it from disintegrating."[83]

Although now committed to bombing in principle, Johnson remained reluctant to move forward without greater stability in Saigon, but the continuing turmoil there made it increasingly clear that no matter how long he waited, the political situation in South Vietnam was unlikely to stabilize.[84] As chaos mounted and morale plummeted in the South, U.S. policymakers came to argue that bombing was needed to stabilize the situation and instill confidence in South Vietnam, rather than stability being a pretext for bombing. As George Herring has put it, "By the end of January 1965 . . . the major argument against escalation had become the most compelling argument for it."[85] With prompting from McNamara and McGeorge Bundy, Johnson decided it was time to act: "Stable government or no stable government . . . we'll do what we have to do. I'm prepared to do that. *We will move strongly.* [Gen. Nguyen] Khanh is our boy."[86] Contrary to the selection effects argument, in other words, "no one

---

81. "Courses of Action in Southeast Asia," p. 665.
82. Taylor, "The Current Situation in South Viet-Nam—November 1964," in *FRUS,* 1964–1968, Vol. 1, doc. 426.
83. Logevall, *Choosing War,* p. 269.
84. On January 27, 1965, South Vietnamese strongman Gen. Nguyen Khanh launched a bloodless coup that ousted the civilian premier, Tran Van Huong.
85. Herring, *America's Longest War,* p. 127.
86. Quoted in Logevall, *Choosing War,* p. 318 (emphasis in original). For Bundy and McNamara's prompting, see the famous "fork in the road" memo, January 27, 1965, in *FRUS,* 1964–1968, Vol. 2, doc. 42.

among them [top U.S. officials] viewed the disintegration in South Vietnam as a reason to de-escalate, to find some means to withdraw from the war."[87] Instead they argued that force should be initiated against Hanoi despite the precarious political and military situation in South Vietnam, the low likelihood that attacking North Vietnam would reverse those conditions, and the risk that Hanoi and its Chinese ally might invade the South, leading to a nontrivial chance that the United States would have to resort to nuclear weapons. As Fredrik Logevall has written, "The most striking thing about the Johnson administration's decision to move to an escalated war was not that it contained contradictions . . . but that it came despite deep pessimism among many senior officials that the new measures would succeed in turning the war around."[88]

Indeed, many Johnson administration officials subscribed to the "good doctor" theory originally articulated by McNaughton, which argued that fighting and losing in Vietnam was better than withdrawing without a fight. According to the theory, to establish the credibility of U.S. commitments, "We must have kept promises, been tough, taken risks, gotten bloodied, and hurt the enemy very badly," even while losing, to be seen as a good doctor whose patient (South Vietnam) died despite the doctor's best efforts.[89] Top U.S. policymakers, in other words, were willing to start a war and lose to bolster U.S. credibility.

Even as Rolling Thunder got under way in early March 1965, and U.S. Marines came ashore at Da Nang, Johnson's closest advisers were not optimistic regarding the ability of the bombing to turn the tide. Arguing in February for the commencement of bombing, McGeorge Bundy warned, "At its very best the struggle in Vietnam will be long," and estimated that his suggested policy of sustained reprisals against North Vietnam had between a 25 and 75 percent chance of success.[90] Bundy, McNamara, and Secretary of State Dean Rusk met on the evening of March 5 to discuss Vietnam, and Bundy conveyed the thrust of their discussion in a memo to the president the next day: "Two of the three of us think that the chances of a turn-around in South Vietnam remain less than even; the brutal fact is that we have been losing ground at an increasing rate in the countryside in January and February. . . . And there is no

---

87. Logevall, *Choosing War,* p. 271.
88. Ibid., p. 271.
89. John McNaughton, "Aims and Options in Southeast Asia," 1st draft, in Gravel, *The Pentagon Papers,* Vol. 3, p. 582.
90. Bundy to Johnson, "The Situation in Vietnam," February 7, 1965. Nor were the service chiefs confident that Operation Rolling Thunder would produce results. See Gravel, *The Pentagon Papers,* Vol. 3, p. 320.

evidence yet that the new government has the necessary will, skill and human resources which a turn-around will require."[91] McNaughton, who had recently returned from South Vietnam, confirmed these fears in mid-March, noting that the "thing is much worse" than anyone in Washington thought.[92]

In short, the United States chose to initiate the bombing of North Vietnam even though most top U.S. officials believed that bombing was unlikely to break Hanoi's will to support the insurgency, significantly degrade the Vietcong's ability to fight and reverse the military situation in South Vietnam, or produce a stable, popular government in Saigon.

THE TURN TO GROUND TROOPS

As Johnson administration officials feared, Rolling Thunder did not improve the political or military situation in South Vietnam. Yet another coup by a group of officers, led by Air Vice Marshal Nguyen Cao Ky and Gen. Nguyen van Thieu, removed Premier Khanh from the scene for good in late February, but this was not necessarily a turn for the better. As William Bundy would later put it, the Ky-Thieu combination "seemed to all of us the bottom of the barrel, absolutely the bottom of the barrel."[93] A variety of U.S. government studies found that Hanoi remained convinced it was winning and that the bombing had done little to curtail infiltration, weaken North Vietnam's military capabilities, or diminish Hanoi's will to continue the struggle.[94] As Chairman of the Joint Chiefs of Staff Gen. Earle Wheeler summed it up on April 1, 1965, "We are losing the war out there."[95]

As it became clear that Rolling Thunder was failing to achieve its objectives, the pressure to send U.S. ground troops to save the situation grew increasingly intense. Army Chief of Staff Gen. Harold Johnson returned from Vietnam in mid-March recommending the insertion of an army division in the central highlands plus a four-division blocking force to cut infiltration routes. In late March, the head of Military Assistance Command, Vietnam (MACV), Gen. William Westmoreland, recommended an increase of 33,000 troops, including

---

91. Bundy to Johnson, Memorandum, March 6, 1965, in *FRUS, 1964–1968*, Vol. 2, doc. 183.
92. Quoted in VanDeMark, *Into the Quagmire*, p. 96.
93. Quoted in Herring, *America's Longest War*, p. 137. Desertion from South Vietnam's army grew to 11,000 per month, and morale failed to improve in the South. See Logevall, *Choosing War*, p. 367.
94. VanDeMark, *Into the Quagmire*, p. 105; Wheeler to McNamara, Memorandum, April 6, 1965, in *FRUS, 1964–1968*, Vol. 2, doc. 241, quoted in Karnow, *Vietnam*, p. 430; and State Department, Division of Intelligence and Research, "The Effects of the Bombings of North Vietnam," June 29, 1965, quoted in George McT. Kahin, *Intervention: How America Became Involved in Vietnam* (Garden City, N.Y.: Anchor, 1987), p. 337. Bombing also cemented closer ties between Hanoi and its more powerful communist allies.
95. Quoted in VanDeMark, *Into the Quagmire*, p. 109.

an army division and an airborne brigade. In mid-April, the Joint Chiefs called for an increase of three divisions; participants in the Honolulu conference on April 20 agreed to raise U.S. troop levels to 82,000. In June, Westmoreland again requested more troops—a lot more: an increase to 175,000 (44 battalions) by the end of 1965, with an additional 95,000 to come in 1966. After a series of meetings in mid-July, Johnson eventually agreed to the 175,000 figure, although he preferred to announce only a 50,000-man increase to 125,000 in public on July 28.

Was the decision to Americanize the ground war in South Vietnam driven by optimism that U.S. troops could defeat the Vietcong and PAVN units in a quick and decisive fashion? Hardly. Pessimism was widespread among the key decisionmakers in Washington and Saigon. When General Johnson returned from South Vietnam in mid-March, for example, he observed that the "much-weakened South Vietnamese armed forces . . . were incapable of handling the Viet Cong, and the United States therefore needed to introduce its own troops into the war." As discouraging as this message was, worse was yet to come: the general warned that "victory could require five years and 500,000 U.S. troops, well beyond what the President and the rest of the civilians had expected."[96] McNamara's recommendation of 82,000 troops after the Honolulu conference in April "was not presented to the president as a prescription for winning the war, but, rather, simply for denying victory to the Viet Cong," which he thought would take "more than six months, perhaps a year or two."[97]

By June the situation in South Vietnam was so bad that U.S. military and political officials agreed that the war, as General Westmoreland reported to the Joint Chiefs early in the month, was on the verge of being lost. Two weeks later, Westmoreland definitively stated that the United States faced a protracted war of attrition: "By way of introduction, the premise behind whatever further actions we may undertake, either in SVN [South Vietnam], or DRV, must be that we are in for the long pull. The struggle has become a war of attrition. Short of [a] decision to introduce nuclear weapons against sources and channels of enemy power, I see no likelihood of achieving a quick, favorable end to the war."[98] Ambassador Taylor in Saigon concurred, commenting that

---

96. Moyar, *Triumph Forsaken*, p. 368.
97. Kahin, *Intervention*, p. 319.
98. Westmoreland to JCS, telegram, June 7, 1965, in *FRUS, 1964–1968*, Vol. 2, doc. 337; and Westmoreland to Wheeler, telegram, June 24, 1965, in *FRUS, 1964–1968*, Vol. 3, doc. 17. As Wheeler would later recall, "In the summer of 1965, it became amply clear that it wasn't a matter of whether the North Vietnamese were going to win the war; it was a question of when they were go-

restraining Hanoi required turning the tide on the ground in the South, which "may take a long time and we should not expect quick results. . . . There is no strategy that can bring about a quick solution."[99] The civilians in Washington were equally downcast. As McNamara confided to the British foreign secretary, "None of us at the centre of things [in U.S. policymaking] talk about winning a victory."[100] Similarly, William Bundy lamented that none of the options on the table—including the deployment of 300,000 U.S. troops—"seems likely to raise the chances of real success from perhaps (by hypothesis) 15 percent without such actions to 25–30 percent with them."[101] George Ball questioned whether victory could be had with 500,000 troops.[102] Even the president admitted privately and publicly in June that the communists were winning the war.[103]

The crucial debates on ground troops took place on July 21–22, 1965.[104] In preparation, McNaughton wrote a paper for McNamara in which he strikingly argued that the probability of success if the United States sought to win with 200,000 to 400,000 or more combat troops rose only to 50 percent by 1968.[105] McNamara's July 20 paper, which formed the basis of discussion in the meetings that followed, painted a similarly grim picture, assessing that "the situation in South Vietnam is worse than a year ago (when it was worse than a year before that)." McNamara noted that communist forces were "hurting ARVN [Army of the Republic of Vietnam] forces badly" and might soon conquer the central highlands; the Ky government probably would not last until the end of the year; pacification efforts were losing ground; interdiction had failed; and the communists "seem to believe that South Vietnam is on the run and near collapse." McNamara recommended increasing the U.S. troop commitment by

---

ing to win it." Quoted in Larry Berman, *Planning a Tragedy: The Americanization of the War in Vietnam* (New York: W.W. Norton, 1982), pp. 69–70.

99. Saigon Embassy to State, telegram, June 3, 1965, in *FRUS, 1964–1968*, Vol. 2, doc. 328.

100. Quoted in Logevall, *Choosing War*, p. 368. In an oral history ten years later, McNamara said he thought that the United States was "on a certain course of defeat," and moreover that "it wasn't clear to me that we could avoid defeat by any action in our power." Quoted in Gareth Porter, *Perils of Dominance: Imbalance of Power and the Road to War in Vietnam* (Berkeley: University of California Press, 2005), p. 222.

101. Quoted in Gibbons, *The U.S. Government and the Vietnam War*, p. 334.

102. Kaiser, *American Tragedy*, p. 448.

103. McGeorge Bundy, "Personal Notes of a Meeting with President Johnson," June 10, 1965, in *FRUS, 1964–1968*, Vol. 2, doc. 343; and VanDeMark, *Into the Quagmire*, pp. 160, 162.

104. Johnson had in fact already decided to approve Westmoreland's request before these discussions took place, but the July 21–22 deliberations are still highly revealing of various officials' estimates of victory. Gibbons, *The U.S. Government and the Vietnam War*, pp. 380–382.

105. John McNaughton, "Analysis and Options for South Vietnam," July 13, 1965, cited in Kahin, *Intervention*, p. 357.

thirty-four battalions to a total of 175,000 men by the end of 1965, with another 100,000 in early 1966 and further deployments possible after that depending on the circumstances. McNamara conceded, however, that "success" might merely mean a return to the VC guerrilla campaign of 1960–64, and warned that there was no obvious way to get U.S. troops out of South Vietnam.[106]

The ensuing discussions revealed that McNamara's evaluation of the consequences of Americanizing the war was far too optimistic. McNamara, for example, acknowledged that pacification would take a minimum of two years and would likely require another 100,000 U.S. troops. Indeed, Westmoreland had informed McNamara that even if MACV's troop request was met in full, "No reasonable assurances can be given that we can prove to the Viet Cong that they cannot win and force them to settle on our terms."[107] McGeorge Bundy supported McNamara's recommendation, yet warned, "There are no early victories in store, although casualties are likely to be heavy." Ball also argued strongly against ground troops, contending that "this war will be long and protracted. The most we can hope for is a messy conclusion." General Wheeler thought there was no chance of winning in a year regardless of the number of troops sent: "We might start to reverse the unfavorable trend in a year and make definite progress in three years," he said. Similarly, in the meeting on July 22, Marine Corps Commandant Gen. Wallace Greene told the president that 500,000 troops would be needed, and the war would last five years.[108]

In short, the United States chose to enter the ground war in South Vietnam despite the widely shared view among the president and his advisers that such a conflict would be a protracted war of attrition, lasting three to five years and requiring perhaps half a million U.S. troops in which victory was far from assured.

EXPLAINING THE DECISION FOR WAR

Many factors undoubtedly contributed to Johnson's decision to fight in Indochina; untangling this knot is impossible here.[109] Two factors are particu-

---

106. Incongruously, though, McNamara concluded that his recommended course of action stood "a good chance of achieving an acceptable outcome within a reasonable time in Vietnam." McNamara to Johnson, Memorandum, July 20, 1965, in *FRUS*, 1964–1968, Vol. 3, doc. 67.

107. Quoted in Gibbons, *The U.S. Government and the Vietnam War*, p. 373.

108. The notes of these meetings are reprinted in Kahin, *Intervention*, pp. 368–386. Quotes are from pp. 376, 374, 377; Greene's estimate is on p. 384.

109. Some argue, for example, that fear of the domino effect in Asia prompted Johnson to hold the line in Vietnam. Moyar, *Triumph Forsaken*, pp. 290, 375–376. Closely related is the view that Johnson fought in Vietnam to deter China. Kaiser, *American Tragedy*, p. 362. Others minimize the impor-

larly relevant to the analysis in this article, however, and thus deserve emphasis. First, as noted above, many of the top civilian policymakers in the administration believed that losing in Vietnam was preferable to pulling out without a fight because going down fighting would bolster U.S. credibility. According to this view, "The United States should strive to project the image of a patient (South Vietnam) who died despite the heroic efforts of the good doctor (the United States)." As McNaughton wrote, even if intervention failed, "it would, by demonstrating U.S. willingness to go to the mat, tend to bolster allied confidence in the U.S. as an ally."[110] Put bluntly, many of those charged with making the decision to go to war and planning military operations believed "it was not necessary to win in Vietnam."[111] Stalemate (and a willingness to take casualties) was sufficient to communicate U.S. credibility. In short, beliefs about how to convey resolve, then current among civilian policymakers, made defeat an acceptable outcome.

Second, many historians have argued—and many of Johnson's statements confirm—that domestic politics affected the decision for war, and that Johnson feared political punishment if he withdrew from Vietnam more than if he engaged in war. Historians largely agree, for example, that protecting the Great Society was the crucial factor that explains why Johnson escalated the war the way he did: slowly, incrementally, and most of all, secretly. According to George Kahin, "The slow pace at which Johnson escalated . . . must in considerable part have reflected his mounting concern that a sharp, clear-cut escalatory move in Vietnam might so rock the boat as to threaten passage of many as-yet-unrealized pieces of Great Society legislation." Similarly, Logevall maintains that Johnson's concerns regarding the Great Society "directly influenced the *way* in which he expanded the war—in particular, they dictated that the escalation be as quiet as possible so as to avoid the need for choosing between the war and the programs, between guns and butter." Participants in the key debates over escalation, such as McGeorge Bundy and members of the

---

tance of credibility concerns and domino beliefs, arguing instead that the United States intervened because of its preponderance of power over the communist bloc. Porter, *Perils of Dominance*. A third view emphasizes the causal power of analogies in explaining the decision for war, and a fourth stresses Johnson's fear of humiliation and loss of personal credibility. See, respectively, Yuen Foong Khong, *Analogies at War: Korea, Munich, Dien Bien Phu, and the Vietnam Decisions of 1965* (Princeton, N.J.: Princeton University Press, 1992); and Logevall, *Choosing War*, pp. 389–393.

110. Quoted in Logevall, *Choosing War*, p. 272. See also H.R. McMaster, *Dereliction of Duty: Lyndon Johnson, Robert McNamara, the Joint Chiefs of Staff, and the Lies That Led to Vietnam* (New York: HarperPerennial, 1997), pp. 184–185.

111. McMaster, *Dereliction of Duty*, p. 332.

Joint Chiefs, sensed Johnson's desire to protect his domestic legislation lurking in the background. Bundy recalled after one meeting that "his [Johnson's] unspoken object was to protect his legislative program." Johnson himself later explained the need to keep his "foreign policy in the wings" by arguing that "the day it [Congress] exploded into a major debate on the war, that day would be the beginning of the end of the Great Society."[112]

Beyond influencing how Johnson escalated the war, however, some argue that his fears over the fate of the Great Society contributed to his decision to go to war in the first place. It was not that the public or Congress was clamoring for war; polls showed that the public was split on the question of escalation, and many of the most powerful senators from the president's own party were skeptical of escalation. Rather, Johnson worried that if he pulled U.S. forces out of Vietnam, Republicans and conservative Democrats would join forces to block the Great Society. This thesis has been restated recently by Francis Bator, who served as deputy national security adviser in the Johnson administration. According to Bator, "The war deprived the Great Society reforms of some executive energy and money. *But Johnson believed—and he knew how to count votes— that had he backed away from Vietnam in 1965, there would have been no Great Society to deprive. It would have been stillborn in Congress.*" Bator argues that Johnson "knew that an honest discussion of the Westmoreland [forty-four battalion] plan would provoke a coalition of budget balancers and small-government Republicans, who balked at the high cost of guns and butter, and Deep South senators, who were determined to block civil rights legislation. They would need only 34 votes out of 100 to block cloture—20 Deep South senators plus 14 conservative Republicans."[113] Only a minority of senators was required to obstruct Johnson's domestic reforms, and Johnson believed that this blocking coalition would be triggered by any attempt to quit Vietnam.

At first glance, this thesis appears implausible given Johnson's landslide victory just a few months earlier and the solid Democratic majorities that resulted in both houses of Congress. As Johnson's new vice president, Hubert

---

112. Kahin, *Intervention*, p. 320; Logevall, *Choosing War*, p. 391 (emphasis in original); Bundy, quoted in Gibbons, *The U.S. Government and the Vietnam War*, p. 426; and Johnson, quoted in Doris Kearns Goodwin, *Lyndon Johnson and the American Dream* (New York: Harper and Row, 1976), pp. 281–282. See also VanDeMark, *Into the Quagmire*, pp. 54, 71, 109–110, 131, 162, 213, 217; and McMaster, *Dereliction of Duty*, pp. 194, 263, 308–309, 326. McMaster goes so far as to argue that Johnson's oft-expressed fear of Chinese or Soviet intervention was a ploy to protect his domestic political agenda (pp. 314, 317).
113. Francis M. Bator, "No Good Choices: LBJ and the Vietnam/Great Society Connection," *Diplomatic History*, Vol. 32, No. 3 (June 2008), pp. 309, 321–322 (emphasis in original).

Humphrey, wrote to him the month after the inauguration, "1965 is the year of minimum political risk for the Johnson Administration."[114] McGeorge Bundy later contended that the president could have used his considerable powers of persuasion to make the case for withdrawal without suffering significant domestic political consequences: "I think if [Johnson] had decided that the right thing to do was to cut our losses, he was quite sufficiently inventive to do that in a way that would not have destroyed the Great Society."[115] Yet there is substantial evidence that Johnson "believed that losing Vietnam in the summer of 1965 would have wrecked his plans for a Great Society" by triggering a destructive debate similar to the one over "who lost China?" that devastated Truman's presidency.[116] Johnson perceived this Vietnam dilemma soon after he inherited the presidency, telling his Senate mentor Richard Russell in May 1964, "Well, they'd impeach a president that would run out, wouldn't they?" Johnson later repeated this view during the height of the debate over ground troops in July 1965: "When I land troops they call me an interventionist . . . and if I do nothing I'll be impeached."[117] As Johnson later told his biographer, Doris Kearns,

I knew from the start . . . that I was bound to be crucified either way I moved. If I left the woman I really loved—the Great Society—in order to get involved with that bitch of a war on the other side of the world, then I would lose everything at home. All my programs. All my hopes to feed the hungry and shelter the homeless. All my dreams to provide education and medical care to the browns and the blacks and the lame and the poor. But if I left that war and let the Communists take over South Vietnam . . . there would follow in this country an endless national debate—a mean and destructive debate—that would shatter my Presidency, kill my administration, and damage our democracy.[118]

Although the evidence is far from conclusive and the debate is still ongoing, there is plausible evidence that the democratic process in this case gave the president incentives to undertake, rather than avoid, a risky war with uncertain prospects for success.

---

114. Quoted in Gibbons, *The U.S. Government and the Vietnam War*, pp. 94–95. Humphrey argued that Johnson should exploit his unique position to cut his losses and extricate himself from Vietnam.
115. Quoted in Logevall, *Choosing War*, p. 391. See also Fredrik Logevall, "Comment on Francis M. Bator's 'No Good Choices: LBJ and the Vietnam/Great Society Connection,'" *Diplomatic History*, Vol. 32, No. 3 (June 2008), p. 357.
116. Larry Berman, "Comment on Francis M. Bator's 'No Good Choices: LBJ and the Vietnam/ Great Society Connection,'" *Diplomatic History*, Vol. 32, No. 3 (June 2008), p. 362. This is also Berman's conclusion in *Planning a Tragedy*, p. 146.
117. Quoted in VanDeMark, *Into the Quagmire*, p. 162.
118. Quoted in Kearns Goodwin, *Lyndon Johnson and the American Dream*, pp. 251–252.

Conclusion

This article sought to reevaluate the argument that democracies are more likely to win wars. The most comprehensive study on this subject to date, Dan Reiter and Allan Stam's *Democracies at War*, offers two explanations for democratic victory. According to the selection effects theory, democratic war initiators are better at choosing winnable wars because leaders fear the electoral costs of failure and benefit from a vigorous marketplace of ideas that weeds out bad ideas and provides leaders with high-quality information to inform policy decisions. The war-fighting theory, by contrast, maintains that democratic culture and the popular nature of democratic regimes produces superior soldiers who are better leaders and fight with greater initiative than soldiers from nondemocratic societies. I reanalyzed the quantitative data offered in support of these two arguments and found that the results were not robust. For example, rather than code all nonwar initiators as targets, it is more accurate to code only states that were attacked as targets and the remainder as joiners. Moreover, given that democratic leaders sometimes suffer politically for wars that end in draws, it seems reasonable to include draws in the analysis. When these two changes are made, democratic war initiators, targets, and joiners are no longer statistically more likely to prevail. These findings raise doubts regarding the quantitative basis for the view that democracies are more likely to win wars.

I also bring qualitative evidence to bear specifically on selection effects. An examination of the Johnson administration's decision to fight in Vietnam revealed that, contrary to the selection effects argument, Lyndon Johnson and his principal foreign policy advisers did not go to war in Vietnam confident of victory. In fact, from late 1963 onward the news out of South Vietnam was largely negative, and U.S. policymakers knew that bombing the North or introducing U.S. ground troops would not in all likelihood reverse the downward trends in the South. George Ball articulated the selection effects mechanism in a June 1965 memo to President Johnson. After detailing the dismal conditions in South Vietnam and the absence of any assurance that Americanizing the war could change things, Ball wrote, "'Good statesmanship' . . . required cutting 'losses when the pursuit of particular courses of action threaten . . . to lead to a costly and indeterminant result.'"[119] Yet Johnson forged ahead into a war his military advisers warned would take at least five years and half a million men.

---

119. Quoted in VanDeMark, *Into the Quagmire*, p. 166.

Moreover, my analysis reveals that domestic political considerations if anything provided an added impetus to credibility and domino theory arguments for going to war. Johnson believed he would face a backlash if he withdrew from Vietnam that would imperil his Great Society programs and ruin his presidency. He worked assiduously to avoid a fractious debate between guns and butter by escalating the United States' military role in Vietnam quietly and gradually. Johnson also apparently believed that any attempt to renege on the U.S. commitment to an independent, noncommunist South Vietnam would have triggered a blocking coalition in the Senate capable of killing his legislative program. In short, democratic politics, rather than helping Johnson choose a short, victorious war, may instead have pushed him to fight a war he knew to be unwinnable because he believed that the domestic consequences of not fighting were worse.

The recent Iraq case suggests a second way that domestic politics in democracies can lead to costly quagmires: leaders may seek to conceal the potential costs of wars of choice from the public in order to build support for going to war in the first place. In 2003 no one doubted that the U.S. military would quickly defeat the Iraqi Army. The major question was what would happen after the Baathist regime was toppled. The potential costs and difficulties of occupying and administering Iraq were well known from the 1991 Gulf War when they deterred the George H.W. Bush administration from overthrowing Saddam Hussein. Many of the same men who argued against going to Baghdad in 1991, including Dick Cheney and Colin Powell, held even more powerful positions in 2003. Yet the George W. Bush administration denied that there would even be an occupation of Iraq, arguing that Americans would be welcomed as liberators who would quickly turn over power to Iraqi exiles such as Ahmed Chalabi. The Bush administration made many other specious arguments for attacking Iraq, but soft-pedaling the costs and avoiding the hard questions about how to govern and rebuild an Iraq riven by ethnoreligious tensions and crippled by thirty years of Saddam Hussein's police state surely helped the administration obtain the war it wanted but also the quagmire that followed.

Additional research is needed to explore the extent to which these two mechanisms undermine the likelihood of democratic victory.[120] It will be im-

---

120. For another argument about how domestic politics can lead to democratic defeat in war, with an application to Vietnam, see Caverley, "Democracies Will Continue to Fight Small Wars . . . Poorly."

portant to consider, for example, how much Johnson's decision to fight in Vietnam to protect the Great Society at home stemmed from idiosyncratic factors, such as his personality, his set of political beliefs, or the dynamics of U.S. politics in the 1960s. A brief perusal of historical cases, however, suggests that Vietnam is not the only instance of a democracy initiating war under adverse circumstances. When democratic Poland attacked the newborn Soviet Union after World War I, for example, Polish leader Joseph Piłsudski was reportedly not confident he would prevail. The British minister to Warsaw reported at the time of Poland's offensive that "there is no doubt that he [Piłsudski] went into the Ukraine with his eyes open, for he had told me some time ago that every army which had entered the Ukraine had had reason to repent having done so."[121] Similarly, Israeli officials were far from unanimous about the likelihood of victory before the Six-Day War in 1967. During a cabinet meeting at the beginning of the crisis, for example, Prime Minister Levi Eshkol asked Yitzhak Rabin, the chief of staff of the Israel Defense Forces (IDF): "What's to stop the Egyptians from taking the south? The Syrians from attacking our settlements?"[122] Former Prime Minister David Ben-Gurion vehemently opposed the war because he did not believe Israel could win. Even Moshe Dayan thought the situation was "graver than in 1956, and far more so than in 1948." Finally, Rabin suffered a breakdown during the crisis and warned that Israeli forces might suffer tens of thousands of dead.[123]

Two recent conflicts—the 1999 Kargil War and the 2008 Russo-Georgian War—were also initiated by democracies despite high risks and low likelihoods of success.[124] Pakistan, for example, initiated the first war between two nuclear-armed states when it attacked India, seemingly contradicting the high degree of caution that infuses democratic decisionmaking.[125] Similarly,

---

121. Quoted in Zygmunt J. Gàsiorowski, "Joseph Piłsudski in the Light of British Reports," *Slavonic and East European Review,* Vol. 50, No. 121 (October 1972), p. 559. See also Desch, *Power and Military Effectiveness,* p. 73.
122. Quoted in Michael B. Oren, *Six Days of War: June 1967 and the Making of the Modern Middle East* (New York: Presidio, 2002), p. 89.
123. Tom Segev, *1967: Israel, the War, and the Year That Transformed the Middle East,* trans. Jessica Cohen (New York: Metropolitan, 2007), pp. 248, 243, 286.
124. At the time of the Kargil incursion, Pakistan scored +7 on the Polity index, and Georgia scored +6 in 2008 (down slightly from +7 the previous year). The Polity Project considers states that score +6 or higher to be democracies. See *Polity IV Project: Political Regime Characteristics and Transitions, 1800–2007,* http://www.systemicpeace.org/polity/polity4.htm. Although technically not an interstate war, the Israeli war against Hezbollah in the summer of 2006 is another recent example of a dubious democratic war choice.
125. Evidence also indicates that Pakistan breached the Line of Control in Kashmir because it was growing ever more pessimistic about the prospects for prying the Muslim-majority province away

Georgia launched a major assault on the breakaway enclave of South Ossetia in August 2008 despite the presence of Russian "peacekeeping" forces in the province, several of whom were killed in the initial attack. The Russian Army struck back, drove the Georgians out, and went on to occupy not only South Ossetia but (temporarily) large chunks of Georgia proper. Finally, democratic states that join ongoing wars can also provide evidence regarding selection effects. It is not clear, for example, that the United States selected easy, winnable wars when it joined the two world wars and the Korean War.

Regarding the second mechanism, in which democratic leaders conceal the long-term costs of war to gain consent for fighting now, the 1982 Israeli invasion of Lebanon is a possible case. Although Israel achieved its short-term objective of eliminating the Palestine Liberation Organization from Lebanon, the invasion spawned a new terrorist organization, Hezbollah, dedicated to violently resisting Israel's presence in the country. The IDF occupied southern Lebanon for eighteen years but was ultimately unable to neutralize this resistance and withdrew in 2000. Michael Desch has already shown how a small faction of the Israeli leadership was able to take the nation to war against the will of the cabinet and the public.[126] They may have also downplayed the potential long-term costs of war to maintain short-term consent.

This article has argued that the relationship between democracy and military outcomes is more complicated than previously thought. Democratic processes can drive national leaders to start or enter wars they are not confident of winning, or get caught in quagmires by failing to confront the possible long-term consequences of a short-term victory. Just as further research is needed to parse out the countervailing effects of democracy, additional work is also needed on other determinants of military effectiveness. Desch, for instance, has restated the case for material power in accounting for war outcomes. Stephen Biddle has challenged the role of material factors, arguing instead that skilled force employment does a better job of explaining victory and defeat.[127] Why some states are better at force employment than

from India. Sumit Ganguly and Devin T. Hagerty, *Fearful Symmetry: India-Pakistan Crises in the Shadow of Nuclear Weapons* (Seattle: University of Washington Press, 2005), p. 158. For more on Kargil, see S. Paul Kapur, *Dangerous Deterrent: Nuclear Weapons Proliferation and Conflict in South Asia* (Stanford, Calif.: Stanford University Press, 2007), pp. 117–131.
126. Desch, *Power and Military Effectiveness*, pp. 100–101.
127. Ibid.; and Stephen Biddle, *Military Power: Explaining Victory and Defeat in Modern Battle* (Princeton, N.J.: Princeton University Press, 2004).

others, however, is left out of Biddle's framework. One promising explanation is civil-military relations, but others have argued that conceptions of collective identity may hold the key.[128] This debate is ongoing, and more work needs to be done to resolve it, but this article suggests that democracy is not the simple answer it has been portrayed to be.

---

128. On civil-military relations, see Risa A. Brooks, *Shaping Strategy: The Civil-Military Politics of Strategic Assessment* (Princeton, N.J.: Princeton University Press, 2008); Weeks, "Leaders, Accountability, and Foreign Policy in Non-Democracies"; Biddle and Long, "Democracy and Military Effectiveness"; and Biddle and Zirkle, "Technology, Civil-Military Relations, and Warfare in the Developing World." On identity, see Castillo, "The Will to Fight"; and Jason M. Lyall, "Paths of Ruin: Why Revisionist States Arise and Die in World Politics," unpublished manuscript, Princeton University, 2006.

Correspondence
Dan Reiter and
Allan C. Stam

Another Skirmish in the Battle over Democracies and War
Alexander B. Downes

To the Editors (Dan Reiter and Allan C. Stam write):

In previous articles and in our 2002 book *Democracies at War,* we argued that democracies are particularly likely to win their wars. Democratic political institutions provide incentives for elected leaders to launch only short, winnable, low-cost wars, so they may avoid domestic political threats to their hold on power. Democracies tend to win the wars they initiate because democratic leaders generally "select" themselves into winnable wars, and they are more likely to win when they are targeted because their armies fight with better initiative and leadership.

Analyzing all interstate wars from 1816 to 1987, we found strong empirical support for our theory.[1] Other scholarship has produced findings supportive of our theory. Elsewhere, two different formal game-theoretic models produced the hypothesis that democracies are especially likely to win the wars they initiate.[2] The empirical results generated to test these and related hypotheses have withstood challenges to data selection and research design.[3]

Using data sets and research designs different from ours, other scholars have uncovered empirical patterns consistent with our theory that democracies are especially likely to win the crises they initiate,[4] that wars and crises are

Dan Reiter is Chair of the Department of Political Science at Emory University. Allan C. Stam is Professor of Political Science at the University of Michigan. The authors would like to thank Michael Horowitz for helpful comments.

Alexander B. Downes is Assistant Professor of Political Science at Duke University. The author thanks Pablo Beramendi, Risa Brooks, Jeffrey Church, Kathryn McNabb Cochran, Matthew Fuhrmann, John Mearsheimer, and Todd Sechser for helpful advice and comments.

1. Dan Reiter and Allan C. Stam, "Democracy, War Initiation, and Victory," *American Political Science Review,* Vol. 92, No. 2 (June 1998), pp. 377–389; Dan Reiter and Allan C. Stam III, "Democracy and Battlefield Military Effectiveness," *Journal of Conflict Resolution,* Vol. 42, No. 3 (June 1998), pp. 259–277; and Dan Reiter and Allan C. Stam, *Democracies at War* (Princeton, N.J.: Princeton University Press, 2002.
2. Darren Filson and Suzanne Werner, "Bargaining and Fighting: The Impact of Regime Type on War Onset, Duration, and Outcomes," *American Journal of Political Science,* Vol. 48, No. 2 (April 2004), pp. 296–313; and Bruce Bueno de Mesquita, Alastair Smith, Randolph M. Siverson, and James D. Morrow, *The Logic of Political Survival* (Cambridge, Mass.: MIT Press, 2003), especially p. 226.
3. David H. Clark and William Reed, "A Unified Model of War Onset and Outcome," *Journal of Politics,* Vol. 65, No. 1 (February 2003), pp. 69–91.
4. Christopher F. Gelpi and Michael Griesdorf, "Winners or Losers? Democracies in International Crises, 1918–94," *American Political Science Review,* Vol. 95, No. 3 (September 2001), pp. 633–647;

International Security, Vol. 34, No. 2 (Fall 2009), pp. 194–204
© 2009 by the President and Fellows of Harvard College and the Massachusetts Institute of Technology.

shorter when democracies and democratic initiators are involved, and that democracies become increasingly likely to initiate wars as their likelihood of victory increases.[5] H.E. Goemans's recent empirical work exploring the relationship among conflict outcome, regime type, and the postwar fate of leaders confirms our theory, noting that his main result "now offers empirical support for some of these theories [of international conflict] (Bueno de Mesquita et al. 1999, 2003; Reiter and Stam 2002)."[6] And, the long-established democratic peace has been explained using our theoretical assumption that variations in domestic political institutions create variations in conflict behavior.[7] Even the research designs of our critics, trivially adjusted, generate supportive results for our theory.[8] Lastly, in recent work, we have extended the data set forward to 2001 and confirmed our earlier results. Notably, in the 1988–2001 period, democratic initiators won five interstate wars, and tied or lost none.[9]

The spring 2009 issue of *International Security* featured an article by Alexander Downes entitled "How Smart and Tough Are Democracies? Reassessing Theories of Democratic Victory in War."[10] This article critiques our empirical finding that democracies are more likely to win their wars. We wel-

---

and Jean Sebastien Rioux, "A Crisis-Based Evaluation of the Democratic Peace Proposition," *Canadian Journal of Political Science*, Vol. 31, No. 2 (June 1998), pp. 263–283.
5. D. Scott Bennett and Allan C. Stam III, "The Declining Advantages of Democracy: A Combined Model of War Outcomes and Duration," *Journal of Conflict Resolution* Vol. 42, No. 3 (June 1998), pp. 344–366; Branislav Slantchev, "How Initiators End Their Wars: The Duration of Warfare and the Terms of Peace," *American Journal of Political Science*, Vol. 48, No. 4 (October 2004), pp. 813–829; Reiter and Stam *Democracies at War*, chap. 2; Bruce Bueno de Mesquita, Michael T. Koch, and Randolph M. Siverson, "Testing Competing Institutional Explanations of the Democratic Peace: The Case of Dispute Duration," *Conflict Management and Peace Science*, Vol. 21, No. 4 (Winter 2004), pp. 255–267; and Bruce Bueno de Mesquita, James D. Morrow, Randolph M. Siverson, and Alastair Smith, "Testing Novel Implications from the Selectorate Theory of War," *World Politics*, Vol. 56, No. 3 (April 2004), pp. 363–388.
6. H.E. Goemans, "Which Way Out? The Manner and Consequences of Losing Office," *Journal of Conflict Resolution*, Vol. 52, No. 6 (December 2008), pp. 771–794, at p. 790.
7. Bruce Russett and John Oneal, *Triangulating Peace: Democracy, Interdependence, and International Organizations* (New York: W.W. Norton, 2001); and David L. Rousseau, Christopher Gelpi, Dan Reiter, and Paul K. Huth, "Assessing the Dyadic Nature of the Democratic Peace, 1918–1988," *American Political Science Review*, Vol. 90, No. 3 (September 1996), pp. 512–533.
8. On Michael C. Desch, "Democracy and Victory: Why Regime Type Hardly Matters," *International Security*, Vol. 27, No. 2 (Fall 2002), pp. 5–47, see Dan Reiter and Allan C. Stam, "Understanding Victory: Why Political Institutions Matter," *International Security*, Vol. 28, No. 1 (Summer 2003), pp. 168–179. On Ajin Choi, "Democratic Synergy and Victory in War, 1816–1992," *International Studies Quarterly*, Vol. 48, No. 3 (September 2004), pp. 663–682, see Dan Reiter and Allan C. Stam, "Democracy and War Outcomes: Extending the Debate," paper presented at the annual meeting of the International Studies Association, Chicago, Illinois, February 28–March 3, 2007.
9. Reiter and Stam, "Democracy and War Outcomes: Extending the Debate."
10. Alexander B. Downes, "How Smart and Tough Are Democracies? Reassessing Theories of Democratic Victory in War," *International Security*, Vol. 33, No. 4 (Spring 2009), pp. 9–51.

Table 1. Initiation, Regime Type, and War Outcome, 1816–1987

	Regime Type and War Outcome								
	Democracy			Mixed Regime			Autocracy		
	Loss	Draw	Win	Loss	Draw	Win	Loss	Draw	Win
Target	8	2	11	30	11	15	26	6	18
	38%	10%	52%	54%	20%	27%	52%	12%	36%
Initiator	1	6	13	15	5	22	15	7	22
	5%	30%	65%	36%	12%	52%	34%	16%	50%
Imult1 Total	9	8	24	45	16	37	41	13	40
	22%	20%	59%	46%	16%	38%	44%	14%	43%

SOURCE: Initiation and outcome codings are from Alexander B. Downes, "How Smart and Tough Are Democracies? Reassessing Theories of Democratic Victory in War," *International Security,* Vol. 33, No. 4 (Spring 2009), pp. 9–51.

NOTE: Democracies are 1–4 on Polity scale, mixed regimes are 5–17, and autocracies are 18–21.

come the opportunity to discuss important issues of research design and address Downes's critiques (there is unfortunately insufficient room here to lay out our critical reaction to Downes's case study of the Vietnam War). Downes offers four main critiques of our work: our dependent variable of war outcomes should include a third "draw" category as well as "win" and "lose" categories; a third category of war belligerent, "joiner," should be added to "initiator" and "target"; we miscoded some initiation outcomes; and we miscoded some war outcomes. We disagree with all of Downes's critiques, but because of limited space we focus here on two errors of theory and substance he made. We demonstrate that correcting those mistakes produces empirical results that, contrary his claims, support our theory.[11]

Inspection of Downes's revised version of our data reveals our hypothesized patterns. Table 1 is a cross tabulation of initiation, regime type, and war outcomes, using Downes's codings. As our theory predicts, democratic initiators win more than either autocratic or mixed initiators. Strikingly, as our previous works also showed, democratic initiators almost never lose (13 wins, 1 loss).

11. Downes's other alterations to our data and research design are questionable. For example, coding the 1969–70 War of Attrition as a draw rather than an Israeli victory is strange, given that Israel achieved its political goal (of not budging from its forward position), Egypt failed to achieve its political goal (of dislodging Israel), and Egypt suffered more than nine times as many as casualties as Israel.

Note that this relationship of democratic initiators winning more than other initiators holds if draws are dropped or included (for more on this issue, see below). That is, as shown in table 1, our hypothesis is supported whether the comparison is win versus lose (democratic initiators win 93 percent of the time, autocratic initiators win 59 percent of the time, mixed regime initiators win 59 percent of the time), or as Downes prefers win versus draw versus lose (democratic initiators win 65 percent, autocratic initiators win 50 percent, mixed regime initiators win 52 percent). Cross tabulations aside, Downes's analysis presents two major concerns.

First, we were puzzled that Downes's statistical analysis used a research design that did not correctly test our central argument and associated hypotheses. In *Democracies at War* and in our 1998 *American Political Science Review* article, we argued (hypothesis 2.2) that the relationship between regime type and war outcomes among war initiators (and not targets) is curvilinear, as democratic initiators are the most likely to win, highly repressive initiators are less likely to win, and moderately repressive (or "mixed regime") initiators are least likely to win. We built this curvilinear hypothesis on mainstream international relations theory that dynamics such as internal political logrolling make mixed regimes even more likely than autocracies to initiate highly risky wars and to misperceive the likelihood of ultimate victory.[12] We then used a simple polynomial transformation to test for this hypothesized curvilinear effect, finding strong empirical support. None of Downes's models employs this or any curvilinear transformation. He instead tests for linear relationships between democracy, initiation, and outcome. It is not surprising that a linear model applied to a curvilinear relationship failed to find statistically significant results.

These results bring us to our second rebuttal of Downes's critique. Both Downes's criticism of our decision to drop draws from the analysis and his subsequent decision to include draws in an ordered probit analysis are misguided. The problem with Downes's approach is that the relationship between initiation, regime type, and draws is more complex than Downes's simplistic portrayal, as Allan Stam argued in his 1996 book *Win, Lose, or Draw,* and as we demonstrated in chapter 7 of our 2002 book.[13] In that chapter, building on a

---

12. See also H.E. Goemans, *War and Punishment: The Causes of War Termination and the First World War* (Princeton, N.J.: Princeton University Press, 2000).
13. Allan C. Stam III, *Win, Lose, or Draw: Domestic Politics and the Crucible of War* (Ann Arbor: University of Michigan Press, 1996).

1998 article by D. Scott Bennett and Stam, entitled "The Declining Advantages of Democracy," we hypothesized and found that in shorter wars, democracies are much more likely to win (and less likely to draw) than nondemocracies. If a war persists past the first year of fighting, however, the relationship changes, as the passage of time in war plays to the advantages of authoritarian regimes, not democracies.[14] As a result, as war duration extends past the first year or two of a war, democracies go from being more likely to win and less likely to draw, to being less likely to win and more likely to settle for draws (some have argued that as war endures, mixed regime leaders are more likely to gamble for resurrection than settle for a result short of decisive victory[15]). This is because publics often experience war fatigue as casualties mount, and democratic leaders feel more pressure than other kinds of leaders to end the war, by draw if necessary. Branislav Slantchev also found support for this conjecture, discovering that democratic initiators are more likely to win because they start wars they are confident they can win quickly, but their advantage disappears if the war drags on.[16] Downes's ordered probit analysis tests a hypothesis that we do not advance, nor do we expect there to be support for: that democracy has a uniform effect on the likelihood of draws across all wars. Hence, it is not surprising that a mismatch between a simple research design and a more complex reality produces statistically insignificant results.

This inconsistent relationship between democracy and draws (sometimes democracies are more likely to fight to a draw, and sometimes less likely) moved us several years ago to simplify matters by dropping draws in the war outcomes analysis in chapter 2 of *Democracies at War* and treating the analysis of the relationship between political institutions and war duration and outcomes in a separate, more nuanced analysis. Tackling the role of draws requires a more sophisticated research design than is allowed for in a data set in which the unit of analysis is a belligerent in a war and the dependent variable is war outcome (the approach we use in chapter 2 of our book and that Downes analyzes in his article). One needs instead a research design that accounts for the length of a war. The 1998 Bennett and Stam article and chapter 7 of our book did this by making the unit of analysis each year of war and by making the dependent variable whether in any particular year the war en-

---

14. See, for example, Stam, *Win, Lose, or Draw*, p. 177.
15. Goemans, *War and Punishment*.
16. Bennett and Stam, "The Declining Advantages of Democracy"; and Slantchev, "How Initiators Win Their Wars."

dured, ended in victory for one side or the other, or ended in a draw. Again, this more theoretically appropriate research design provided support for our theory, demonstrating that war outcomes are shaped by the public opinion constraints imposed on elected leaders in democracies.

The relatively high frequency of democracies experiencing draws supports this conjecture (see table 1). Democracies do better than other states at picking winnable wars (hence, when democracies initiate wars, they win more often than they draw or lose as compared to other kinds of regimes). As we and others argue, when democracies start wars, they plan on fighting short, victorious wars. Democracies are significantly more likely to experience short, victorious wars than other kinds of states. As the war endures and the possibility of a short, victorious war fades, democracies become more likely to seek a negotiated settlement or a draw rather than fighting on. Other kinds of governments, facing different internal pressures, more commonly continue to fight on in the hopes of experiencing a decisive victory. That is, democracies are less likely than other states to make the mistake of starting a war that does not end in swift victory. When democracies do make the mistake of starting a war that does not result in swift victory, however, the eventual war outcome is more often a draw than it is when other kinds of states make that mistake.

Although we take issue with Downes's characterization of several of our data-coding decisions, due to space constraints we do not argue those points here. Instead we demonstrate that our findings are robust to minor changes in data codings, changes that Downes erroneously claimed alter our previous results. We reanalyzed his data, using his initiation codings, his initiation/joiner/target distinctions, and his war outcome codings. We corrected the two errors discussed above, employing the fractional polynomial terms we used in our previous work to test for a curvilinear effect and dropping draws. The results of this analysis are provided in table 2. As we found in our previous work, the initiation term and the democracy-initiation terms for the curvilinear functional form are all substantively and statistically significant. We should note that the presence of several interaction terms and the presence of fractional polynomials make it difficult to interpret easily the substantive effects of these variables by viewing the table. Our 1998 article and 2002 book provide detailed discussion and presentation of similar statistical results. Further analysis revealed that the net substantive effects of regime type and initiation on war outcomes are as predicted, as democratic initiators are more likely to win their wars (90 percent chance, holding other values at their means and modes) than autocratic initiators (78 percent chance), which in turn are more likely to

Table 2. Exploring Curvilinear Relationships among Democracy, Initiation, and War Outcomes

Politics × Target	0.0135
	(0.0193)
Initiation	6.10***
	(1.76)
Poly Pol 1 (first curvilinear term)	−4.74**
	(1.66)
Poly Pol 2 (second curvilinear term)	−4.71**
	(1.68)
Capabilities	3.90***
	(0.739)
Allies	4.79***
	(1.10)
Troop Quality	0.0538
	(0.0374)
Terrain	−13.7***
	(3.71)
Strategy × Terrain	4.37***
	(1.21)
Strategy 1	9.52**
	(3.29)
Strategy 2	4.43*
	(2.50)
Strategy 3	4.51**
	(1.61)
Strategy 4	3.68**
	(1.28)
Constant	−6.83***
	(1.83)

SOURCE: Alexander B. Downes's data on initiation and war outcomes. Downes, "How Smart and Tough Are Democracies? Reassessing Theories of Democratic Victory in War," *International Security*, Vol. 33, No. 4 (Spring 2009), pp. 9–51.
N = 196. Probit analysis; robust standard errors in parentheses. Pseudo R^2 = 0.53; log-likelihood = 63.588738.
*significant at 0.05; **significant at 0.01; ***significant at 0.001. All tests are one-tailed.

win their wars than mixed regime initiators (57 percent chance). In these data, democratic targets are not more likely to win their wars.

Our work has moved forward, but not ended, the debate on the relationships between democracy and war. Much new and exciting research is ongoing, exploring for example possible determinants of public support during war, such as casualties, perceptions of success, stakes, and elite discourse. There is in particular robust debate about whether casualties, perception of success, or some combination of the two affected public attitudes during the 2003 Iraq War and its insurgency aftermath.[17] There is substantial theoretical and empirical research that explores the effects of political institutions on conflict by using the leader as the unit of analysis.[18] But the empirical record still strongly supports our fundamental proposition that democracies are significantly more likely to win the wars they fight, and in particular the wars they initiate.

—*Dan Reiter*
Atlanta, Georgia

—*Allan C. Stam*
Ann Arbor, Michigan

Alexander B. Downes Replies:

In my article "How Smart and Tough Are Democracies? Reassessing Theories of Democratic Victory in War," I questioned the widespread view that democracies are more militarily effective than states with other types of political regimes.[1] Focusing on the most prominent example of this view, Dan Reiter and Allan Stam's book *Democracies at War*, I advanced two arguments.[2] First, Reiter and Stam's empirical analysis of the relationship between democracy and victory inappropriately omitted draws (an undesirable war outcome) and lumped all war participants into two categories—initiators and targets—when there are really three—initiators, targets, and joiners (states that entered into wars after they started). When draws and joiners are incorporated into

17. Christopher Gelpi, Peter Feaver, and Jason Reifler, *Paying the Human Costs of War: American Public Opinion and Casualties in Military Conflicts* (Princeton, N.J.: Princeton University Press, 2009); Adam Berinsky, *In Time of War: Understanding Public Opinion, From World War II to Iraq* (Chicago: University of Chicago Press, 2009); Scott Sigmund Gartner, "The Multiple Effects of Casualties on Public Support for War: An Experimental Approach," *American Political Science Review*, Vol. 102, No. 1 (February 2008), pp. 95–106; and Dan Reiter, *How Wars End* (Princeton, N.J.: Princeton University Press, 2009).
18. Goemans, *War and Punishment*; Goemans, "Which Way Out?"; and Scott Wolford, "The Turnover Trap: New Leaders, Reputation, and International Conflict," *American Journal of Political Science*, Vol. 51, No. 4 (October 2007), pp. 772–788.
1. Alexander B. Downes, "How Smart and Tough Are Democracies? Reassessing Theories of Democratic Victory in War," *International Security*, Vol. 33, No. 4 (Spring 2009), pp. 9–51.
2. Dan Reiter and Allan C. Stam, *Democracies at War* (Princeton, N.J.: Princeton University Press, 2002).

the analysis, democratic initiators, targets, and joiners are not significantly more likely than nondemocracies to win interstate wars. Second, focusing on the democratic selection effects argument, I suggested that under certain conditions domestic politics might actually cause democratic failure. I demonstrated, for example, that President Lyndon Johnson escalated U.S. involvement in Vietnam despite believing that victory was unlikely and the costs would be high, and that this decision was caused in large part by his fear that not fighting in Vietnam would derail his domestic legislative agenda. In short, I found that democratic leaders may knowingly decide to fight losing wars to protect their domestic priorities.

I thank Dan Reiter and Allan Stam for their thoughtful reply to my article, and I welcome the opportunity to respond to their critique. Reiter and Stam offer two rebuttals to my arguments. First, they contend that I did not test their "central argument," which is that a curvilinear relationship exists between regime type and war outcomes for war initiators: democracies are most effective, followed by dictatorships, and mixed regimes are worst. Reiter and Stam suggest that "it is not surprising that a linear model applied to a curvilinear relationship failed to find statistically significant results." Second, Reiter and Stam reject my decision to incorporate into the analysis wars that end in draws. They argue that "the relationship between initiation, regime type, and draws is more complex" than the "simplistic portrayal" I offered in my article.

I am not persuaded by either of these rebuttals. Reiter and Stam's main argument is that democracies are significantly more likely than nondemocracies to win wars; the curvilinear hypothesis is secondary. Moreover, when properly analyzed, the data (with or without draws) reveal no significant evidence of a curvilinear association between democracy and victory. Therefore, the analysis in my article, which treated regime type linearly and found that democracies are not significantly more likely than nondemocracies to win, is valid. Finally, including draws is necessary because forward-looking, risk-averse democratic leaders should seek to avoid all undesirable war outcomes, not only outright defeats. Omitting draws thus generates a bias in favor of finding a positive relationship between democracy and victory.

A CURVILINEAR RELATIONSHIP?
In my article, I tested what I understood to be the principal finding from the literature on democracy and victory, namely, that democracies are more likely to emerge victorious in wartime than nondemocracies. Although Reiter and Stam claim in their response that their "central argument" concerns a curvilinear relationship between regime type and victory, this contention is difficult

to sustain. In the introduction to *Democracies at War,* Reiter and Stam state their thesis clearly: "Our central argument is that democracies win wars because of the off-shoots of public consent and leaders' accountability to the voters."[3] Reiter and Stam develop two arguments rooted in democracy for why democratic war initiators and targets prevail; they test these arguments against alternative explanations for democratic victory. With regard to democratic initiators, Reiter and Stam spend several pages carefully laying out their selection effects theory for why democracies are particularly likely to win wars they start, but they spend two paragraphs briefly outlining the logic for a curvilinear relationship between regime type and war outcomes.[4] Moreover, in previous debates with their critics, Reiter and Stam have not characterized this curvilinear relationship as being their main argument.[5] Reiter and Stam are thus primarily concerned with differences between democracies and nondemocracies, and secondarily interested in variation between different kinds of nondemocracies. Hence I tested their central argument in my article and found that democracies do not have a significant advantage over nondemocracies in winning wars when all wars are included in the analysis.

Nevertheless, Reiter and Stam are correct that I did not test their curvilinear hypothesis. In their rebuttal, they offer two pieces of evidence to support this hypothesis: a cross tabulation of initiation, regime type, and war outcomes (table 1), and a probit model that employs two fractional polynomials (FPs), which are designed to detect nonlinear effects of continuous independent variables (table 2). Neither of these pieces of evidence, however, actually supports their argument. Table 1, for example, shows that democratic initiators win 65 percent of the time when draws are included, a steep drop from the 93 percent figure reported by Reiter and Stam in *Democracies at War.* The table also shows a substantial narrowing of the gap in winning percentages among the three regime types: when draws are excluded, the disparity between democracies and autocracies is 33 percentage points, and between democracies and mixed regimes it is 35 points. When all war outcomes are considered, however, these differences are cut by nearly two-thirds for mixed regimes and more than half for autocracies. Mixed regimes also prevail at a slightly higher rate

3. Ibid., p. 3.
4. Ibid., pp. 19–25.
5. See, for example, Dan Reiter and Allan C. Stam, "Understanding Victory: Why Political Institutions Matter," *International Security,* Vol. 28, No. 1 (Summer 2003), pp. 168–179, written in response to Michael C. Desch, "Democracy and Victory: Why Regime Type Hardly Matters," *International Security,* Vol. 27, No. 2 (Fall 2002), pp. 5–47.

than autocracies—52 versus 50 percent—contrary to the proposed curvilinear hypothesis.

Cross tabulations, however, are relatively weak tests: not only are the relationships examined bivariate, failing to control for other possible causes of war outcomes, but this particular type of cross tabulation (among three variables) does not permit an assessment of the statistical significance of the correlations. Democratic initiators appear to be somewhat more likely to win than the two types of autocratic initiators, but it is impossible to draw any firm conclusions.

In table 2 of their rebuttal, Reiter and Stam present what initially appears to be more convincing evidence for the curvilinearity hypothesis. Using my data and codings—but excluding draws as a war outcome—Reiter and Stam replace the linear regime type variables (the Polity index, ranging in this case from 1 to 21, and Polity × initiator) with two fractional polynomials.[6] Each of these terms is statistically significant, which Reiter and Stam claim supports the proposition that a curvilinear relationship exists between regime type and victory for war initiators.

This interpretation is incorrect. The key piece of evidence that must be examined when testing for nonlinear effects of independent variables is whether the curvilinear specification fits the data better than a linear specification.[7] Although it is true that the FPs in Reiter and Stam's model are statistically significant, this does not mean that treating democracy as curvilinear rather than linear explains more of the variance in war outcomes. In fact, the model that includes the FPs explains slightly less of the variance. The log-likelihood of the model in Reiter and Stam's response, for example, is −63.59. The log-likelihood of the same model but with the linear variables for regime type (Polity and Polity × initiator) instead of the FPs is −62.40.[8] Because a smaller difference from zero indicates better model fit, the linear specification explains more of the variation in war outcomes.[9] In other words, the FP results indicate that FPs are not needed—even when the dependent variable excludes draws—because a curvilinear specification offers no improvement over a linear one.[10]

---

6. For details on FP models, see Patrick Royston and Willi Sauerbrei, *Multivariable Model-Building: A Pragmatic Approach to Regression Analysis Based on Fractional Polynomials for Modelling Continuous Covariates* (Chichester, U.K.: John Wiley, 2008).

7. Ibid., pp. 82–83.

8. The regression tables are available on my website, http://www.duke.edu/~downes/publications.htm.

9. The difference in model fit is probably not statistically significant, but the point is that the curvilinear specification does not fit the data as well as the linear one.

10. This criticism also applies to the FP results in *Democracies at War*, where the linear and curvilinear versions of democracy produce models with identical log-likelihoods. Reiter and Stam, *Democracies at War*, p. 45, models 4 and 5. Contrary to the claim in their letter, however, the FPs

Do these results hold when all war outcomes are examined (as I argued in my article should be the case) rather than wins and losses only? The answer is yes: comparing the fit of the linear versus curvilinear models run on the data that include draws yields log-likelihood statistics of -168.46 and -169.15, respectively. Again, the linear specification outperforms the curvilinear one (the FPs in this model are also statistically insignificant).

These tests demonstrate that the relationship between democracy and victory for initiators is not curvilinear, as Reiter and Stam have suggested, but rather is best treated as linear, as I did in my article. The argument that my results are insignificant because I neglected a curvilinear relationship is thus without foundation. Moreover, as I demonstrated in my article, when democracy is treated as linear, democratic initiators, targets, and joiners are not significantly more likely to prevail in war.

THE CONTENTIOUS DRAWS

Reiter and Stam also claim that my decision to incorporate draws into the analysis is misguided because democracies' proclivity to settle for draws changes as wars continue. Democracies rarely accept draws in a war's early stages, but by the fifth year the probability that they do so reaches 0.7.[11] Reiter and Stam suggest that selection effects cause democratic leaders to choose "wars that are short, low-casualty, and victorious," but "when they guess wrong . . . and the war does not end quickly," democracies "seek a draw in order to exit sooner rather than later."[12]

This explanation faces two difficulties. First, as I pointed out in my article, democratic leaders do not simply guess wrong: sometimes they knowingly select their countries into wars that are likely to be costly and inconclusive, which defies the selection effects logic. In 1965, for example, President Johnson and his top advisers clearly understood that conditions in South Vietnam were grim, and that escalation—either attacking Hanoi or sending large U.S. ground forces to the South—stood little chance of reversing the situation, but they still

---

used in this earlier work and in their correspondence are not the same. The formulas for the FPs in the letter are $x - \frac{1}{2} \times$ initiator and $x - \frac{1}{2}(\ln(x)) \times$ initiator, with x consisting of Polity $+ 11$ (email communication with Dan Reiter, June 15, 2009). The formulas in the book are $x - \frac{1}{2}$ and $x - \frac{1}{2}(\ln(x))$, with x consisting of (Polity $\times$ initiator $+ 11)/10$. Ibid., p. 41. It is easy to see that the second formula results in war targets being included in the analysis, assigning them a value of 1.1—(Polity $\times 0 + 11)/10 = 1.1$—the same score that an initiator with a Polity score of zero receives. Because the FPs were constructed incorrectly to include targets, these results cannot be used as evidence to support the hypothesis of a curvilinear effect of democracy on the likelihood of victory for war initiators.

11. Reiter and Stam, *Democracies at War*, p. 171.
12. Ibid., p. 178.

decided to do both. Reiter and Stam support this view in *Democracies at War*, writing that when Johnson was making his fateful decisions to escalate, "the outlook for the conflict was not promising even at this early stage," and "the American leadership did not in 1965 foresee an imminent victory."[13] In short, democratic leaders sometimes choose to go to war knowing that a costly and indecisive result—such as a draw—may ensue. Excluding these cases biases democratic selection arguments toward finding a statistical relationship between democracy and victory.

Second, given Reiter and Stam's argument that democratic leaders aim to stay in office but face an increased likelihood of removal if they settle a war on less-than-victorious terms, it is unclear why they would ever settle for draws. Democratic leaders need to deliver policy success to remain in power, but draws are typically perceived as failures, meaning that leaders who preside over them do so at their peril. Logically, therefore, democratic leaders should try to avoid draws just as they do defeats; and if stuck in a quagmire, they should be reluctant to accept a draw and instead "gamble for resurrection" in the hope that they can somehow obtain victory.[14] Reiter and Stam, however, attribute the propensity to gamble to autocratic leaders, who they claim "choose to risk outright defeat in hopes of victory." Yet Reiter and Stam also state that authoritarian rulers are relatively immune from being removed from office by their constituents, and thus have little to fear from settling a war short of victory.[15] According to this logic, autocrats have no incentive to gamble on decisive victory because they can repress domestic protest against the war's outcome, whereas democrats are fairly certain to be removed for failing to win. In short, the logic of Reiter and Stam's argument contradicts their empirical result.[16]

CONCLUSION

The aim of my article was to foster further debate and inquiry on the determinants of military effectiveness by demonstrating that the influence of democracy was more equivocal than previously believed, in some circumstances contributing to poor outcomes. It was not my aim to debunk democracy as an

---

13. Quotations are from ibid., pp. 173–174. For further evidence, see also ibid., pp. 13, 174.
14. For this argument, see George W. Downs and David M. Rocke, "Conflict, Agency, and Gambling for Resurrection: The Principal-Agent Problem Goes to War," *American Journal of Political Science*, Vol. 38, No. 2 (May 1994), pp. 362–380.
15. Reiter and Stam, *Democracies at War*, p. 168.
16. For an argument that mixed regimes rather than democracies gamble for resurrection, see H.E. Goemans, *War and Punishment: The Causes of War Termination and the First World War* (Princeton, N.J.: Princeton University Press, 2000).

independent variable in general, only to question its explanatory power in this specific empirical domain. Indeed, although material power surely plays a role, I endorse Stephen Biddle's argument that the causes of states' military effectiveness are predominantly nonmaterial and found primarily at the domestic level.[17] I suggested a pair of mechanisms whereby domestic politics can cause democracies to choose poorly, but other scholars are exploring how state-level variables other than regime type affect how states perform in wartime.[18] I hope that my article and this exchange contribute to this growing literature.

—Alexander B. Downes
Durham, North Carolina

17. Stephen Biddle, *Military Power: Explaining Victory and Defeat in Modern Battle* (Princeton, N.J.: Princeton University Press, 2004).
18. For exemplary citations, see Downes, "How Smart and Tough Are Democracies?" p. 51 n. 128.

The Deception Dividend

John M. Schuessler

FDR's Undeclared War

When do leaders resort to deception to sell wars to their publics?[1] Dan Reiter and Allan Stam, drawing on the liberal tradition, have advanced a "selection effects" explanation for why democracies tend to win their wars, especially those they initiate: leaders, because they must secure public consent first, "select" into those wars they expect to win handily.[2] This article highlights, in contrast, those cases where the selection effect breaks down—cases where leaders, for realist reasons, are drawn toward wars where an easy victory is anything but assured. Leaders resort to deception in these cases to preempt what is sure to be a contentious debate over whether the use of force is justified by shifting blame for hostilities onto the adversary. The article assesses the plausibility of these claims by revisiting the events surrounding the United States' entry into World War II. The case constitutes a "tough test" of the argument insofar as there is considerable disagreement among historians as to whether Franklin Roosevelt's administration wanted war in 1941 and what lengths it was willing to go to sway public opinion. The goal is to develop a plausible case that President Roosevelt welcomed U.S. entry into the war by the fall of 1941 and attempted to manufacture events accordingly. Specifically, both the "unde-

John M. Schuessler is Assistant Professor of Strategy and International Security in the Department of Leadership and Strategy at the Air War College.

The author is grateful to John Mearsheimer for his guidance and support throughout the time this article was written. He wishes to thank Jonathan Caverley, Alexander Downes, Benjamin Fordham, Charles Glaser, Michael Glosny, Michael Horowitz, Nuno Monteiro, Michelle Murray, Robert Pape, Negeen Pegahi, Jeffrey Record, Sebastian Rosato, Lora Viola, and the anonymous reviewers for helpful comments and suggestions. Seminar participants at Harvard University's Belfer Center for Science and International Affairs and at the University of Chicago also provided useful feedback. For financial and institutional support, he thanks the Smith Richardson Foundation (by way of the University of Chicago's Program on International Security Policy), the Belfer Center's International Security Program, and the Eisenhower Institute. The views expressed in this article are those of the author and do not reflect the official policy or position of the U.S. government or the Department of Defense.

1. I define "deception" as deliberate attempts by leaders to mislead the public about the thrust of official thinking, in this case about the decision to go to war. Deception can take different forms. Lying, where a leader makes a knowingly false statement, is a particularly blatant form of deception, but for that reason is also less common than other, subtler forms such as spinning, where a leader uses exaggerated rhetoric, and concealment, where a leader withholds vital information. Deception campaigns invariably involve all three types of deception. For more on the different kinds of deception, see John J. Mearsheimer, *Why Leaders Lie: The Truth about Lying in International Politics* (New York: Oxford University Press, forthcoming).
2. Dan Reiter and Allan C. Stam, *Democracies at War* (Princeton, N.J.: Princeton University Press, 2002).

International Security, Vol. 34, No. 4 (Spring 2010), pp. 249–165

clared war" in the Atlantic and the oil embargo on Japan should be understood as intended, at least partly, to manufacture an "incident" that could be used to justify hostilities. An important implication of this argument is that deception may sometimes be in the national interest.

The rest of the article proceeds as follows. First, I briefly summarize the long-standing debate between liberalism and realism over the difference that democracy makes for international relations, highlighting the role that deception plays in it. Second, I develop the main argument, elaborating the conditions under which the selection effect might break down and explaining why leaders would be tempted to resort to deception in these cases. Third, I conduct a plausibility probe by revisiting the events surrounding the United States' entry into World War II. I conclude with several theoretical and policy implications that follow from the analysis.

The Debate about Democracy

The debate between liberalism and realism over the influence of democracy in international relations has been long-standing. From Immanuel Kant onward, liberals have argued that democracies are more prudent than nondemocracies in the conduct of their external relations, because leaders are accountable to publics that will endorse the use of force only when the benefits clearly outweigh the costs. The public's sensitivity to costs is one of the factors most commonly cited to explain the democratic peace, the finding that mature democracies rarely, if ever, fight each other.[3] Public opinion, from the liberal perspective, exerts a moderating influence on leaders, ensuring that they resort to war only when it is prudent to do so. Deception, in turn, has little to recommend it, insofar as it erodes the constraints that keep leaders honest.[4] When leaders can manufacture consent, they no longer have to take public opinion into account and can pick fights as casually as their nondemocratic counterparts, increasing the risk of calamity.

Realists, for their part, argue that the central fact of international life is anarchy and the condition of insecurity that follows from it.[5] Given that states can

---

3. There are two basic logics underpinning the democratic peace: a normative one and an institutional one. Public constraint is one of the causal mechanisms making up the institutional logic. See Sebastian Rosato, "The Flawed Logic of Democratic Peace Theory," *American Political Science Review,* Vol. 97, No. 4 (November 2003), pp. 585–602.
4. Miroslav Nincic, *Democracy and Foreign Policy: The Fallacy of Political Realism* (New York: Columbia University Press, 1992), chap. 5.
5. For foundational treatments of realism, see Kenneth N. Waltz, *Theory of International Politics* (New York: McGraw-Hill, 1979); and John J. Mearsheimer, *The Tragedy of Great Power Politics* (New York: W.W. Norton, 2001).

never be certain of each other's intentions, they have no choice but to adopt a "self-help" posture, which entails arming, alliances, and when all else fails, war. For realists, whatever peace exists among democracies has less to do with democracy per se than with the balance of power and interests among them.[6] When realists do discuss the inner workings of democracy, it is to explain how leaders overcome "disruptions from below"—usually from a moralistic, short-sighted, and emotional public—to pursue realpolitik.[7] Thomas Christensen, for example, seeks to explain the hostile relationship between the United States and China in the opening decade of the Cold War, when one might have expected them to ally against the Soviet Union. His argument is that U.S. leaders could not sell containment in balance of power terms because of various difficulties with the public, most notably limited information, short time horizons, and ideological rigidity. Thus, they resorted to anticommunist crusading, which led to conflict with China.[8] From the realist perspective, democracy is less a moderating influence than a distorting one. Deception is thus justified in cases where public opinion deviates from what realists consider to be in the national interest. Hans Morgenthau sums up the realist position succinctly: "A tragic choice often confronts those responsible for the conduct of foreign affairs. They must either sacrifice what they consider good policy upon the altar of public opinion, or by devious means gain popular support for policies whose true nature they conceal from the public."[9]

Explaining Deception

Historically, realists have been more forgiving of deception than liberals, primarily because they have a less sanguine view of the public. Both camps, however, have left an important question unaddressed: When do leaders choose deception to convince their publics of the need for war? My main argument, which I develop below, is twofold. First, leaders are sometimes drawn toward wars where an easy victory is anything but assured—that is, the selection effect sometimes breaks down. Second, leaders resort to deception in these cases to preempt debate over whether the use of force is justified by shifting blame for hostilities onto the adversary.

---

6. Christopher Layne, "Kant or Cant: The Myth of the Democratic Peace," *International Security*, Vol. 19, No. 2 (Fall 1994), pp. 5–49.
7. Nincic, *Democracy and Foreign Policy*, pp. 6–11.
8. Thomas J. Christensen, *Useful Adversaries: Grand Strategy, Domestic Mobilization, and Sino-American Conflict, 1947–1958* (Princeton, N.J.: Princeton University Press, 1996).
9. Hans J. Morgenthau, *In Defense of the National Interest: A Critical Examination of American Foreign Policy* (New York: Alfred A. Knopf, 1951), pp. 223–224.

THE BREAKDOWN OF THE SELECTION EFFECT

The current debate between liberals and realists is over why democracies tend to win their wars, especially those they initiate. Reiter and Stam, in *Democracies at War,* examined a data set of interstate wars from 1816 to 1990 and found that democracies won 93 percent of the wars they initiated, compared with 60 percent for dictatorships and 58 percent for mixed regimes.[10] Although less successful than democratic initiators, democratic targets were also more likely to win than other types of targets, prevailing 63 percent of the time compared with 34 percent for dictatorships and 40 percent for oligarchs. Reiter and Stam conducted a battery of statistical tests to control for other possible explanations of why states win wars, including military-industrial capabilities, troop quality, military strategy, terrain, distance, and alliances. Even when controlling for these factors, Reiter and Stam found that democracies were more likely to win their wars than nondemocracies, whether they were initiators or targets.[11]

Why do democracies tend to win their wars? Reiter and Stam provide two answers. The first argument is that democracies are more selective about the wars they choose to fight than nondemocracies, initiating only those they are likely to win.[12] The logic underpinning this selection effect is twofold. First, leaders in democracies have powerful incentives not to start losing wars, because they can be deposed from office for doing so, whereas their counterparts in nondemocracies lack this electoral incentive.[13] Second, leaders in democracies must make a public case for war if they are to secure domestic consent for it. When they do so, opposition parties, the press, and independent experts can scrutinize their claims as part of a process of collective deliberation.[14] This vetting process provides democracies with the information necessary to discriminate between wars that can be won quickly and decisively and those that cannot. Nondemocracies benefit from no such marketplace of ideas, the result being that democracies produce better estimates of the probability of victory than nondemocracies do.[15] Second, Reiter and Stam argue that democratic ar-

---

10. Reiter and Stam, *Democracies at War,* p. 29.

11. Ibid., p. 30. Alexander B. Downes points out that the statistical significance of Reiter and Stam's results depends on two choices: their decision to equate war targets and joiners and their decision to exclude draws. See Downes, "How Smart and Tough Are Democracies? Reassessing Theories of Democratic Victory in War," *International Security,* Vol. 33, No. 4 (Spring 2009), pp. 9–51.

12. Reiter and Stam, *Democracies at War,* chaps. 2, 6.

13. Bruce Bueno de Mesquita and Randolph M. Siverson, "War and the Survival of Political Leaders: A Comparative Study of Regime Types and Political Accountability," *American Political Science Review,* Vol. 89, No. 4 (December 1995), pp. 841–885.

14. On the process of collective deliberation, see Benjamin I. Page and Robert Y. Shapiro, *The Rational Public: Fifty Years of Trends in Americans' Policy Preferences* (Chicago: University of Chicago Press, 1992), pp. 362–366.

15. Reiter and Stam, *Democracies at War,* p. 23. The marketplace of ideas argument is made most fa-

mies outperform nondemocratic armies on the battlefield because their soldiers, products of a political culture that emphasizes individualism, fight with greater initiative and exhibit better leadership than their nondemocratic counterparts do.[16] Here, Reiter and Stam part ways with other proponents of the "democratic victory" thesis, such as David Lake, who argue that democracies win their wars by bringing a preponderance of power to bear against their opponents.[17]

The most comprehensive critique of the democratic victory thesis is Michael Desch's *Power and Military Effectiveness*.[18] In the book, Desch makes the case that regime type "hardly matters" for explaining who wins and loses wars. Desch's critique of the selection effect, in particular, is worth elaborating. He makes three points. First, compared with other variables, such as material capabilities, democracy has a relatively small substantive effect on the probability of victory.[19] Second, there are logical reasons to think that leaders are relatively unconstrained by public opinion and that caution about starting a war should not be unique to democracies, as authoritarian leaders who lose wars are frequently exiled, imprisoned, or put to death.[20] Finally, Desch argues that the case study method, rather than statistical modeling, is the optimal way to test for the selection effect and that there is scant evidence of it at work in six of the ten cases that should support the theory.[21] In the lead-up to the Mexican War (1846–48), the Greco-Turkish War (1897), the Russo-Polish War (1919–20), the Sinai War (1956), the Six-Day War (1967), and the Lebanon War (1982), leaders stifled debate and initiated wars that lacked broad public support. Nor were they punished politically for doing so. Even if the democratic side won in the end, it is difficult to credit the outcome to the selection effect.

Desch, in short, dismisses the selection effect on logical and empirical grounds. One suspects, however, that Reiter and Stam, and other proponents of the democratic victory thesis, will not be entirely swayed by his critiques.

---

mously by Jack Snyder. See Snyder, *Myths of Empire: Domestic Politics and International Ambition* (Ithaca, N.Y.: Cornell University Press, 1991), pp. 31–55.

16. Reiter and Stam, *Democracies at War*, chap. 3.

17. David A. Lake, "Powerful Pacifists: Democratic States and War," *American Political Science Review*, Vol. 86, No. 1 (March 1992), pp. 24–37.

18. Michael C. Desch, *Power and Military Effectiveness: The Fallacy of Democratic Triumphalism* (Baltimore, Md.: Johns Hopkins University Press, 2008).

19. Ibid., pp. 39–43.

20. Ibid., pp. 43–45. On the impact of regime type on leaders' postwar fate, see H.E. Goemans, *War and Punishment: The Causes of War Termination and the First World War* (Princeton, N.J.: Princeton University Press, 2000).

21. Desch, *Power and Military Effectiveness*, pp. 49–51. Desch drops six cases from consideration because the democracies in question did not actually initiate war. These include the Boxer Rebellion (1900), the Czech-Hungarian War (1919), and World War II (1941–45).

First, the fact remains that democracies win the large majority of the wars they initiate. How are scholars to account for this finding? As a reviewer has noted, one wishes that Desch were able to offer a convincing theoretical argument that explains the correlation between democracy and victory in war.[22] Second, it is not particularly damning for the selection effect that democracy has a small substantive effect on the probability of victory relative to other variables such as material capabilities. The argument, after all, is not that democracy is more decisive for victory than material capabilities, but that democracies are likely to select into wars where they enjoy decisive advantages in material capabilities over their opponents.[23] Third, the notion that leaders in democracies are no more accountable for their performance in war than their counterparts in nondemocracies is belied by the fact that democracies fight so few losing wars relative to nondemocracies.[24] Finally, Desch leverages a considerable amount of case evidence to suggest that the selection effect cannot account for why democracies win the wars they initiate. Yet, even he finds the mechanism at work in four cases: the Spanish-American War (1898), the First Balkan War (1912–13), the Bangladesh War (1971), and the Turkey/Cyprus Invasion (1974).[25] The case evidence, in other words, is hardly conclusive one way or the other.

Rather than discredit the selection effect, my goal is to specify the conditions under which it breaks down. When, in other words, might leaders be drawn toward wars where an easy victory is anything but assured? Reiter and Stam address the question of when democracies start wars in chapter 6 of their book. Their basic argument is that democracies go to war when the public consents to it. This begs the question, of course, of when such consent is likely to be forthcoming. Reiter and Stam's answer is open-ended: "The central factor determining whether a public will consent to war is its definition of the national interest."[26] A narrow definition of the national interest implies a short list of circumstances under which the public will consent to war. An expansive

22. David M. Edelstein, "Review of *Power and Military Effectiveness: The Fallacy of Democratic Triumphalism*," *Perspectives on Politics*, Vol. 6, No. 4 (December 2008), pp. 863–864.

23. Dan Reiter and Allan C. Stam, "Understanding Victory: Why Political Institutions Matter," *International Security*, Vol. 28, No. 1 (Summer 2003), pp. 168–179. As Michael C. Desch points out, Reiter and Stam are inconsistent on this point. In their book, they go to some effort to dispel the notion that democracies win their wars because they are richer or better able to extract resources from their economies. See Desch, "Democracy and Victory: Fair Fights or Food Fights?" *International Security*, Vol. 28, No. 1 (Summer 2003), pp. 185–186.

24. David Kinsella, "No Rest for the Democratic Peace," *American Political Science Review*, Vol. 99, No. 3 (August 2005), p. 455; and Branislav L. Slantchev, Anna Alexandrova, and Erik Gartzke, "Probabilistic Causality, Selection Bias, and the Logic of the Democratic Peace," *American Political Science Review*, Vol. 99, No. 3 (August 2005), p. 461.

25. Desch, *Power and Military Effectiveness*, p. 50.

26. Reiter and Stam, *Democracies at War*, p. 148.

definition of the national interest implies a longer list of such circumstances. The public's sense of the national interest, in turn, is a function of the state's power, the level of external threat it faces, and its past history. Reiter and Stam are reluctant to make more determinate claims, "as there is great variability among democracies in different eras as to the conditions that are sufficient to generate public consent for the use of force against some other state."[27]

Democratic publics allow for the initiation of war to advance a variety of national interests. If the selection effect is to hold, however, these national interests cannot be so vital as to justify excessive risk taking. Generally, the public should consent to war only insofar as a quick and decisive victory is assured. In their work on casualty sensitivity, Christopher Gelpi, Peter Feaver, and Jason Reifler find that the public is indeed "defeat phobic," that is, beliefs about a war's likely success matter most in determining the public's willingness to tolerate casualties. When the public is confident of eventual victory, casualties have little effect on popular support. But when the public comes to believe that victory is unlikely, even small numbers of casualties erode support. Gelpi, Feaver, and Reifler marshal a wealth of evidence to test their claims, reanalyzing polling data from a number of conflicts dating back to the Korean War and conducting original surveys designed explicitly to tap into public attitudes on casualties.[28] The implication of their work is that the public is going to be reluctant to pay the human cost of war when victory is uncertain, especially when that cost promises to be high.

This reluctance will be all the more evident when threats are diffuse or indeterminate. Christensen, among others, has argued that "the public will judge the importance of international challenges by how immediately and clearly they seem to compromise national security."[29] By this standard, public consent for war should be most forthcoming when an adversary appears ready to launch an attack, either on the homeland or on a key ally, and least forthcoming when no such attack is on the horizon. For example, in the event that the adversary is geographically distant and does not have much in the way of an offensive capability, the public should be reluctant to pick a fight with it.[30] Normative considerations should reinforce this tendency, as force

---

27. Ibid., p. 148.
28. Christopher Gelpi, Peter D. Feaver, and Jason Reifler, *Paying the Human Costs of War: American Public Opinion and Casualties in Military Conflicts* (Princeton, N.J.: Princeton University Press, 2009). Gelpi, Feaver, and Reifler seek to qualify the less contingent claim, associated with John E. Mueller, that public support for war declines inexorably with mounting casualties. See Mueller, "Trends in Support for the Wars in Korea or Vietnam," *American Political Science Review*, Vol. 65, No. 2 (June 1971), pp. 358–375.
29. Christensen, *Useful Adversaries*, p. 27.
30. On the sources of threat, see Stephen M. Walt, *The Origins of Alliances* (Ithaca, N.Y.: Cornell University Press, 1987), pp. 21–26.

is generally viewed as legitimate only as a last resort and only for defensive purposes.[31]

If the public is indeed "defeat phobic," the more so when the threat is less than imminent, then why might leaders be drawn toward wars where an easy victory is anything but assured? Realism provides the beginnings of an answer. Democratic decisions for war, from a realist perspective, are determined and constrained not only by public consent but also by the imperatives of self-help in an anarchic environment. Democracies are not always free to pick on weak and vulnerable opponents but must, from time to time, take on formidable foes because their territorial integrity or political independence would be threatened otherwise. The goal is either to amass power at a rival's expense or to prevent that rival from doing the same. Granted, war is most likely to start when the potential attacker envisions a quick victory—either because of an imbalance in military capability, the relative ease of attack versus defense, or a disparity in skill—but the political consequences of continued peace are sometimes so unacceptable as to lead decisionmakers to assume significant risk.[32]

This will be the case, for example, when a power shift is under way and the declining state faces a shrinking window of opportunity or growing window of vulnerability. Windows increase the risk of war in a number of ways, most directly by tempting the declining state to launch a preventive war.[33] As Jack Levy defines it, "The preventive motivation for war arises from the perception that one's military power and potential are declining relative to that of a rising adversary, and from the fear of the consequences of that decline. . . . The temptation is to fight a war under relatively favorable circumstances *now* in order to block or retard the further rise of an adversary and to avoid both the worsening of the status quo over time and the risk of war under less favorable circumstances later."[34] Dale Copeland, consistent with this logic, has found that major wars are typically initiated by dominant military powers that fear significant decline.[35] The stronger the preventive motivation, the more willing the leadership should be to gamble on a war that holds out some hope of forestalling decline. The public, however, is likely to be reluctant to absorb the costs of a major war when the threat is more potential than actual. This leads to a gulf, as Randall Schweller argues, between what realpolitik requires of the leadership

---

31. Martha Finnemore, *The Purpose of Intervention: Changing Beliefs about the Use of Force* (Ithaca, N.Y.: Cornell University Press, 2003), p. 19.
32. John J. Mearsheimer, *Conventional Deterrence* (Ithaca, N.Y.: Cornell University Press, 1983), chap. 2, especially pp. 62–63.
33. Stephen Van Evera, *Causes of War: Power and the Roots of Conflict* (Ithaca, N.Y.: Cornell University Press, 1999), chap. 4, especially pp. 76–79.
34. Jack S. Levy, "Declining Power and the Preventive Motivation for War," *World Politics,* Vol. 40, No. 1 (October 1987), p. 87 (emphasis in original).
35. Dale C. Copeland, *The Origins of Major War* (Ithaca, N.Y.: Cornell University Press, 2000).

and what the public is willing to sanction.[36] Schweller's original finding, that democracies do not wage preventive wars, has since been qualified.[37] The logic of his argument remains compelling, however, and suggests that the leadership should have some difficulty securing public consent for war in the event that it is preventively motivated.

Another scenario where leaders may have difficulty securing public consent for war is offshore balancing, which entails intervening in a distant region to forestall the rise of a powerful adversary when local states prove unequal to the task. As John Mearsheimer argues, not only do great powers aim to dominate their own region, but they also strive to prevent rivals in other regions from gaining hegemony.[38] If there are two or more great powers in other regions of the world, they are likely to spend most of their time competing with each other, rather than threatening distant states. Given the "stopping power of water"—or the ways in which large bodies of water limit the power projection capabilities of land forces—what kind of threat can a regional hegemon pose to an offshore balancer?[39] First, it may be able to offer material support to a hostile coalition in the offshore balancer's backyard. The United States had to deal with this problem three times during the twentieth century: German involvement in Mexico during World War I, German designs on South America during World War II, and the Soviet Union's alliance with Cuba during the Cold War.[40] Alternatively, a regional hegemon could isolate the offshore balancer economically and politically, threatening it with strangulation and forcing it to adopt the trappings of a "garrison state."[41] Robert Art, for example, has argued that the United States could have remained secure over the long term had it not entered World War II and had it allowed Germany and Japan to win. Its standard of living and its way of life, however, would most likely have suffered, which was the real reason to have entered the war.[42]

36. Randall L. Schweller, "Domestic Structure and Preventive War: Are Democracies More Pacific?" *World Politics*, Vol. 44, No. 2 (January 1992), p. 248.

37. Jack S. Levy and Joseph R. Gochal, "Democracy and Preventive War: Israel and the 1956 Sinai Campaign," *Security Studies*, Vol. 11, No. 2 (Winter 2001/02), pp. 1–49; Marc Trachtenberg, "Preventive War and U.S. Foreign Policy," *Security Studies*, Vol. 16, No. 1 (January–March 2007), pp. 1–31; Scott A. Silverstone, *Preventive War and American Democracy* (New York: Routledge, 2007); and Jack S. Levy, "Preventive War and Democratic Politics," *International Studies Quarterly*, Vol. 52, No. 1 (March 2008), pp. 1–24.

38. Mearsheimer, *The Tragedy of Great Power Politics*, pp. 140–143.

39. On the "stopping power of water," see ibid., pp. 114–128.

40. Michael C. Desch, *When the Third World Matters: Latin America and United States Grand Strategy* (Baltimore, Md.: Johns Hopkins University Press, 1993).

41. Aaron L. Friedberg, *In the Shadow of the Garrison State: America's Anti-Statism and Its Cold War Grand Strategy* (Princeton, N.J.: Princeton University Press, 2000).

42. Robert J. Art, "The United States, the Balance of Power, and World War II: Was Spykman Right?" *Security Studies*, Vol. 14, No. 3 (July–September 2005), pp. 365–406.

When a potential hegemon emerges, the offshore balancer's first preference is to pass the buck, to stand aside and allow local states to check the threat.[43] If local states prove unequal to the task, however, the offshore balancer may intervene to slow the potential hegemon's momentum and, if possible, reestablish a rough balance of power in the region. The political problem confronting leaders is that, in the event war is necessary, the fighting is likely to be costly and protracted, as the offshore balancer attempts to roll back a potential hegemon at the peak of its strength. The public can thus be expected to insist on evidence of direct provocation before it consents to a fight to the finish on the adversary's turf. This is likely part of the explanation for why offshore balancers prefer to pass the buck and intervene only at the last moment, as the United States and Great Britain have done repeatedly.[44]

In sum, the selection effect is most likely to break down when leaders, for realist reasons, are drawn toward wars where an easy victory is anything but assured.[45] My hypothesis is that leaders will be tempted to resort to deception in these cases to preempt what can only be a contentious debate over whether the use of force is justified.

THE RESORT TO DECEPTION

From Kant onward, liberals have argued that a crucial difference between democracies and nondemocracies is that democratic leaders must enlist broad support when going to war.[46] Reiter and Stam's central claim, consistent with this tradition, is that democratic decisions for war are determined and constrained by public consent.[47] Rather than dispute this claim, I ask how leaders might go about manufacturing consent in cases where the selection effect breaks down and there is the prospect of significant public dissent.[48] In cases where victory might come at a steep price or threats are ambiguous, an open declaration of war is likely to trigger a contentious debate between the leadership and the political opposition over whether the use of force is justified. As Kenneth Schultz argues, "The opposition generally has greater political incentives to support the threat of force the better it expects the outcome to be. On

---

43. On buck-passing, see Mearsheimer, *The Tragedy of Great Power Politics*, pp. 157–158.
44. Ibid., chap. 7.
45. Downes's treatment of the Vietnam War suggests that domestic politics may also contribute to leaders' decisions to initiate war despite poor odds of victory. See Downes, "How Smart and Tough Are Democracies?"
46. Michael Doyle, "Liberalism and World Politics," *American Political Science Review,* Vol. 80, No. 4 (December 1986), pp. 1151–1169.
47. Reiter and Stam, *Democracies at War,* p. 144.
48. I agree with Reiter and Stam that democratic leaders want to generate public consent before entering into war. Unlike them, however, I highlight the ways in which leaders can manufacture such consent, by way of deception.

the other hand, when the opposition expects that the threat or use of force has weak support in the electorate, it has incentives to capitalize on that fact by giving voice to those concerns."[49] The more contentious the debate at the elite level, in turn, the more polarized the public is likely to become as people respond to cues from fellow partisans.[50]

Rather than press their case in the marketplace of ideas under such unfavorable conditions, leaders will be tempted to preempt debate by shifting blame for hostilities onto the adversary. The trick is to prepare domestic opinion for a possible, and even probable, war while providing firm assurances that it will come only as a last resort and only when the other side forces the issue. This entails two types of deception: (1) leaders must obscure the fact that they are open to war; and (2) they must wait for a crisis of sufficient magnitude to justify escalation.

First, if war threatens to be divisive, leaders are likely to go to some lengths to obscure the fact that they are considering it. This can entail a number of political and military distortions. For example, if alliance commitments have been made that threaten to entangle the state in a brewing conflict, leaders may downplay those commitments or deny that they are binding. Leaders can also negotiate in bad faith, professing a willingness to resolve matters peacefully while conducting talks with the adversary in such a way that they are likely to break down. In the modern context, this can involve channeling coercive diplomacy through international organizations, in the expectation that such diplomacy will ultimately fail but that the resulting war will be seen as more legitimate for having made the effort.[51] To preempt charges of warmongering, leaders may even renounce the use of force altogether—for example, if election-year politics demand it. Militarily, the key is to mobilize in a way that does not betray aggressive intent. One option is to escalate incrementally, deploying forces in a piecemeal fashion so as not to arouse suspicion. In cases where larger, and more dramatic, buildups are required, justifications can be offered that the forces are necessary for defensive purposes or for coercive diplomacy, but are not evidence of a war footing. Throughout, leaders are likely to shroud their war planning in secrecy.

---

49. Kenneth A. Schultz, *Democracy and Coercive Diplomacy* (Cambridge: Cambridge University Press, 2001), pp. 82–83.
50. On the "polarization effect," see John R. Zaller, *The Nature and Origins of Mass Opinion* (Cambridge: Cambridge University Press, 1992), pp. 100–113.
51. Alexander Thompson argues that when powerful coercers work through international organizations (IOs), they do so strategically to lower the international political costs of coercion. See Thompson, "Coercion through IO's: The Security Council and the Logic of Information Transmission," *International Organization*, Vol. 60, No. 1 (January 2006), p. 9. An additional benefit would be the enhanced domestic support that comes from giving diplomacy a chance, however halfheartedly.

Second, to justify escalation to war, leaders must wait for a suitable pretext, usually some hostile act on the part of the adversary that leaders can point to as evidence of aggressive intent. These "justification-of-hostility crises," as Ned Lebow calls them, are unique in that leaders of the initiating state make a decision for war before the crisis commences: "The purpose of the crisis is not to force an accommodation but to provide a *casus belli* for war. Initiators of such crises invariably attempt to make their adversary appear responsible for the war. By doing so they attempt to mobilize support for themselves, both at home and abroad and to undercut support for their adversary."[52] Justification-of-hostility crises tend to follow a standard script: leaders exploit a provocation, real or imagined, to inflame public opinion; make unacceptable demands upon the adversary in response; legitimize those demands with reference to generally accepted international principles; publicly deny or understate their real objectives in the crisis; and then point to rejection of their demands as a casus belli.[53] In extreme cases, leaders may even try to provoke the adversary into striking first, although this strategy is difficult to achieve when the other side is intent on avoiding war.[54] Either way, the goal is to preempt a contentious debate over whether the use of force is justified by exploiting a crisis and the "rally-round-the-flag effect" that follows it.[55] Under emergency conditions, public consent is sure to be more forthcoming than if leaders had declared war outright. In the next section, I assess the plausibility of these claims by revisiting the events surrounding the United States' entry into World War II.

FDR's Undeclared War

In the realm of national mythology, Americans remember World War II as "the good war." According to the standard narrative, the United States desired only to be left alone but was forced to fight in the face of German and Japanese aggression. When one takes a closer look at the historical record, though, it becomes clear that World War II was hardly forced on the United States. Well before the Japanese attack on Pearl Harbor on December 7, 1941, President Roosevelt came to the conclusion that the United States would have to act as a balancer of last resort in Europe, but he understood that public support for a declaration of war was unlikely to be forthcoming in the absence of a major

---

52. Richard Ned Lebow, *Between Peace and War: The Nature of International Crisis* (Baltimore, Md.: Johns Hopkins University Press, 1981), p. 25.
53. Ibid., p. 29.
54. Ibid., pp. 37–40.
55. John E. Mueller, *War, Presidents, and Public Opinion* (New York: John Wiley and Sons, 1973), pp. 208–213.

provocation. In this light, both the "undeclared war" in the Atlantic and the oil embargo on Japan should be understood as designed, at least in part, to invite an "incident" that could be used to justify hostilities. While controversial, a plausible case can be made that Roosevelt welcomed U.S. entry into the war by the fall of 1941 and manufactured events accordingly.[56]

I chose the World War II case for two reasons. First, it is a potential outlier for the selection effects argument.[57] On the one hand, consistent with the argument's logic, Roosevelt was sensitive to the domestic mood and waited until public consent was forthcoming to ask for an official declaration of war. On the other hand, there is compelling evidence that Roosevelt settled on a war policy well before Pearl Harbor and that in the interim he engaged in a significant amount of deception, maneuvering the country in the direction of open hostilities while assuring a wary public that the United States would remain at peace. Second, the case is a "tough test" of the argument insofar as there is considerable disagreement among historians as to whether the Roosevelt administration wanted war in 1941 and what lengths it was willing to go to sway public opinion.[58] Admittedly, some uncertainty about Roosevelt's intentions is irreducible, but I attempt to build a strong circumstantial case that he misled the public about his thinking. The role of deception in the World War II case remains controversial, but it is worth exploring for exactly this reason.

I expand on the argument in the rest of the section. First, I outline the grand strategy that motivated U.S. entry into the war. I also address the debate among historians about whether Roosevelt sought full-scale intervention prior to Pearl Harbor. Second, I discuss the anti-interventionism that Roosevelt had to contend with as he expanded U.S. involvement. Third, I detail the deceptions that he used to counteract this anti-interventionism. Finally, I discuss the consequences of these deceptions.

THE OFFSHORE BALANCING STRATEGY

During World War II, the United States essentially pursued a grand strategy of offshore balancing, acting as a balancer of last resort in Europe and Asia after its allies failed to contain potential hegemons in those regions. The primary

---

56. FDR's deceptions were a prominent theme in the first wave of revisionist scholarship on U.S. entry into World War II. See, for example, Charles A. Beard, *President Roosevelt and the Coming of War, 1941: Appearances and Realities* (New Haven, Conn.: Yale University Press, 1948).
57. On the heuristic value of outliers, see Alexander L. George and Andrew Bennett, *Case Studies and Theory Development in the Social Sciences* (Cambridge, Mass.: MIT Press, 2005), p. 75; and John Gerring, *Case Study Research: Principles and Practices* (Cambridge: Cambridge University Press, 2007), pp. 105–108.
58. On tough tests, see George and Bennett, *Case Studies and Theory Development in the Social Sciences*, pp. 120–123; and Gerring, *Case Study Research*, pp. 115–122.

challenge to U.S. interests was Nazi Germany, which threatened to overrun the European continent. The Nazis came perilously close to doing so with the fall of France and the invasion of the Soviet Union. U.S. policymakers feared that if Adolf Hitler were able to dominate Europe he would pose an intolerable military, economic, and ideological threat to the Western Hemisphere. To check the Nazis, Roosevelt extended assistance to European powers resisting Hitler, first Great Britain and then the Soviet Union, in the form of lend-lease aid and convoy protection. As the allies' situation grew more precarious over the course of 1941, however, FDR reluctantly concluded that stronger measures would be required and began to seek out opportunities to escalate U.S. involvement in the conflict. In Asia the United States faced a secondary but still serious threat from Japan, which was embroiled in a protracted war with China and menaced the Soviet Union to the north and the British supply line to the south. As Japanese forces expanded into the Southwest Pacific, U.S. policymakers moved to contain them. Official policy was to deter the Japanese while not provoking them, so as not to divert scarce resources from the primary theater in Europe.

Below I elaborate on these points. First, I discuss the Nazi threat and the offshore balancing strategy that the United States adopted in response. Second, I argue that this offshore balancing strategy implied U.S. entry into the war, while acknowledging disagreement among historians on the issue. I conclude by discussing the Japanese threat and initial U.S. attempts at containment.

THE NAZI THREAT IN EUROPE. By late 1938, in the wake of Germany's forcible annexation of the Sudetenland, Roosevelt was convinced that Hitler wanted nothing less than world domination and that after subduing the European continent he was sure to turn his sights on the United States.[59] Because of the nature of modern warfare, with aircraft capable of attacking at great speeds over long distances, FDR felt that the United States could no longer afford to retreat within its borders and rely on the oceanic barriers for safety. Instead, he envisaged the European democracies as the United States' front line. As long as France and Great Britain resisted Nazi advances, the United States was afforded a degree of protection from the Nazi threat. If France and Great Britain were to succumb to Nazi rule, though, Germany would pose an intolerable threat to the Western Hemisphere. Roosevelt's primary fear was that Hitler would exploit political unrest in Latin America to in-

---

59. For the impact of the Munich crisis on FDR's thinking, see Barbara Rearden Farnham, *Roosevelt and the Munich Crisis: A Study of Political Decision-Making* (Princeton, N.J.: Princeton University Press, 1997).

stall pro-Nazi regimes, which would provide staging areas for direct attacks on the United States.[60]

Roosevelt's intention to pass the buck to the European democracies suffered a major blow with the fall of France in June 1940.[61] Thereafter, the survival of Great Britain became of paramount importance. As FDR warned in a fireside chat at the close of the year, "If Great Britain goes down, the Axis powers will control the continents of Europe, Asia, Africa, Australasia, and the high seas—and they will be in a position to bring enormous military and naval resources against this hemisphere. It is no exaggeration to say that all of us, in all the Americas, would be living at the point of a gun—a gun loaded with explosive bullets, economic as well as military."[62] To avert this outcome, it was vital that the United States extend whatever assistance was necessary to sustain Great Britain in its fight against the Nazis, even if that meant courting war.

By the fall of 1940, FDR's military advisers were thinking along similar lines.[63] Indeed, some were willing to go further. In early November, Harold Stark, chief of naval operations, circulated one of the more explicit statements of the offshore balancing logic that was to motivate U.S. entry into the war a little more than a year later, commonly known as the Plan Dog memorandum. In it, Stark warned, "Should Britain lose the war, the military consequences to the United States would be serious." He further cautioned that Britain's mere survival would be insufficient. Rather, "To win, she must finally be able to effect the complete, or, at least, the partial collapse of the German Reich." This would require an invasion of Axis territory, with U.S. help. In Starks's estimation, "The United States, in addition to sending naval assistance, would also need to send large air and land forces to Europe or Africa, or both, and to participate strongly in this land offensive." At the same time, Stark understood that "account must be taken of the possible unwillingness of the people of the United States to support land operations of this character." Overall, Stark recommended "an eventual strong offensive in the Atlantic as an ally of the British, and a defensive in the Pacific."[64]

---

60. On the perceived Nazi threat to the Western Hemisphere, see Desch, *When the Third World Matters*, chap. 3.

61. On the impact of the fall of France on U.S. thinking, see David Reynolds, *The Creation of the Anglo-American Alliance, 1937–41: A Study in Competitive Cooperation* (Chapel Hill: University of North Carolina Press, 1981), pp. 108–113.

62. Franklin D. Roosevelt, *The Public Papers and Addresses of Franklin D. Roosevelt*, comp. Samuel Rosenman, Vol. 9 (for 1940) (New York: Russell and Russell, 1941), p. 635.

63. Initially, military planners recommended a posture of hemispheric defense in response to Hitler's victories in Europe, but that consensus was short lived. See Mark A. Stoler, *Allies and Adversaries: The Joint Chiefs of Staff, the Grand Alliance, and U.S. Strategy in World War II* (Chapel Hill: University of North Carolina Press, 2000), pp. 24–29.

64. The Plan Dog memorandum is reprinted in Steven T. Ross, ed., *American War Plans, 1919–1941,*

At the time, Roosevelt was not ready to admit that British victory would require a U.S. declaration of war and an invasion of Europe, so he avoided any direct approval of the Plan Dog memorandum.[65] He subscribed to its basic thinking, however, so military planning proceeded under the assumption that, in the event of war, the United States would pursue a Germany-first approach in conjunction with Great Britain while maintaining a defensive posture against Japan. This assumption underpinned the joint planning done by Anglo-American officers in the spring of 1941 and is reflected even more clearly in the Victory Program. The latter, prepared at Roosevelt's request in the wake of Germany's invasion of the Soviet Union in late June 1941, outlined production requirements in the event of a global war. Echoing Plan Dog, it reiterated that "if our European enemies are to be defeated, it will be necessary for the United States to enter the war, and to employ a part of its armed forces offensively in the Eastern Atlantic and in Europe or Africa." Moreover, it stressed that "naval and air forces seldom, if ever, win important wars" and that "it should be recognized as an almost invariable rule that only land armies can finally win wars."[66] The logic, as Mark Stoler argues, was unassailable: "If British survival and German defeat were essential to U.S. security and if this required ground operations beyond British capabilities or competence, then Washington would have to create and deploy the enormous army needed for such operations."[67] Vitally, it would have to do so by July 1943, at which point Germany would have consolidated its gains.

U.S. ENTRY INTO THE WAR? As evidenced by Plan Dog and the Victory Program, Roosevelt's military planners were in agreement that Germany's defeat would require U.S. entry into the war, including an invasion of Europe. Historians, however, remain divided on the issue of whether the president himself sought full-scale intervention. As one review of the literature concludes, "After a half century of research, it must be noted that there is still no scholarly consensus as to whether the president sought full-scale intervention in the European war."[68] David Reynolds, for example, argues that FDR's "de-

Vol. 3: *Plans to Meet the Axis Threat, 1939–1940* (New York: Garland, 1992), pp. 225–250. Direct quotes are taken from pp. 229, 241, and 247.
65. For FDR's reaction to Plan Dog, see James R. Leutze, *Bargaining for Supremacy: Anglo-American Naval Collaboration, 1937–1941* (Chapel Hill: University of North Carolina Press, 1977), pp. 202–205, 219; and Stoler, *Allies and Adversaries*, pp. 33–37.
66. Quoted in Robert E. Sherwood, *Roosevelt and Hopkins: An Intimate History*, rev. ed. (New York: Harper and Brothers, 1950), pp. 412, 415.
67. Stoler, *Allies and Adversaries*, p. 49. In a separate estimate, the U.S. Army concluded that Germany's defeat would require the creation of a ground force of 215 divisions and nearly 9 million men.
68. Justus D. Doenecke, "Historiography: U.S. Policy and the European War, 1939–1941," *Diplomatic History*, Vol. 19, No. 4 (Fall 1995), p. 696.

sire in 1941 was that America's contribution to the war would be in arms not armies—acting as the arsenal of democracy and the guardian of the oceans but not involved in another major land war in Europe."[69] Roosevelt had good reasons for avoiding formal entry into the war, according to Reynolds: public opinion would demand that rearmament take priority over lend-lease aid, with potentially disastrous consequences for the allies; and war with Germany would inevitably mean war with Japan, given the Tripartite Pact.[70] Stoler, along similar lines, argues that FDR viewed the Russian and British armies as substitutes for U.S. ground forces.[71] As late as the fall of 1941, he points out, Roosevelt deferred army expansion so as to release supplies for the Allies.

Roosevelt undeniably had deep misgivings about entering the war, given its expected cost. He was known to say privately that it would be politically impossible to send another American Expeditionary Force to Europe.[72] At the same time, the president's military advisers had made it clear to him that an invasion of Europe would be necessary to defeat Hitler. To the extent that Roosevelt was committed to that goal, it follows that he should have been at least open to full belligerency. Robert Dallek, who has written the most comprehensive account of Roosevelt's foreign policy, concludes as much, arguing that the president considered U.S. entry into the war inevitable by the summer of 1941.[73] By this time, as Steven Casey argues, FDR was "under pressure from both advisers and allies to plunge America directly into the war"; and if he had been free to act, he "would probably have favored a balanced response to the Nazi danger, based primarily on the use of U.S. land power."[74] Because of pressing domestic constraints, however, Roosevelt was reluctant to push for a formal declaration of war or the creation of an American Expeditionary Force. Instead, I argue, he maneuvered the country in the direction of open hostilities while assuring a wary public that the United States would remain at peace.[75]

CONTAINING JAPAN. If Roosevelt intended to balance against Germany, even at the risk of war, what were the implications for U.S. policy toward Japan?

---

69. Reynolds, *The Creation of the Anglo-American Alliance,* p. 288.
70. Ibid., pp. 218–219.
71. Stoler, *Allies and Adversaries,* pp. 56–57.
72. Reynolds, *The Creation of the Anglo-American Alliance,* p. 212.
73. Robert Dallek, *Franklin D. Roosevelt and American Foreign Policy, 1932–1945* (Oxford: Oxford University Press, 1995), pp. 265, 285.
74. Steven Casey, *Cautious Crusade: Franklin D. Roosevelt, American Public Opinion, and the War against Nazi Germany* (Oxford: Oxford University Press, 2001), p. 42.
75. Waldo Heinrichs concludes that, from the Atlantic Conference on, FDR was not seeking war but was knowingly risking it to forward supplies to the allies and contain Japan. He admits, though, that "it is hard to believe that he did not understand that sooner or later, one way or the other, this course of action would lead to war." See Heinrichs, *Threshold of War: Franklin D. Roosevelt and American Entry into World War II* (Oxford: Oxford University Press, 1988), p. 159.

First, it meant a clear recognition that the Japanese threat was of secondary importance and that the United States should not court war in the Pacific while it deepened its involvement in the Atlantic. Roosevelt himself, as Casey points out, "strongly believed that Nazi Germany posed the most powerful threat to U.S. interests" and that "wherever possible, the conflict in Asia had to be contained and dampened so that scarce U.S. resources would not be diverted away from Europe."[76] As I have already shown, this was a sentiment that was shared by the military. War planners agreed that in any war against the Axis the United States should pursue a Germany-first approach while maintaining a defensive posture against Japan.

At the same time, there was widespread agreement within Roosevelt's administration that the United States should make some effort to contain Japanese expansion, which had already consumed parts of China and, by the summer of 1941, menaced both the Soviet Union and the British supply line in the Southwest Pacific. Waldo Heinrichs summarizes the administration's thinking nicely: "Whether Japan went north or south it threatened to upset the improving balance of forces. This careening expansionism must be stopped. Japan must be boxed in, contained, immobilized."[77] To this end, the United States reinforced the Philippines with land and air forces, expanded military aid to China, and, most important, applied increasingly stringent economic sanctions on Japan. The dilemma, as FDR understood, was how to deter Japanese aggression without provoking a war in the process.[78] One of the enduring puzzles of the period is why the United States failed in this regard and ended up at war with Japan in December 1941. This is a subject I return to later.

ANTI-INTERVENTIONISM

Even as the Axis threat intensified, Roosevelt had to contend with domestic opposition to expanded U.S. involvement in the war. Most prominently, a diverse anti-interventionist movement, with significant representation in Congress, mobilized to challenge the president's initiatives. This movement fed off a deeply ambivalent public, which came to support aid to the Allies but was strongly opposed to a declaration of war or a commitment of ground forces. I discuss the anti-interventionist movement, Congress, and public opinion in turn.

---

76. Casey, *Cautious Crusade*, p. 13.
77. Heinrichs, *Threshold of War*, p. 145. Heinrichs represents the dominant view among historians when he describes U.S. policy in the Pacific in containment terms.
78. Scott D. Sagan, "From Deterrence to Coercion to War: The Road to Pearl Harbor," in Alexander L. George and William E. Simons, eds., *The Limits of Coercive Diplomacy* (Boulder, Colo.: Westview, 1994), p. 61.

"THE MANY MANSIONS OF ANTI-INTERVENTIONISM." The anti-interventionists of 1939–41 were a highly diverse coalition, one that included communists, fascists, and pacifists. As Justus Doenecke characterizes the movement, "The house of anti-interventionism contained many mansions."[79] At the same time, the anti-interventionists were by no means restricted to the fringes of society. Indeed, their ranks included former presidents (Herbert Hoover), influential members of Congress (Hamilton Fish and Robert Taft), powerful elements of the press (William Randolph Hearst), celebrities (Charles Lindbergh), radio personalities (Charles Coughlin), prominent business and labor labors (Gen. Robert E. Wood and John L. Lewis), and even members of the Roosevelt administration (Joseph Kennedy and Adolf Berle). The organization that came to embody these diverse tendencies was the America First Committee (AFC), which established itself during the 1940 presidential election as the primary anti-interventionist organization in the United States. By the time it disbanded on December 11, 1941, the AFC had more than 250,000 members. Its unifying message was that the traditional philosophy of hemisphere defense was still viable.[80]

With the AFC leading the way, the anti-interventionists protested any steps, such as lend-lease, that they thought would result in U.S. involvement in the war. To bolster their case, leading spokesmen hammered home a number of points in speeches and articles: that the Axis posed neither a military nor an economic threat to the United States; that the Soviet Union would emerge as the conflict's primary beneficiary; that Great Britain was an imperial power and unfit ally; that aid to the Allies would come at the expense of national defenses; that overseas involvement would lead to domestic ruin; and that a negotiated peace could be secured.[81] In the end, the anti-interventionists were defeated at every point, but in the short term, they complicated Roosevelt's task with Congress and the public.

THE NEUTRALITY ACTS. During his time in office, Roosevelt enjoyed his share of legislative triumphs, most famously with the New Deal. Congress, however, proved less cooperative when it came to measures that might embroil the country in another European war. By the mid-1930s, as Dallek reminds us, "Americans generally believed that involvement in World War I had been a mistake, that Wilson's freedom to take unneutral steps had pushed the country into the fighting, and that only strict limitations on presidential discre-

79. Justus D. Doenecke, *Storm on the Horizon: The Challenge to American Intervention, 1939–1941* (Lanham, Md.: Rowman and Littlefield, 2000), p. 8.
80. David Reynolds, *From Munich to Pearl Harbor: Roosevelt's America and the Origins of the Second World War* (Chicago: Ivan R. Dee, 2001), p. 94.
81. Doenecke expands on all these themes in *Storm on the Horizon*.

tion could keep this from happening again."[82] The result was a series of Neutrality Acts. The first, passed in August 1935, imposed an embargo on the sale of arms to belligerents. The second, passed in February 1936, added a ban on loans. The final Neutrality Act, passed in May 1937, forbade travel on belligerent ships. Collectively, the Neutrality Acts were designed "to minimize the economic entanglements and naval incidents that had supposedly drawn America into the Great War."[83]

Initially, Roosevelt was not opposed to the idea of a neutrality act, so long as it permitted him to exercise a degree of executive discretion. The legislation that resulted strongly tied the president's hands, however, and as the situation in Europe deteriorated in the late 1930s, he began to chafe at the constraints it imposed. Roosevelt wanted the freedom of maneuver to sell armaments to Britain and France and later to finance and transport them, but he faced an uphill battle in Congress when it came to revision of the legislation. This is not to say that he met only with failure. With the outbreak of war in Europe, Congress repealed the arms embargo and placed trade with belligerents on a "cash-and-carry" basis. Right up until Pearl Harbor, however, the president remained reluctant to ask for full repeal of the Neutrality Acts. He understood that there were limits to congressional interventionism. Indeed, in the latter half of 1941, Congress only narrowly revised the acts to allow the arming of merchant ships. Close votes such as these convinced Roosevelt that winning a declaration of war would be impossible in the absence of a substantial provocation from abroad.[84]

AN AMBIVALENT PUBLIC. In its ambivalence toward the war, Congress reflected public opinion. On the one hand, overwhelming majorities detested the Nazi regime and wanted the Allies to win. As German victories piled up in 1940 and 1941, more and more Americans came to favor a moderate interventionist stance that included military preparedness and aid to the Allies.[85] By May 1941, for example, almost 70 percent of the public was convinced that Hitler sought to dominate the United States; the same percentage believed it was more important to help Britain than to keep out of the war.[86] Public opinion was even more bellicose toward Japan, with persistent calls for the liberation of China well before the administration made it an issue.[87] On the other

---

82. Dallek, *Franklin D. Roosevelt and American Foreign Policy*, p. 109.
83. Reynolds, *The Creation of the Anglo-American Alliance*, p. 54.
84. Dallek, *Franklin D. Roosevelt and American Foreign Policy*, p. 292.
85. On public attitudes toward Nazi Germany and World War II, see Casey, *Cautious Crusade*, pp. 19–30.
86. Dallek, *Franklin D. Roosevelt and American Foreign Policy*, p. 267.
87. Paul W. Schroeder, *The Axis Alliance and Japanese-American Relations* (Ithaca, N.Y.: Cornell University Press, 1958), pp. 182–199.

hand, as Casey points out, Americans "remained adamantly opposed to a direct, unlimited, and formal involvement in the conflict."[88] No more than a third of the public ever supported a declaration of war. Roosevelt's perpetual problem, as he put it to the British ambassador to the United States, was "to steer a course between the two factors represented by: (1) The wish of 70 percent of Americans to keep out of war" and "(2) The wish of 70 percent of Americans to do everything to break Hitler, even if it means war."[89] It was in this context, I argue, that Roosevelt resorted to deception.

THE UNDECLARED WAR

President Roosevelt was famously sensitive to the domestic mood. He firmly believed that an effective policy abroad first required a consensus at home.[90] Given the persistence of anti-interventionism, such a consensus was bound to be elusive in the case of a declaration of war. FDR thus maneuvered the country in the direction of open hostilities while assuring a wary public that the United States would remain at peace. I now outline the deceptions that this strategy entailed. First, I discuss Roosevelt's efforts to obscure the belligerent drift of U.S. policy. Second, I argue that by mid-1941 he was searching for pretexts in the Atlantic and Pacific that would justify open hostilities.

THE GUISE OF NONBELLIGERENCY. As predicted, Roosevelt went to some lengths to obscure the belligerent drift of U.S. policy. He did so in three ways. First, he insisted that the United States was a nonbelligerent even as he waged a proxy war on behalf of the Allies. This pattern is evident as far back as the congressional debate over revision of the Neutrality Acts in the fall of 1939. In that debate, the Roosevelt administration presented "cash-and-carry" as a peace measure even though it was primarily intended to benefit Britain and France, which had the "cash" to pay for arms and the ships to "carry" them.[91] The administration, as Robert Divine argues, "carried on the elaborate pretense that the sale of arms to the Allies was but the accidental byproduct of a program designed solely to keep the United States clear of war."[92] The gap between rhetoric and reality only widened after the fall of France, when the administration moved to bolster the British. During that period, Roosevelt defended such unneutral acts as destroyers-for-bases, lend-lease, and interven-

---

88. Casey, *Cautious Crusade*, p. 30.
89. Quoted in Dallek, *Franklin D. Roosevelt and American Foreign Policy*, p. 289.
90. Ibid., p. 39. FDR regularly consulted a variety of sources to stay abreast of popular attitudes. These included newspapers, media surveys, gossip, mail, and increasingly opinion polls. See Casey, *Cautious Crusade*, pp. 16–19.
91. Reynolds, *From Munich to Pearl Harbor*, pp. 63–68.
92. Robert A. Divine, *The Illusion of Neutrality* (Chicago: University of Chicago Press, 1962), p. 297.

tion in the Battle of the Atlantic as intended to secure the Western Hemisphere. For example, he provocatively titled the lend-lease legislation "An Act to Promote the Defense of the United States." Roosevelt justified expanded naval patrolling in similar terms, as entailed by a series of "Western Hemisphere Defense Plans." Roosevelt was not being entirely disingenuous in this regard; his fears for the security of the Western Hemisphere did intensify as Britain's position worsened.[93] The president could have been more forthcoming, however, that a "common-law" alliance was developing with Britain, given that such a policy entailed a serious risk of war with Germany and Japan.[94]

Second, and related, Roosevelt pledged to keep the country out of the fighting, even as official thinking came around to the conclusion that entry into the war might eventually be required. The most prominent examples come from the 1940 election campaign, when Roosevelt's opponent, Wendell Willkie, charged him with warmongering. To rebut these charges, the president felt compelled to predict that the country would not become involved in fighting of any kind. Most famously, he promised at a campaign stop in Boston, "I have said this before, but I shall say it again and again and again: Your boys are not going to be sent into any foreign wars."[95] After some thought, he left out the qualifying phrase in the Democratic platform, "except in case of attack." Nor did FDR's pledges cease with the election. In a fireside chat at the end of December 1940, the president reassured the public that "the people of Europe who are defending themselves do not ask us to do their fighting" and that "our national policy is not directed toward war."[96] Roosevelt, again, may have shared some of these sentiments, but he also must have understood that such unqualified assurances of peace were at odds with the central tendency of U.S. policy.

Finally, Roosevelt was careful to shroud military planning in secrecy. He not only kept internal exercises such as Plan Dog and the Victory Program secret, but he also authorized secret talks between U.S. planners and their British counterparts, which were held from the end of January to the end of March 1941.[97] Known as the "ABC talks," the consultations revolved around contingency plans in case the United States entered the war. They culminated in agreement "that Germany was the primary threat to the security of both countries, that the defeat of Germany and Italy was the priority, that the Atlantic

93. Heinrichs, *Threshold of War*, p. 82.
94. On the "common-law" alliance between the United States and Britain, see Reynolds, *The Creation of the Anglo-American Alliance*, pt. 3.
95. Quoted in Dallek, *Franklin D. Roosevelt and American Foreign Policy*, p. 250.
96. Roosevelt, *The Public Papers and Addresses of Franklin D. Roosevelt*, Vol. 9, p. 640.
97. On FDR's authorization of the talks, see Leutze, *Bargaining for Supremacy*, pp. 202–205.

lifeline between Britain and the U.S.A. must be secured and that a purely defensive, deterrent policy should be maintained against Japan."[98] The participants even produced a war plan to that effect in case the United States entered the conflict. One informed observer has written that these staff talks "provided the highest degree of strategic preparedness that the United States or probably any another non-aggressor nation has ever had before entry into war."[99] Such detailed planning belied FDR's assurances that the United States would definitely remain out of the fighting, which explains why the president went to such lengths to keep the talks secret. British planners, for example, dressed in civilian clothes and were officially described as technical advisers to the British Purchasing Commission.[100]

LOOKING FOR AN "INCIDENT" IN THE ATLANTIC. As the ABC talks were coming to a close, the Germans intensified their war on British shipping and communications. By March 1941, Britain was losing ships at the rate of more than 500,000 tons a month; and because its shipyards were under air attack, those losses were becoming increasingly difficult to replace.[101] Meanwhile, Nazi armies were advancing through the Balkans, overrunning Yugoslavia and Greece in April. Uncertainty abounded as to where they would move next: southeast toward the Suez Canal, across the English Channel, or southwest toward Dakar, which was within aircraft range of Brazil.[102] Ominously, Hitler was massing his forces for a possible invasion of the Soviet Union. With the Axis threat intensifying, Roosevelt began to conclude that the United States would ultimately have to join the fighting, but he was reluctant to ask for a declaration of war in the absence of a provocation from abroad. As Dallek summarizes the president's thinking, "While he believed that the public would strongly line up behind intervention if a major incident demonstrated the need to fight, he did not feel that he could evoke this response simply by what he said or did."[103] The problem was how to court such an incident when Hitler was going out of his way to avoid confrontation with the United States. With the outcome of the war in Europe still undecided, Hitler was hardly ready to invite U.S. belligerency. Roosevelt, however, saw opportunity in the "undeclared war" that was about to be waged in the Atlantic.[104]

FDR first edged into the Battle of the Atlantic on March 15, 1941, when he

98. Reynolds, *The Creation of the Anglo-American Alliance*, p. 184.
99. Sherwood, *Roosevelt and Hopkins*, p. 273.
100. Reynolds, *From Munich to Pearl Harbor*, p. 117.
101. Heinrichs, *Threshold of War*, p. 30.
102. Ibid., p. 26.
103. Dallek, *Franklin D. Roosevelt and American Foreign Policy*, p. 267.
104. William L. Langer and S. Everett Gleason, *The Undeclared War, 1940–1941* (New York: Harper and Brothers, 1953).

ordered the Atlantic Fleet to prepare for active duty. Roosevelt's principal concern was to ensure that lend-lease supplies arrived safely in Britain. Over the course of the summer, however, as forces became available, the president adopted an increasingly confrontational posture. In April he extended the neutrality zone to include Greenland and the Azores. This allowed the U.S. Navy to patrol the western Atlantic and broadcast the location of German ships to the British. In July he moved to occupy Iceland, which he envisioned as an eventual transshipment point for Atlantic convoys. In August he met with Churchill to discuss war aims, culminating in the Atlantic Charter. In September the navy began escorting convoys as far as Iceland, with orders to "shoot on sight" in the event that they encountered Axis warships. By this time, the United States was at war with Germany in the western Atlantic in all but name only.

In his public appearances, FDR defended U.S. intervention in the Atlantic in hemispheric security terms. For example, in a May 27 radio address, he warned listeners that "unless the advance of Hitlerism is forcibly checked now, the Western Hemisphere will be within range of the Nazi weapons of destruction." To counter this threat, the president promised to "actively resist wherever necessary, and with all our resources, every attempt by Hitler to extend his Nazi domination to the Western Hemisphere."[105] Roosevelt came to have a more provocative end in mind, though: to force an "incident" that would justify open hostilities. His conversations with Winston Churchill are especially revealing in this regard. At the Atlantic Conference in August 1941, he explained to the prime minister that "he was skating on pretty thin ice in his relations with Congress" and that "if he were to put the issue of peace and war to Congress, they would debate it for three months." Instead, FDR indicated that "he would wage war, but not declare it, and that he would become more and more provocative." As Churchill understood it, "Everything was to be done to force an 'incident' . . . which would justify him in opening hostilities."[106] This was hardly an isolated remark. FDR used similar language with his advisers throughout the summer, leading Henry Stimson, his secretary of war, to speculate, "The President shows evidence of waiting for the accidental shot of some irresponsible captain on either side to be the occasion of his going to war."[107]

---

105. Franklin D. Roosevelt, *The Public Papers and Addresses of Franklin D. Roosevelt*, comp. Samuel Rosenman, Vol. 10 (for 1941) (New York: Russell and Russell, 1950), pp. 181, 190.
106. Quoted in Reynolds, *The Creation of the Anglo-American Alliance*, pp. 214–215. Reynolds suggests that such bellicose rhetoric was meant to boost British morale and should not be taken at face value. See ibid., pp. 66–67, 149.
107. Quoted in Dallek, *Franklin D. Roosevelt and American Foreign Policy*, p. 265. Stimson did not intend this remark as a compliment. He, along with other cabinet "hawks," grew increasingly frustrated with FDR for his indecisiveness during the spring and summer of 1941, as well as for his

In the weeks following his meeting with Churchill, FDR used just such an "incident" to announce his policy of escorting and "shoot on sight" in the Atlantic. The details are as follows: on September 4, a U.S. destroyer, the *Greer*, exchanged fire with a German submarine in the North Atlantic. The submarine fired on the *Greer* only after being pursued for several hours and evading depth charges from a British plane. As U.S. Navy officials reported to the president on September 9, "Submarine was not seen by *Greer* hence there is no positive evidence that the submarine knew nationality of ship at which it was firing."[108] In a fireside chat delivered on September 11, FDR deliberately distorted the details of the incident. He claimed that the *Greer*'s identity as a U.S. ship was unmistakable and that the German submarine fired first without warning. He went on to outline a Nazi design to abolish the freedom of the seas as a prelude to domination of the United States. Reminding his audience that he had sought no shooting war with Hitler, FDR insisted nonetheless that "when you see a rattlesnake poised to strike, you do not wait until he has struck before you crush him."[109] The president then announced that the U.S. Navy would protect all merchant ships engaged in commerce in the North Atlantic and that Axis ships would enter U.S. waters at their peril. The navy began escorting convoys shortly thereafter.

Although the *Greer* incident certainly facilitated U.S. intervention in the Atlantic, it was hardly the casus belli that FDR had promised Churchill in August. As subsequent events made clear, naval skirmishes would not suffice to generate the kind of consensus for war that the president wanted. In the latter half of October, German submarines claimed 126 American lives in attacks on the destroyers *Kearny* and *Reuben James*. Nevertheless, Congress barely passed revisions to the Neutrality Act arming U.S. merchant ships and allowing them to enter combat zones. FDR reluctantly concluded that it would take a more dramatic event than any sinking in the Atlantic to draw the United States into the war.[110] After all, Hitler was determined to postpone hostilities with the United States until victory in Russia was assured

---

lack of candor with the public. Their fear, at least, was that FDR would not intervene soon enough to stave off disaster.

108. On the *Greer* incident, see ibid., pp. 287–288; Reynolds, *The Creation of the Anglo-American Alliance*, p. 216; and Heinrichs, *Threshold of War*, pp. 166–168.

109. Roosevelt, *The Public Papers and Addresses of Franklin D. Roosevelt*, Vol. 10, pp. 384, 386–387, 390.

110. Dallek, *Franklin D. Roosevelt and American Foreign Policy*, p. 292. Gerhard L. Weinberg has argued that the Roosevelt administration exploited intercepted German naval messages to avoid rather than provoke incidents during this period. See Weinberg, *A World at Arms: A Global History of World War II* (Cambridge: Cambridge University Press, 1994), pp. 240–241. Marc Trachtenberg rebuts these claims. See Trachtenberg, *The Craft of International History: A Guide to Method* (Princeton, N.J.: Princeton University Press, 2006), pp. 84–87.

and had issued orders to his navy to that effect.[111] In the event, he declared war on the United States only after the Japanese attack on Pearl Harbor.

PEARL HARBOR: BACK DOOR TO WAR? Even today, conspiracy theories persist suggesting that Roosevelt deliberately allowed the Pearl Harbor attack to bring a unified country into the war. Among serious scholars, though, such arguments have been discredited.[112] David Kennedy points out that "despite decades of investigation, no credible evidence has ever been adduced to support the charge that Roosevelt deliberately exposed the fleet at Pearl Harbor to attack in order to precipitate war."[113] Rather, the scholarly consensus is that the administration did not anticipate a Japanese attack on Pearl Harbor, preoccupied as it was with developments in Southeast Asia. "The evidence," as Reynolds concludes, "points to confusion and complacency, not conspiracy, in Washington."[114]

Although such conspiracies have been found wanting, the notion that Roosevelt allowed matters to come to a head with Japan so that the United States could have a "back door" into the European war remains defensible.[115] The argument has recently been revived, for example, by historian Marc Trachtenberg.[116] As he points out, official policy was to deter the Japanese while not provoking them, so as not to divert scarce resources from the primary theater in Europe. Given these limited aims, it is puzzling that the president adopted such an aggressive line in the summer and fall of 1941, imposing an oil embargo and then insisting that the Japanese withdraw from China before supplies would be resumed. When negotiations predictably unraveled and Japanese action was imminent, Roosevelt talked openly of maneuvering Japan into firing the first shot rather than working for a diplomatic compromise. In these ways, Pearl Harbor was neither a complete surprise to the president nor entirely unwelcome. I elaborate on these points in the remainder of the section.

In late June 1941, Japanese leaders debated how they could best exploit Hitler's invasion of the Soviet Union. On July 2 they decided to move forward with the Southern Advance, which promised control of the resources of Southeast Asia and the encirclement of China. Japanese troops made their first

111. Heinrichs, *Threshold of War,* p. 109.
112. Roberta Wohlstetter's account remains definitive. See Wohlstetter, *Pearl Harbor: Warning and Decision* (Stanford, Calif.: Stanford University Press, 1962).
113. David M. Kennedy, *The American People in World War II: Freedom from Fear,* pt. 2 (Oxford: Oxford University Press, 1999), p. 90.
114. Reynolds, *From Munich to Pearl Harbor,* p. 163.
115. The "back door" argument is made most famously by Charles Callan Tansill. See Tansill, *Back Door to War: The Roosevelt Foreign Policy, 1933–1941* (Chicago: Regnery, 1952).
116. Trachtenberg, *The Craft of International History,* chap. 4.

moves in this direction on July 24, occupying southern Indochina. At the time, the leadership did not expect a war with the United States to result.[117] To its surprise, the Roosevelt administration responded by imposing an oil embargo, if only a de facto one. "The United States," as Waldo Heinrichs puts it, "had imposed an embargo without saying so."[118] As was widely recognized in Washington, this was a provocative move given that Japan was heavily dependent on oil imports to supply its war effort in China. Roosevelt himself understood that an embargo could well lead to a Japanese attack on the Dutch East Indies to secure alternative sources of supply and thus to war with Great Britain and the United States.[119] Admittedly, his participation in the embargo was indirect: he issued an order freezing Japanese assets and then failed to intervene when funds were not released for exports. This has led some observers to conclude that midlevel officials hijacked the policy process and engineered an embargo against the president's wishes.[120] Other evidence, however, suggests that the oil embargo almost certainly met with Roosevelt's approval. At the Atlantic Conference, for example, Roosevelt assured Churchill that he would maintain economic measures against Japan in full force. It is hard to escape the conclusion, which Trachtenberg reaches, that the president "deliberately opted for a policy which he knew would in all probability lead to war with Japan."[121]

If that was Roosevelt's intent, the embargo had the desired effect. It accelerated preparations for a general war that Japanese leaders had hoped to avoid. Faced with looming shortages, Japanese policymakers decided that they had no choice but to secure the raw materials of Southeast Asia by force, even if that entailed war with Great Britain and the United States. The moderates among them, however, still held out hope for a diplomatic settlement that would allow oil shipments to resume. To secure an agreement, Japanese leaders were willing to forgo further expansion, to withdraw from territory Japan had occupied as part of the Southern Advance, and even to work toward an

---

117. Heinrichs, *Threshold of War*, pp. 118–121; and Sagan, "From Deterrence to Coercion to War," pp. 66–67.
118. Heinrichs, *Threshold of War*, p. 177.
119. Dallek, *Franklin D. Roosevelt and American Foreign Policy*, pp. 273–274; and Trachtenberg, *The Craft of International History*, p. 96.
120. See, for example, Jonathan G. Utley, *Going to War with Japan, 1937–1941* (Knoxville: University of Tennessee Press, 1985).
121. Trachtenberg, *The Craft of International History*, p. 100. Trachtenberg rebuts Jonathan Utley's argument that Roosevelt had lost control of policy. He cites Waldo Heinrichs, who agrees that Roosevelt had some hand in the oil embargo but argues that it was intended to deter a Japanese attack on the Soviet Union (see *Threshold of War*, pp. 141–142). What is puzzling, from this perspective, is why the embargo remained in effect even after it was clear that a Japanese attack on the Soviet Union was unlikely and why the United States insisted on a Japanese withdrawal from China as a condition of its removal.

accommodation on the China issue. As Paul Schroeder argues, "From the beginning of August till late November the Japanese government bent all its efforts toward securing some kind of agreement with the United States which would relax the economic pressure now grown intolerable, without sacrificing all of Japan's gains."[122] U.S. negotiators, however, insisted on a complete withdrawal from China, a demand that the Japanese army found hard to accept after four years of intense fighting there. Coupled with the oil embargo, U.S. demands for the liberation of China were more consistent with a policy of rollback than containment.[123] They were certainly at odds with the thrust of U.S. grand strategy, which was to remain on the defensive in Asia until Nazi Germany was defeated. The president's military advisers, in particular, were opposed to this shift from deterrence to coercion. From their perspective, the president seemed to be "insanely willing" to provoke a second war in the Pacific when hostilities were escalating in the Atlantic.[124]

Rather than play for time as the military was recommending, Roosevelt rebuffed Japan's final diplomatic overtures. In early October, he refused to agree to a Leaders' Conference with Prince Konoe, the Japanese premier, because a comprehensive settlement could not be reached beforehand, ostensibly the purpose of the meeting. At the end of November, he abandoned the search for a modus vivendi, a temporary agreement that would have postponed hostilities, in the face of Chinese and British resistance. Roosevelt may have had the solidarity of the Allied coalition in mind when he gave up on negotiations, but that coalition had become important exactly because war was near. Signal intercepts indicated that Japanese action was imminent, most likely in Southeast Asia. At this point, the discussion turned to preparations for war and how to justify war to the public. U.S. negotiators, to this end, reintroduced the issue of the Tripartite Pact into the final round of talks with the Japanese. The thinking was that it would be useful to link Japanese expansion with Hitler's program of world conquest in the minds of Americans.[125] The only remaining question, as FDR remarked to his cabinet on November 25, "was how we should maneuver them [the Japanese] into the position of firing the first shot without allowing too much danger to ourselves."[126] With Japanese forces moving into position, all the public knew was that talks were stalled but not broken.

Japanese planes attacked Pearl Harbor shortly thereafter, on December 7. Expecting an invasion of Thailand, Malaya, or the Dutch East Indies, Roosevelt

---

122. Schroeder, *The Axis Alliance and Japanese-American Relations*, p. 54.
123. Trachtenberg, *The Craft of International History*, p. 91.
124. Stoler, *Allies and Adversaries*, p. 58.
125. Schroeder, *The Axis Alliance and Japanese-American Relations*, pp. 100–101.
126. Quoted in Dallek, *Franklin D. Roosevelt and American Foreign Policy*, p. 307.

was taken aback that the Japanese had targeted the fleet at Hawaii. He was also relieved, however, that they had not limited their attacks to British and Dutch possessions, potentially depriving him of the popular backing needed to declare war. Harry Hopkins, a top Roosevelt adviser, says this explicitly when recounting the events of the day: "I recall talking to the President many times in the past year and it always disturbed him because he really thought that the tactics of the Japanese would be to avoid a conflict with us; that they would not attack either the Philippines or Hawaii but would move on Thailand, French Indo-China, make further inroads on China itself and possibly attack the Malay Straits. This would have left the President with the very difficult problem of protecting our interests. Hence his great relief at the method that Japan used. In spite of the disaster at Pearl Harbor and the blitz-warfare with the Japanese during the first few weeks, it completely solidified the American people and made the war upon Japan inevitable."[127] By December 1941, Roosevelt and his advisers wanted such a war because it promised entry into a European conflict that demanded U.S. intervention. Again, Hopkins makes this explicit when describing the mood at the White House following the Pearl Harbor attacks: "The conference met in not too tense an atmosphere because I think that all of us believed that in the last analysis the enemy was Hitler and that he could never be defeated without force of arms; that sooner or later we were bound to be in the war and that Japan had given us an opportunity."[128]

What reason did Hopkins have to believe that Japanese military action would facilitate U.S. entry into the European war? It is important to point out in this context that it had been official policy since early 1941 that in the event of a conflict with Japan, the United States would go to war with Germany and Italy at once.[129] As it happens, FDR was not forced to declare war first, as he had intelligence in hand suggesting that Hitler would declare war on the United States in the event of a Japanese attack.[130] Even if Hitler had refused to declare war, however, Roosevelt would likely have been able to channel the popular anger surrounding the Pearl Harbor attacks into a declaration of war against Germany. The idea that the Japanese were in league with Hitler had

---

127. Quoted in Sherwood, *Roosevelt and Hopkins*, p. 428.
128. Quoted in ibid., p. 431. On the sense of relief surrounding Pearl Harbor, see Dallek, *Franklin D. Roosevelt and American Foreign Policy*, p. 311; Casey, *Cautious Crusade*, p. 47; and Trachtenberg, *The Craft of International History*, p. 129.
129. Trachtenberg, *The Craft of International History*, p. 126; and Leutze, *Bargaining for Supremacy*, pp. 225, 242.
130. Trachtenberg, *The Craft of International History*, p. 125. See, for example, F.H. Hinsley, *British Intelligence in the Second World War: Its Influence on Strategy and Operations*, Vol. 2 (New York: Cambridge University Press, 1981), p. 75.

been in place since the signing of the Tripartite Pact, and for that reason Pearl Harbor was widely blamed on the Axis as a whole.[131] According to a Gallup poll taken on December 10, after Pearl Harbor but before Hitler's declaration of war, 90 percent of respondents favored an immediate declaration of war on Germany.[132] U.S. leaders simply took it for granted, right after the Pearl Harbor attacks, that war with Germany was imminent.

In summary, a plausible reading of the evidence suggests that if there was a strategy underpinning Roosevelt's actions in the latter half of 1941, it was almost certainly that of the "back door." Since taking office, the president had been unwilling to take a hard line with Japan over its territorial ambitions in Asia, but he allowed matters to come to a head when the situation in Europe was deteriorating. He did so even though the thrust of official thinking suggested that he appease Japan and even though he was providing public assurances that war was not at hand. One of the few logical explanations is that he was taking advantage of the East Asian situation to bring the United States into the European war, a war that the public had been reluctant to embrace.

DID DECEPTION PAY?

Roosevelt understood that a declaration of war was bound to be divisive. Thus, he maneuvered the United States in the direction of open hostilities while providing misleading assurances to the contrary. In the process, he concealed the belligerent drift of U.S. policy and sought out pretexts that would justify escalation. In the wake of the Pearl Harbor attack, Roosevelt got the result he wanted, entry into the European war. Even anti-interventionists dropped their opposition to U.S. involvement, if only temporarily. That said, FDR's maneuvering and indirection were hardly without consequence. Most important, the United States had to fight an essentially unnecessary second war in the Pacific because Roosevelt could not bring himself to declare war on Germany as an act of policy.[133] Indeed, Pearl Harbor yielded a groundswell of support for revenge against Japan, which interfered with the administration's "Germany-first" strategy. Roosevelt worked throughout the war to counter such sentiment, downplaying Japanese intentions and capabilities. Such concerns contributed to his decision to invade North Africa in November 1942: FDR wanted to get U.S. troops into action as quickly as possible against the Nazis, so as to preempt demands for more effort in the Pacific.[134] More indirectly, FDR's arbitrary use of executive power paved the way for the imperial

---

131. Trachtenberg, *The Craft of International History*, p. 127.
132. Richard F. Hill, *Hitler Attacks Pearl Harbor: Why the United States Declared War on Germany* (Boulder, Colo.: Lynne Rienner, 2003), p. 209 n. 37.
133. Trachtenberg, "Preventive War and U.S. Foreign Policy," p. 28.
134. Christensen, *Useful Adversaries*, p. 251; and Casey, *Cautious Crusade*, chap. 3.

presidency of the Vietnam era, enabling Lyndon Johnson to wage a massive, undeclared war in Southeast Asia. Like Roosevelt, Johnson concealed his intentions and exploited pretexts such as the Gulf of Tonkin "incident," all in the service of a cause that has come to be considered more dubious than the fight against the Nazis.[135] What this particular legacy suggests is that deception always comes with a cost, even when otherwise most justified.

Conclusion

Leaders resort to deception to sell wars to their publics for two reasons. First, leaders are sometimes drawn toward wars where an easy victory is anything but assured—that is, the selection effect can break down. Second, leaders resort to deception in these cases to preempt debate over whether the use of force is justified by shifting blame for hostilities onto the adversary. Empirically, one can make a compelling case that this is exactly what the Roosevelt administration did in the lead-up to World War II, bringing the country into the war by way of the "back door."

One must be careful, of course, about deriving implications from a single case, especially one as controversial as World War II. That said, three are worth developing, although they must remain speculative at this point. First, the selection effect may break down when systemic pressures are severe. Contrary to the argument put forth by Reiter and Stam, democracies are not always free to pick on weak and vulnerable opponents but must, from time to time, take on formidable foes because their territorial integrity or political independence would be threatened otherwise. Roosevelt, for example, recognized that public consent was not forthcoming for a protracted war against Germany, but he felt that the United States had no choice but to intervene as a balancer of last resort to forestall the Nazi threat to the Western Hemisphere. He was willing, therefore, to court "incidents" in the Atlantic and Pacific that could be used as pretexts to justify hostilities. Democracies are obviously reluctant to fight open-ended wars and hence prone to pass the buck.[136] When systemic pressures are severe, however, their leaders are likely to do what is necessary to balance against external threats, including resorting to deception. This provides some vindication for realism.

Second, in cases where the selection effect does break down, alternative ex-

---

135. Senator J. William Fulbright famously asserted in 1971, "FDR's deviousness in a good cause made it easier for LBJ to practice the same kind of deviousness in a bad cause." Quoted in Dallek, *Franklin D. Roosevelt and American Foreign Policy*, p. 289.
136. Randall L. Schweller argues that democracies will be particularly slow to balance against threats. See Schweller, *Unanswered Threats: Political Constraints on the Balance of Power* (Princeton, N.J.: Princeton University Press, 2006), p. 48.

planations for why the democratic side ultimately prevailed are required. In an influential study of the outcome of World War II, Richard Overy finds that the Allied victory resulted from more than an overwhelming advantage in manpower and resources. Other elements—technological quality, combat prowess, organization and leadership, and fighting a moral war—were also important.[137] Overy's findings are consistent with trends in the scholarly literature on the sources of military effectiveness. The point of departure for that literature is that the creation of military power is only partially dependent on states' basic material and human assets. Cultural and societal factors, political institutions, and pressure from the international arena all affect how states organize and prepare for war, and ultimately influence their effectiveness in battle.[138] It remains to be seen how important regime type is as a source of military power relative to these other factors.

Finally, and perhaps most important, deception cannot be ruled out a priori as contrary to the national interest. If he had not resorted to deception, Roosevelt may not have been able to generate public consent for entry into World War II; and if the United States had not entered the war, then Hitler may have come to dominate Europe and to threaten the Western Hemisphere.[139] More generally, in the event that the leadership and the public disagree over the necessity of using force, deception may be justified when the leadership's assessment of the threat environment is sounder than the public's. This is not to deny that deception is costly. It is simply to say that deceiving may sometimes be less costly than not deceiving.

Allow me to conclude with a few words about the Iraq War. There is now relative consensus that the George W. Bush administration engaged in threat inflation to secure domestic support for deposing Saddam Hussein, and the focus of the academic debate has been on explaining why the marketplace of ideas failed to expose its distortions.[140] Less remarked upon, and in stark contrast to the World War II case, is that leading officials were able to be forthcoming about their belligerent intentions in the lead-up to the invasion. Indeed, if the Bush administration had wanted to be coy about its thinking on Iraq, it probably would not have elevated preventive war to the level of doctrine.[141]

---

137. Richard Overy, *Why the Allies Won* (New York: W.W. Norton, 1995).

138. Risa A. Brooks and Elizabeth A. Stanley, eds., *Creating Military Power: The Sources of Military Effectiveness* (Stanford, Calif.: Stanford University Press, 2007).

139. For some skepticism on this point, see Bruce M. Russett, *No Clear and Present Danger: A Skeptical View of the U.S. Entry into World War II* (New York: Harper and Row, 1972).

140. On threat inflation in the context of the Iraq case, see A. Trevor Thrall and Jane K. Cramer, eds., *American Foreign Policy and the Politics of Fear: Threat Inflation since 9/11* (London: Routledge, 2009).

141. George W. Bush, *The National Security Strategy of the United States of America* (Washington, D.C.: White House, September 2002).

Ironically, the selection effects argument provides a straightforward explanation for why the Bush administration was able to be so transparent: the public is likely to support the use of force when an easy victory is seemingly assured. Debate will then be truncated, because the political opposition has little incentive to present an alternative to a policy that is popular or widely regarded as successful. Under these conditions, leaders can be more forthcoming about their belligerent intentions. They may still engage in threat inflation, to convince the public that there is some danger that warrants the use of force, but whatever deceptions are entailed are unlikely to come under serious scrutiny in the marketplace of ideas. An initial reading of the evidence suggests that this is what happened in the Iraq case.[142] No firm conclusions can be drawn, however, until the documentary record is more complete.

---

142. John M. Schuessler, "Deception and the Iraq War," Air War College, 2009.

Correspondence

John M. Schuessler

FDR, U.S. Entry into World War II, and Selection Effects Theory

To the Editors (Dan Reiter writes):

In "The Deception Dividend: FDR's Undeclared War," John Schuessler argues that in 1941 President Franklin Roosevelt took actions against Germany and Japan to increase the likelihood that the United States would enter World War II, despite public hesitance to do so.[1] Schuessler writes, "The 'undeclared war' in the Atlantic and the oil embargo on Japan should be understood as designed, at least in part, to invite an 'incident' that could be used to justify hostilities" (p. 145). The broader implication is that if elected leaders can manipulate their countries into war despite public hesitation, then public opinion does not affect democratic foreign policy choices. More specifically, the ability of elected leaders to circumvent public opinion would damage the selection effects explanation of why democracies win their wars. The selection effects argument, which Allan Stam and I have made, claims that democracies win the wars they initiate because public opinion constrains them to initiate only those wars they are highly confident they will go on to win, and the marketplace of ideas better informs their policy choices.[2]

In this letter, I make three main points. First, even if one concedes the entirety of Schuessler's historical interpretation, the United States in World War II is not a case of a democracy initiating a war it went on to lose, and therefore not evidence against the proposition that democracies are likely to win the wars they initiate. Second, the evidence does not clearly indicate that Roosevelt sought to provoke war, and there is evidence to the contrary. Third, it is hard to think of a better demonstration of selection effects theory than Roosevelt's actions in 1941. He was deeply aware of public opinion and recognized that major foreign policy actions must be popular. Despite his grave con-

Dan Reiter is Chair of the Department of Political Science at Emory University.

John M. Schuessler is Assistant Professor of Strategy and International Security at the Air War College. He thanks Michelle Murray, Douglas Peifer, and Sebastian Rosato for helpful advice and comments. The views expressed in this letter are those of the author and do not reflect the official policy or position of the U.S. government or the Department of Defense.

1. John M. Schuessler, "The Deception Dividend: FDR's Undeclared War," *International Security*, Vol. 34, No. 4 (Spring 2010), pp. 133–165. Further references to this article appear parenthetically in the text.
2. Dan Reiter and Allan C. Stam, *Democracies at War* (Princeton, N.J.: Princeton University Press, 2002).

International Security, Vol. 35, No. 2 (Fall 2010), pp. 176–185

cern about the imperative of U.S. aid to Britain, Roosevelt was careful not to get ahead of public opinion, taking only those actions the public supported and no more. He worked within the marketplace of ideas to influence public opinion through persuasion and public speeches. His public claims about the seriousness of the Nazi and Japanese threat and the importance that the Soviet Union and Britain not lose the war reflected his genuine beliefs, and in hindsight appear to be right. The marketplace of ideas corrected his largest deviation from the facts, his biased portrayal of the *Greer* incident.

Schuessler's empirical argument does not provide evidence against selection effects theory. His argument would damage the theory if Roosevelt duped the public into supporting the initiation of a war that the United States went on to lose, but the United States defeated the Axis. And if one holds that war with Japan and Germany was avoidable and occurred because of U.S. provocation, then this is a case of a democracy choosing war and winning, the prediction of selection effects theory.

Some scholars might claim that the very attempt to deceive is evidence against selection effects theory. The theoretical assumption that elected leaders wish to manipulate public opinion to give them more foreign policy latitude is, however, not inconsistent with selection effects theory. Elected leaders pay attention to public opinion not because they share a normative commitment to democratic processes, but because they worry about the electoral consequences of unpopular policies. Stam and I argued that elected leaders engage in actions inconsistent with normative visions of democratic foreign policymaking, such as covertly undermining elected governments.[3] Credibly blaming the other side for initiation of the war certainly affects public opinion. I noted fifteen years ago in this journal that one reason preemptive wars are so rare is that leaders understand the political advantages of being attacked rather than of attacking.[4]

The historical record indicates that public opinion is not endogenous to the machinations of elected leaders. The ability of elected leaders to form public opinion is highly limited, and the constraint of public opinion dissuades them from initiating wars they will lose. The quantitative evidence demonstrating different implications of the assumption that public opinion constrains the foreign policy of elected leaders, including that democratic initiators are especially likely to win their wars, is deep, broad, and compelling. Domestic

3. Ibid., chap. 6.
4. Dan Reiter, "Exploding the Powder Keg Myth: Preemptive Wars Almost Never Happen," *International Security*, Vol. 20, No. 2 (Fall 1995), pp. 5–34.

political institutions affect international crisis behavior, war initiation, war duration, war outcome, and the postwar fates of leaders.[5]

Consider Schuessler's Germany case. Schuessler frets that Roosevelt "could have been more forthcoming" that U.S. actions before Pearl Harbor to aid Britain, including the destroyers-for-bases and lend-lease programs, risked war with Germany and Japan (p. 154). Schuessler notes that as late as December 1940, Roosevelt pledged to keep the United States out of war. Yet Roosevelt's pre–Pearl Harbor speeches were quite bellicose, focusing on the threat posed by the Axis to the United States and the need for action. They did not reassure Americans of the protection of an umbrella of neutrality. For example, in a speech on March 15, 1941, Roosevelt declared that Nazi Germany was "far worse" than the autocratic Prussian government the United States fought in World War I. The president stated, "They [Nazi Germany] openly seek the destruction of all elective systems of government on every continent—including our own."[6]

More important, Roosevelt addressed this question of U.S. involvement in the war directly. He argued that economic assistance to the Allies would decrease the risks of U.S. involvement in the war not because such actions were risk free, but because failure to support the Allies would ensure Axis victory in Europe. This would make U.S. involvement in the war inevitable, because it would encourage German aggression in the Western Hemisphere and against the United States itself. In his December 1940 fireside chat, Roosevelt said, "Thinking in terms of today and tomorrow, I make the direct statement to the American people that there is far less chance of the United States getting into war if we do all we can now to support the nations defending themselves against attack by the Axis than if we acquiesce in their defeat, submit tamely to an Axis victory, and wait our turn to be the object of attack in another war later on." Further, Roosevelt shied away from offering assurances that aid to Britain was without risks: "If we are to be completely honest with ourselves, we must admit that there is risk in any course we may take."[7]

Schuessler expresses concerns about secret Anglo-American war planning. Of course, such war planning reflected the realistic recognition that U.S. entry into the war against the Axis was a growing likelihood, the secrecy of such planning was demanded by obvious security considerations, and

---

5. Dan Reiter and Allan C. Stam, "Correspondence: Another Skirmish in the Battle over Democracies and War," *International Security*, Vol. 34, No. 2 (Fall 2009), pp. 194–204.
6. See Franklin D. Roosevelt, "On Lend Lease (March 15, 1941)," Presidential Speech Archive, http://millercenter.org/scripps/archive/speeches/detail/3322.
7. Ibid.

such war planning had no effect on increasing the likelihood of U.S. participation in war.

Schuessler next discusses the growing belligerency of U.S. maritime policy in the Atlantic in 1941, arguing that Roosevelt hoped to cause a naval incident that might increase public support for war. Schuessler exaggerates Roosevelt's desire to create such incidents. Indeed, the U.S. Navy used intercepted German communications to avoid rather than create naval incidents.[8] Roosevelt averted a major incident when, on June 11, a German submarine sank the *Robin Moor*, an American freighter, in a non-war zone. Although Roosevelt could have used this episode to justify an expansion of the U.S. presence in the Atlantic, he elected not to call this a direct attack on the United States or start a policy of escorting U.S. commercial vessels.[9]

Regardless, the aggressive U.S. naval stance in the Atlantic that Roosevelt eventually embraced was popular, meaning he was not (secretly) engaging in unpopular policies intended to drag an unwilling nation into war. Schuessler argues that Roosevelt hoped that the *Greer* incident in September 1941 would increase U.S. support for war, and that Roosevelt distorted the details to make Germany look more guilty of aggression. Certainly, the public supported Roosevelt's reaction to the *Greer* incident, as his new "shoot on sight" policy against all German and Italian shipping enjoyed 62 percent approval.[10] When Senate isolationists publicly exposed Roosevelt's distortions, however, there was no congressional or public backlash against Roosevelt or the shoot-on-sight policy.[11] Ultimately, Schuessler himself notes that the *Greer* incident "was hardly the casus belli that FDR had promised Churchill in August. As subsequent events made clear, naval skirmishes would not suffice to generate the kind of consensus for war that the president wanted" (p. 157).

The Japan case is more complex. Schuessler argues that Roosevelt took actions in 1941, especially the July imposition of oil sanctions, that he knew would cause Japan to attack. Schuessler cites an essay by Marc Trachtenberg, who makes a similar argument. There is not space to give Trachtenberg's argument fair treatment, but his views have been critiqued by other historians, and Trachtenberg himself concedes regarding his own interpretation that "few historians would go that far."[12]

---

8. Gerhard Weinberg, *A World at Arms,* 2d ed. (Cambridge: Cambridge University Press, 2005), p. xix
9. Robert Dallek, *Franklin D. Roosevelt and American Foreign Policy, 1932–1945* (New York: Oxford University Press, 1995), p. 268.
10. Ibid., p. 288.
11. Ibid.
12. Patrick Finney, ed., "Roundtable Review, *The Craft of International History: A Guide to Method,*" article review of "The Deception Dividend: FDR's Undeclared War," H-Diplo Roundtables, Vol. 8,

Roosevelt certainly recognized that in the event of war, it would be easier to rally public support if Japan attacked first, but Schuessler presents no evidence that the president actually wanted to cause a Japanese attack, conceding that his own evidence on Roosevelt's motives is "circumstantial" (p. 145). Schuessler asserts that "Roosevelt himself understood that an embargo could well lead to a Japanese attack," citing Robert Dallek's book (p. 159). On the pages cited by Schuessler, however, Dallek paints a far different picture of Roosevelt, remarking: "In June and July, despite his undiminished desire to avoid greater involvement in the Pacific, pressures beyond his control pushed Roosevelt toward a confrontation with Japan. . . . Though giving [his cabinet] 'quite a lecture' [on July 18] against a total oil embargo, which would be a goad to war in the Pacific, he agreed to [sanctions] . . . Roosevelt still had no intention of closing off all oil to Japan."[13]

Indeed, Roosevelt hoped that the sanctions would coerce Japan into curbing its belligerent moves, specifically deterring a Japanese attack on Soviet forces or in Southeast Asia and avert war with the United States. Waldo Heinrichs expressed it this way: "The central dynamic of [Roosevelt's] policies was the conviction that the survival of the Soviet Union was essential for the defeat of Germany and that the defeat of Germany was essential for American security. This more than any other concern, to his mind, required the immobilization of Japan."[14]

In fact, Roosevelt tried to make U.S. actions toward Japan less provocative. In the July 18 debate on oil sanctions, Roosevelt advocated reducing rather than ending oil exports to Japan, stating that he believed that "to cut off oil altogether at this time would probably precipitate an outbreak of war in the Pacific and endanger British communications with Australia and New Zealand."[15] About a week later, Roosevelt made the same point in a cabinet meeting, indicating that he "was still unwilling to draw the noose tight. He thought that it might be better to slip the noose around Japan's neck and give it a jerk now and then."[16] Heinrichs notes that "there were reasons for believing that an embargo might not precipitate a Japanese attack southward. [Maxwell] Hamilton [chief of the division of Far Eastern Affairs] pointed out

No. 16 (December 2007), pp. 1–35; and Marc Trachtenberg, "H-Diplo ISSF Article Review," No. 3 (April 2010), p. 4.

13. Dallek, *Franklin D. Roosevelt and American Foreign Policy,* pp. 273–274.

14. Waldo Heinrichs, *Threshold of War: Franklin D. Roosevelt and American Entry into World War II* (New York: Oxford University Press, 1988), p. 179.

15. Quoted in Irvine H. Anderson Jr., *The Standard-Vacuum Oil Company and United States East Asian Policy, 1933–1941* (Princeton, N.J.: Princeton University Press, 1975), p. 175.

16. Harold L. Ickes, *The Secret Diary of Harold L. Ickes,* Vol. 3: *The Lowering of the Clouds, 1939–1941* (New York: Simon and Schuster, 1954), p. 588.

on July 31 that Japan was weaker economically and now open to attack from all sides. . . . Roosevelt wired [British Prime Minister Winston] Churchill in satisfaction that their concurrent action seemed to be 'bearing fruit'. . . . A policy of maximizing Japanese uncertainty and insecurity seemed to be having a useful effect."[17] Roosevelt reiterated his preference for avoiding a severe provocation of Japan in his August meeting with Churchill at Argentia. Indeed, only with great reluctance did he agree to give Japan a severe warning. And upon returning to Washington, he backtracked and watered down the language of the message to Japan, to reduce the provocative effects.[18]

Schuessler states that Roosevelt's policy of provoking Japan into war was inconsistent with the U.S. policy of defeating Germany first before turning to Asia. Yet it was this very Germany-first orientation that both encouraged the oil sanctions—because the sanctions might prevent a Japanese attack on the Soviet Union, which in turn would help the Soviets survive the German invasion—and made Roosevelt prefer delaying U.S. entry into war in the Pacific to better deal with the German threat. On the latter point, he told Secretary of the Interior Harold Ickes on July 1, "[I]t is terribly important for the control of the Atlantic for us to help keep peace in the Pacific. I simply have not got enough Navy to go round—and every little episode in the Pacific means fewer ships in the Atlantic."[19] Schuessler also quotes Mark Stoler's remark that "from a military perspective the president and the State Department seemed to be insanely willing to provoke a second war in the Pacific," implying that the sanctions policy did not accord with U.S. military thinking at the time (p. 160). In his next paragraph, however, Stoler notes that the U.S. Army leadership supported the hard line on Japan, to prevent a Japanese attack on the Soviet Union.[20]

Beyond the oil embargo, there is the question of U.S. diplomacy with Japan in the fall of 1941. Schuessler argues that, to provoke Japan to attack, Roosevelt refused to discuss Japanese peace overtures. Schuessler's portrayal of Roosevelt's stonewalling in the fall of 1941 is exaggerated and undersupported. Schuessler cites a book by Paul Schroeder, and the critical paragraph in Schuessler's argument (p. 160) cites two pages in Schroeder's book that discuss only the role of the Tripartite Pact in linking Germany and Japan, not any

---

17. Heinrichs, *Threshold of War*, p. 142.
18. Ibid., p. 154; and David Reynolds, *The Creation of the Anglo-American Alliance, 1937–1941: A Study in Competitive Co-operation* (Chapel Hill: University of North Carolina Press, 1982), p. 239.
19. Ickes, *The Secret Diary of Harold Ickes*, Vol. 3, p. 567.
20. Mark A. Stoler, *Allies and Adversaries: The Joint Chiefs of Staff, the Grand Alliance, and U.S. Strategy in World War II* (Chapel Hill: University of North Carolina Press, 2000), p. 58.

alleged stonewalling by Roosevelt.[21] Schuessler writes that Roosevelt imposed prohibitive conditions on a possible meeting with Prince Konoe. However, Roosevelt himself was "prepared, even eager, to go," but the idea of the meeting died because of Secretary of State Cordell Hull's vehement opposition and demand for strict conditions.[22] It is also unclear whether negotiations would have succeeded with greater U.S. concessions, as Japan's true interest in peace was doubtful. In September Japan demanded U.S. trade concessions in East Asia, a commitment to halt all U.S. military preparations in East Asia, and the restoration of trade with Japan before Japan withdrew from Indochina or reached a peace agreement with China.[23] In mid-November Roosevelt circulated a peace plan proposal within the U.S. government and among U.S. allies that included a six-month agreement to freeze arms on both sides and permit Sino-Japanese peace talks.[24]

Schuessler notes that by the end of November Roosevelt had abandoned the idea of a temporary peace plan. This decision, however, came about not because of a desire to provoke the Japanese to attack first, but for other reasons, including recognition that the plan would certainly be rejected (especially given the November 22 Japanese message that Japan would not extend a deadline for successful negotiations past November 29), concern that such an agreement would deliver a severe blow to Chinese morale and resistance, and distrust that the Japanese would adhere to an agreement, as Roosevelt interpreted news of Japanese troop movements as evidence that Japan was negotiating in bad faith.[25] Schuessler quotes Harry Hopkins's early 1942 description of Roosevelt's "relief" that Japan attacked first, but Hopkins also reports that on December 7 Roosevelt privately "discussed at some length his efforts to keep the country out of the war and his earnest desire to complete his administration without war."[26]

Beyond whether or not Roosevelt sought war, his hard-line policy against Japan was popular. He was not out of step with public opinion by engaging in

---

21. See Paul W. Schroeder, *The Axis Alliance and Japanese-American Relations, 1941* (Ithaca, N.Y.: Cornell University Press, 1958), pp. 100–101, quoted in Schuessler, "The Deception Dividend," p. 160 n. 125.

22. Jonathan G. Utley, *Going to War with Japan, 1937–1941* (Knoxville: University of Tennessee Press, 1985), pp. 159–161.

23. Heinrichs, *Threshold of War*, p. 187. One slight softening of the Japanese position was the apparent signal that Japan's alliance with Germany would not affect any U.S.-Japan deal.

24. Dallek, *Franklin D. Roosevelt and American Foreign Policy*, p. 307; Utley, *Going to War with Japan*, p. 167 (see also p. 157); and Heinrichs, *Threshold of War*, p. 209.

25. Dallek, *Franklin D. Roosevelt and American Foreign Policy*, pp. 307–308.

26. Robert E. Sherwood, *Roosevelt and Hopkins: An Intimate History* (New York: Harper and Brothers, 1948), p. 431.

potentially provocative foreign policies, and therefore not engaged in deception. A State Department study in mid-1941 found that American editorials were presenting an "almost unanimous and very insistent demand for a firmer stand in the Far East."[27] A December 5 poll indicated that 69 percent of respondents supported U.S. efforts to prevent Japan from becoming more powerful, even if it meant risking war.[28]

Roosevelt did not drag the United States into World War II through deception or any other means. Consistent with selection effect theory, Roosevelt's foreign policy was in step with, constrained by, and not manipulative of public opinion.

—*Dan Reiter*
Atlanta, Georgia

John M. Schuessler Replies:

I would like to thank Dan Reiter for his spirited rebuttal of my article.[1] Reiter makes three points in his letter. First, even if one accepts my rendering of the World War II case, this does not undermine the selection effects argument. Second, my reading of the evidence is flawed: it is not clear that President Franklin Roosevelt wanted to provoke war with Germany and Japan. Finally, the policy that Roosevelt pursued—a policy of holding the line in the Atlantic and the Pacific—is actually an excellent demonstration of the selection effect. I address each point in turn.

THE SELECTION EFFECT AND THE WORLD WAR II CASE

Reiter suggests that my goal in "The Deception Dividend" was to damage the selection effects argument. As I make clear in the article, however, my goal was not to discredit the selection effect, but rather to specify the conditions under which it breaks down.[2] The conclusion I reached—that the selection effect breaks down when systemic pressures are severe—was borne out by Roosevelt's behavior in the lead-up to World War II.

Reiter contends that the World War II case cannot be counted against the selection effect because the United States went on to defeat the Axis. The out-

---

27. Quoted in Heinrichs, *Threshold of War*, p. 142.
28. Dallek, *Franklin D. Roosevelt and American Foreign Policy*, p. 310.
1. John M. Schuessler, "The Deception Dividend: FDR's Undeclared War," *International Security*, Vol. 34, No. 4 (Spring 2010), pp. 133–165.
2. Ibid., p. 138.

come, in other words, is consistent with the theory. The selection effects argument does more than predict an outcome, however. It is underpinned by a causal logic suggesting that because democratic leaders must secure public consent for war through an open process, they are constrained from starting unduly costly or difficult wars, leading to a high likelihood that democracies will win the wars they start.[3]

My analysis of the World War II case suggests that Roosevelt maneuvered the United States into a costly and protracted war, one whose outcome was by no means foreordained, and one that was controversial enough politically that he never felt able to ask for a declaration of war from Congress. The United States may have ultimately prevailed over the Axis, but Reiter's effort to credit that outcome to the selection effect is a stretch. After all, if the selection effect allows for such maneuvering, then how constraining could it possibly be?

Reiter's fallback position is that the selection effect does not preclude attempts at deception. Surely, though, it should preclude successful attempts, especially when the stakes are as high as they were in the World War II case. Reiter and his coauthor, Allan Stam, concede as much in *Democracies at War*: "An important assumption of this perspective is that consent cannot be easily manufactured by democratic leaders. If democratic leaders could manipulate public opinion into supporting military ventures, then of course public opinion would provide little constraint on democratic foreign policy, as it could be actively molded to support the foreign policy aims of the leadership."[4] If Roosevelt was able to deceive the public about his intentions, the World War II case would be problematic for the selection effects argument, a point Reiter tacitly admits by contesting whether Roosevelt engaged in much deception at all.

ROOSEVELT'S LEVEL OF DECEPTION
Reiter devotes the bulk of his letter to challenging the particulars of my argument that President Roosevelt sought to provoke war with Germany and Japan. He contends that Roosevelt was content to hold the line in the Atlantic and Pacific—ferrying supplies to Britain while containing Japan—a policy that was in line with public opinion. Here I reiterate my case that Roosevelt was well ahead of public opinion on the issue of whether to intervene in the European war and was thus forced to bring the United States into the conflict

---

3. My reading of the selection effects argument is based primarily on Dan Reiter and Allan C. Stam, *Democracies at War* (Princeton, N.J.: Princeton University Press, 2002), chaps. 2, 6, 7.
4. Ibid., p. 146.

through the "back door." In the process, I address as many of Reiter's specific claims as possible.

THE ATLANTIC. After pledging unequivocally during the 1940 presidential campaign that the United States would not enter the European war, Roosevelt became more forceful in 1941 about describing Nazi Germany as a clear and immediate danger to national security.[5] That spring, he prevailed in the "great debate" with the anti-interventionists over whether the United States should remain strictly neutral or actively aid the British war effort. The results were the lend-lease program and the "undeclared war" in the Atlantic. As Reiter alludes to, increasingly large majorities of the American public supported the latter, moderate interventionist, stance, which explains why FDR was able to embrace it in his public appearances.[6]

Reiter, however, fails to address evidence that Roosevelt wanted the United States to do considerably more than act as the "arsenal of democracy." By the fall of 1941, the president's advisers were in agreement that the United States would have to enter the war in force to defeat Nazi Germany.[7] The military had made it clear that this would entail raising a large army to fight on the European continent.[8] Winston Churchill and Joseph Stalin, understandably, were also pressing the president to declare war.

Pinning down Roosevelt's thinking on the issue is difficult, but the evidence suggests that he came to the same conclusion as his advisers and allies: the United States would have to enter the war as a full belligerent. Indeed, Roosevelt said as much in deliberations with them. Most famously, he explained to Churchill at the Atlantic Conference that "he was skating on pretty thin ice in his relations with Congress" and that "if he were to put the issue of peace and war to Congress, they would debate it for three months." Instead, Roosevelt assured Churchill that "he would wage war, but not declare it, and that he would become more and more provocative. If the Germans did not like it, they could attack American forces." In this way, "Everything was to be done to force an 'incident' . . . which would justify him in opening hostilities."[9] One

5. Steven Casey, *Cautious Crusade: Franklin D. Roosevelt, American Public Opinion, and the War against Nazi Germany* (Oxford: Oxford University Press, 2001), pp. 38–39.
6. Ibid., p. 26
7. Ibid., pp. 41–42.
8. Mark A. Stoler, *Allies and Adversaries: The Joint Chiefs of Staff, the Grand Alliance, and U.S. Strategy in World War II* (Chapel Hill: University of North Carolina Press, 2000), chaps. 2, 3.
9. David Reynolds, *The Creation of the Anglo-American Alliance, 1937–41: A Study in Competitive Co-operation* (Chapel Hill: University of North Carolina Press, 1981), pp. 214–215. Beginning in late March 1941, Roosevelt regularly mused to aides that he would welcome an incident with Germany. See ibid., p. 347 n. 38.

can only imagine the public reaction if these comments, or others like them, had been leaked.

Reiter suggests that Roosevelt was not as determined to create an "incident" as his bellicose rhetoric would imply. For support, he cites Gerhard Weinberg, who has argued that the U.S. Navy, with British help, used intercepted German communications to avoid contact with U-boat formations.[10] The primary source upon which Weinberg relies, however, supports a different conclusion than the one he reaches: the naval authorities may have exploited the available intelligence to steer convoys around German submarine concentrations—their primary responsibility, after all, was to ferry supplies safely to Britain—but that does not mean they were trying to avoid confrontations with German warships.[11] After all, the British, who were routing the convoys, had every incentive to maximize tensions in the Atlantic.

The "incidents" that did occur, such as U-boat attacks on the destroyers *Kearny* and *Reuben James,* did not generate the consensus for war that Roosevelt wanted. In November 1941, Congress only narrowly revised the Neutrality Act, convincing Roosevelt that winning a declaration of war would require a substantial provocation from abroad.[12] That provocation came in the form of Japan's attack on Pearl Harbor.

THE PACIFIC. The "back door" thesis—the argument that Roosevelt courted war with Japan so that the United States could enter the war with Germany— is controversial among historians. Reiter, therefore, is more than justified in challenging key elements of it. Why, then, does this thesis deserve serious consideration? The primary reason is that the conventional wisdom, shared by Reiter, that Roosevelt was simply trying to contain Japan in the summer and fall of 1941 suffers from critical flaws.

First, if Roosevelt's goal was to immobilize Japan, why did he agree to an oil embargo, a move he had resisted in the past because it would in all probability have led to a Japanese attack on the Dutch East Indies and thus war with the United States?[13] Reiter suggests that Roosevelt's intent was to impose a partial embargo, one that he could relax in the event Japan backed down. The embargo put in place, however, was total, and mounting evidence suggests that

---

10. Gerhard L. Weinberg, *A World at Arms: A Global History of World War II* (Cambridge: Cambridge University Press, 1994), p. 240.

11. Marc Trachtenberg, *The Craft of International History: A Guide to Method* (Princeton, N.J.: Princeton University Press, 2006), pp. 84–87.

12. Robert Dallek, *Franklin D. Roosevelt and American Foreign Policy, 1932–1945* (New York: Oxford University Press, 1995), p. 292.

13. Ibid., pp. 273–275.

Roosevelt was not only aware of the extent of the embargo but approved of the policy.[14] Reiter asks why Roosevelt packaged the embargo in a relatively mild way, for example, by refusing to attach an ultimatum to it. If Roosevelt's goal was to put the United States on a collision course with Japan without attracting unwanted attention, an ultimatum would have been counterproductive.

Second, if Roosevelt's goal was to contain Japan, why was he not more responsive to Japan's signals that it would refrain from entering the war against the Soviet Union, that it would stall the southern advance in its tracks, and that it would even withdraw from parts of Indochina? Roosevelt was not averse to talks, as Reiter points out: at the very least, the president needed to appear to be negotiating in good faith. The key point, however, is that U.S. negotiators came to insist on a set of demands that Japan was sure to reject, most importantly, Japan's withdrawal from China.[15] Jeffrey Record makes the point best: "The United States was, in effect, demanding that Japan renounce its status as an aspiring great power and consign itself to permanent strategic dependency on a hostile Washington. Such a choice would have been unacceptable to any great power."[16] It is difficult to explain why Roosevelt would confront Japan with such a choice unless he was seeking war.

As far as public attitudes in the United States on the issue, the hard-line policy against Japan was popular enough, as Reiter alludes to, but that does not mean the public was ready to go to war over China.[17] Indeed, the public was generally less concerned with events in Asia than in Europe, which explains why Roosevelt had the freedom of maneuver he needed to back Japan into a corner.

HARD QUESTIONS FOR THE SELECTION EFFECT

To summarize, Roosevelt was certainly constrained by public opinion in the lead-up to World War II. Otherwise, he would have had no reason to resort to all the maneuverings he did to get the United States into the war. The success of those maneuverings, however, should lead scholars to ask some hard questions about the selection effect: Why, in some cases, are leaders not deterred from seeking war by the prospect of punishing costs and the political rancor

---

14. Trachtenberg, *The Craft of International History*, pp. 99–100; and Waldo Heinrichs, *Threshold of War: Franklin D. Roosevelt and American Entry into World War II* (Oxford: Oxford University Press, 1988), pp. 141–142, 246–247 n. 68.

15. As Paul W. Schroeder highlighted at an early point, "There is no longer any doubt that the war came about over China." See Schroeder, *The Axis Alliance and Japanese-American Relations, 1941* (Ithaca, N.Y.: Cornell University Press, 1958), p. 200.

16. Jeffrey Record, *Japan's Decision for War in 1941: Some Enduring Lessons* (Carlisle, Pa.: Strategic Studies Institute, U.S. Army War College, 2009), p. 21.

17. Casey, *Cautious Crusade*, pp. 29–30.

that might result? What explains the deep reluctance of leaders to put the question of war and peace before the public, and how are they able to avoid doing so? That is, how are they able to bypass the marketplace of ideas? In cases where the democratic process is compromised in this way, are the results uniformly negative? Or is deception sometimes justified by the stakes and the ultimate outcome of the conflict? Not to pursue these questions is to foreclose valuable lines of inquiry.

—*John M. Schuessler*
Maxwell Air Force Base, Alabama

International Security

The Robert and Renée Belfer Center for
Science and International Affairs
John F. Kennedy School of Government
Harvard University

Most of the articles in this reader were previously published in **International Security,** a quarterly journal sponsored and edited by the Robert and Renée Belfer Center for Science and International Affairs at Harvard University's John F. Kennedy School of Government, and published by MIT Press Journals. To receive journal subscription information or to find out more about other readers in our series, please contact MIT Press Journals at 55 Hayward Street, Cambridge, MA 02142, or on the web at http://mitpress.mit.edu.